THE INFLUENCE OF MAN
ON
ANIMAL LIFE IN SCOTLAND
A STUDY IN FAUNAL EVOLUTION

BY

JAMES RITCHIE

M.A., D.Sc. (Aberdon.), F.R.S.E.

Assistant-Keeper in the Natural History Department of the
Royal Scottish Museum
(Scottish Education Department)

CAMBRIDGE
AT THE UNIVERSITY PRESS
1920

Quhare I misknaw myne errour, quho it findis
For charité amend it, gentil wicht,
Syne perdoun me sat safer in my lycht ;
And I sal help to smore your falt, leif brother,
Thus vailye quod vailye, ilk gude dede helpis vthir.

     GAWAINE DOUGLAS, Bishop of Dunkeld.

CAMBRIDGE
UNIVERSITY PRESS

University Printing House, Cambridge CB2 8BS, United Kingdom

Cambridge University Press is part of the University of Cambridge.

It furthers the University's mission by disseminating knowledge in the pursuit of education, learning and research at the highest international levels of excellence.

www.cambridge.org
Information on this title: www.cambridge.org/9781107512030

© Cambridge University Press 1920

This publication is in copyright. Subject to statutory exception and to the provisions of relevant collective licensing agreements, no reproduction of any part may take place without the written permission of Cambridge University Press.

First published 1920
First paperback edition 2015

A catalogue record for this publication is available from the British Library

ISBN 978-1-107-51203-0 Paperback

Cambridge University Press has no responsibility for the persistence or accuracy of URLs for external or third-party internet websites referred to in this publication, and does not guarantee that any content on such websites is, or will remain, accurate or appropriate.

THE INFLUENCE OF MAN
ON
ANIMAL LIFE IN SCOTLAND

Frontispiece

SCOTTISH RED DEER IN PREHISTORIC TIMES AND TO-DAY

The Stag to the left represents the prehistoric Red Deer of the Scottish peat mosses, which lived in low-country forests even to the sea coast; to the right is placed for comparison a modern Royal Stag—the difference in size can be traced to the interference of man (see p. 333 *et seq.*).

PREFACE

THE animal life or fauna of a country is no fixed unit of occupation, established and unchanging, but, endowed with the plasticity of life, it carries in itself the imprints of many influences which have played upon it throughout the ages. The lectures contained in the following pages were planned to unravel one important set of such influences—those which radiate from the acts of Man—so that it might be possible to trace the different ways in which Man's power has worked and is working, and to realize to what degree a fauna of to-day owes its character and composition to his interference.

With this end in view it was necessary to select a particular fauna of manageable compass, where the inquisition into Man's influence could be pushed to the furthest limits; and several facts pointed to the fauna of Scotland as best suited for the purpose. Nevertheless, I have not hesitated to refer to examples of Man's influence in other countries, wherever particular types have been strikingly illustrated, or where influences are seen at work which help to explain effects of causes long lost to sight in Scotland, or where, as in the case of counter-pests, modern science has created new kinds of interference which sooner or later are likely to be adopted in this country.

A result of this enquiry has been to emphasize the instability and changefulness of a fauna, and a word may be said as to the general place of Man's influence in the sum of change. Two types of changefulness affect a country's animals—one temporary in incidence and local in effect, a function of circumstance; the other persistent and general, a function of time. Within itself a fauna is in a constant state of uneasy restlessness, an assemblage of creatures which in its parts ebbs and flows as one local influence or another plays upon it. It may be that a succession of favourable seasons breeds many field-voles, and the tide of the field-vole race flows to its high water-mark of numbers. But this new food-supply brings to the feast hungry owls, hawks, stoats and others, and as the tide of the beasts and birds of prey flows, that of the voles ebbs. Yet no sooner is the ebb apparent than the carnivores themselves decline for lack of food; and eventually the dead level is reached again. So the story goes on—there is a constant ebb and flow of parts within the whole, a fauna is in unstable equilibrium, the "balance of nature" is never quite struck.

But while the parts fluctuate, the fauna as a whole follows a path of its own. As well as internal tides which swing to and fro about an average level, there is a drift which carries the fauna bodily along an irretraceable course. While the former adjustments depend on temporary influences, such as adverse or favourable seasons or variations in the amount of foodstuffs, the latter is a secular phenomenon, due it may be to climatic changes or to the ordinary processes of organic evolution, and leaving a slowly marked but permanent imprint on the sum total of the fauna. The extinct animals and lost faunas of past ages illustrate the reality of the faunal drift.

Now, part of Man's influence, where it is inconstant in tendency, is of no more import in the long run than the internal tides of the fauna; but it is strikingly true that the greater part of his influence ranks with the great secular changes. For his interference tends to persist in fixed directions, and so impels individuals in the fauna and the fauna as a whole upon a definite path along which there is no return. So sweeping are the changes wrought by Man and so swift are they in their action that they obscure and almost submerge the slow march of the other processes of nature, and this difference in degree, associated with Man's purposefulness, almost inevitably leads to a sharp distinction being drawn between nature and man. Where, however, this distinction has been emphasized in the following pages the contrast is relative and not absolute, it lies between wild nature and nature man-controlled; in our land the old order of nature has been all but superseded by the new order of mankind, but Man himself is still "Nature's insurgent son."

This book has been made possible by the labours of many, and principally of Scottish naturalists, travellers, historians and lawmakers; their records are the bricks of which the structure is built. I recognize and would acknowledge my debt to all, instancing as of special value to the student of faunal evolution such contributions as were made by the late Dr J. A. Harvie-Brown to the past histories of several Scottish animals. For myself in particular I am indebted to Dr W. Eagle Clarke for hints which led to fruitful investigation, and to Mr Oliver H. Wild for several apt illustrations.

Permission to use figures from published papers was granted by Prof. J. Cossar Ewart, F.R.S., Dr R. Stewart MacDougall and Mr A. Henderson Bishop; blocks were generously lent by the Councils of the Society of Antiquaries of Scotland (Figs. 1, 10, 12, 26, 35, 57, 60), and of the Highland and Agricultural Society (Figs. 5-9, 13-19, 21, 72-74, 76, 81-83), by Mr Bruce Campbell (Fig. 23), Mrs Comyns Lewer, Editor of *The Feathered World* (Fig. 24), the Council of the Zoological Society of Scotland (Fig. 58), and the Trustees of the British Museum (Fig. 68). Full reference to the sources of these blocks is made in the "List of Illustrations." The Director of the Geological Survey of Great Britain allowed the reproduction of an unpublished photograph from the Survey Collections (Fig. 56). It is a pleasure to record my gratitude to one and all of these.

PREFACE

As for the remainder of the illustrations, they are from my own photographs and drawings, those of animals being based for the most part on specimens in the collections of the Royal Scottish Museum. The reproduction of this extensive series of illustrations was made possible by a very generous grant from the Trustees of the Carnegie Trust for the Universities of Scotland.

Most of all, the reader and myself are indebted to my wife, Jessie J. Elliot, who has been associated with this work from its beginning. She shared in the reading of old books and records, has constantly been consulted during the development of the theme, read a proof, and is responsible for the full index.

It ought to be added that the material in the pages which follow was presented in lecture form to general audiences in Aberdeen in December 1917, as a course under the Thomson Lectureship in Natural Science in the United Free Church College.

J. R.

EDINBURGH,
March 1920.

CONTENTS

CHAPTER I

INTRODUCTORY 1   PAGE

I. MAN AND NATURE 1
Scotland particularly fitted for our study—Methods of enquiry—Main directions of man's influence.

II. SCOTLAND AS MAN FOUND IT 6
The arrival of man—Physical condition of Scotland—Climate and vegetation—Animal life.

PART I

MAN'S DELIBERATE INTERFERENCE WITH ANIMAL LIFE 23

CHAPTER II

THE DOMESTICATION OF ANIMALS . . 27
General effects of domestication—Lines of argument—The beginnings of domestication.

I. SHEEP IN SCOTLAND 35
The wild ancestors of domestic sheep—Primitive Scottish sheep, the sheep of Soay and the peat or turbary sheep of Shetland, their early recorded histories and primitive characteristics—Modern breeds, as illustrating changes induced by man—Improvement of wool in Scotland.

II. CATTLE IN SCOTLAND 49
Native wild cattle: the Urus, its distribution, characteristics and domestication—Earliest domesticated cattle: the Celtic Shorthorn, its introduction, characteristics and domestic status—Modern Scottish cattle, as exemplifying the influence of man—The "Wild White Cattle," an offshoot from a domesticated race.

III. THE HORSE IN SCOTLAND 68
Native horses—Domesticated horses in prehistoric and early historic Scotland—Influences which have modified the native race—Ponies of the Hebrides—Shetland "shelties"—Horses of the mainland: Norse influence; "Wild horses"; Breeding and interbreeding in the Middle Ages and later—The modern Clydesdale.

CONTENTS

		PAGE
IV.	SOME LESSER DOMESTICATED ANIMALS . . .	. 86

The dog, a Neolithic introduction—Scottish dogs of the sixteenth century—The wild boar turned swine: Evidences of the wild boar in Scotland; Domesticated pigs—The pigeon, its ancestry and stages of domestication in Scotland—The barnyard fowl, its Indian ancestry; "Scots Greys" and the Scottish "Dumpie"—The domesticated goose: Evidences of the domestication of the wild Grey Lag in Scotland—Sundry other domestics.

CHAPTER III

DELIBERATE DESTRUCTION OF ANIMAL LIFE . 108

Destruction of animal life a primitive necessity—its wastefulness a development of civilization.

I. DESTRUCTION FOR SAFETY OF MAN AND HIS DOMESTIC STOCK 111

Extermination of the Lynx, the Brown Bear, and the Wolf—Decline of the Fox, the Wild Cat and lesser beasts of prey—War against Golden and Sea Eagles, Kites, Ospreys, lesser Hawks and "ravenois foullis"—An indirect result of destruction.

II. DESTRUCTION FOR FOOD 138

Geese and some lesser Birds—Cormorants, Rock-Doves and Kittiwakes—The tragedy of the Garefowl—The Gannet and the Fulmar—"Bird-Butter" and Birds' Eggs—Fisheries—Shell-fisheries.

III. DESTRUCTION FOR SKINS AND OIL 155

Scottish skin exports—Fur Fairs—Extermination of the Beaver—Decrease of Marten and Polecat—Rabbits and Hares—The Fox, Otter and Badger—Destruction of Seals and Whales for Oil.

IV. DESTRUCTION OF VERMIN AND PESTS 176

Rooks and Choughs—Rats, Moles and Sparrows—Rabbits and Hares—Red Deer, Squirrels and others—The Dipper.

V. DESTRUCTION FOR SPORT 184

Disappearance of the Wild Boar—The Badger—Game beasts and birds.

VI. DESTRUCTION FOR PLEASURE OR LUXURY . . . 189

The sins of "collectors"—Bird-catching—Scottish Pearl fishing.

CHAPTER IV

PROTECTION OF ANIMAL LIFE . . . 197

General influences—Protection by law and its effects.

I. PROTECTION OF ANIMALS FOR SPORT 199

Hawking and Hawks—The Quarry—"Wylde Foulys" in general—Modern game birds—The Deer Forest: Red Deer in the Lowlands; Modes of Deer Protection; Effects of Deer legislation; Deer protection at the present day; Roe and Fallow Deer—Lesser Game, Hares.

CONTENTS

		PAGE
II.	PROTECTION OF ANIMALS FOR UTILITY	216

Food animals: Beasts and birds; Fishes of fresh waters; Sea fisheries; Mollusca and Crustacea—Protection of Fur-bearing animals: The Rabbit; Seals—Protection of animals as scavengers: Beasts and birds of carrion—Protection of the farmer's friends.

III. PROTECTION OF ANIMALS FOR AESTHETIC REASONS. 229

Birds "attractive in appearance or cheerful in song."

IV. PROTECTION OF ANIMALS THROUGH POPULAR FAVOUR AND SUPERSTITION 233

Favoured mammals—The poets and birds—Insects—Animal sanctuaries—Superstition as a protector: Sacred animals; Scottish superstitions protecting birds and lesser creatures.

CHAPTER V

THE DELIBERATE INTRODUCTION OF NEW ANIMALS . 241

The significance of Introductions—Some general results.

I. ANIMALS INTRODUCED FOR THE SAKE OF UTILITY . 246

Appearance and spread of the Rabbit—Effects of its introduction—Failure to acclimatize foreign deer—Destructiveness of introductions—Balance of Nature upset—Success depends on an even balance; some foreign examples—Some introduced fishes—The Medicinal Leech—Counter-pests.

II. ANIMALS INTRODUCED FOR THE SAKE OF SPORT . 264

The story of the Pheasant in Scotland—Introduction and spread of the Capercaillie—The Great Bustard—The Red-legged Partridge—Incidental game-birds—Some sporting fishes—Transportations: fishes, game-birds, the Mountain Hare and others.

III. ANIMALS INTRODUCED FOR AMENITY 283

The Peacock—Birds of bright plumage—Gold and Silver Fishes—The histories of the Fallow and other Deer—The American Grey Squirrel—The return and spread of the Common Red Squirrel—The Beaver—Transported Mollusca—"Escapes"—A new motive and its warning.

PART II

MAN'S INDIRECT INTERFERENCE WITH ANIMAL LIFE 301

CHAPTER VI

THE DESTRUCTION OF THE FOREST . . . 304

I. SCOTTISH FORESTS IN TIMES PAST 305

The Lower and Upper Forests of the Peat—Nature or Man the destroyer—Early historical forests—Forests in the fifteenth century and after.

PAGE

II. CAUSES OF THE DESTRUCTION OF THE FOREST . 315
The needs of the household: Fuel and Housebuilding—Incidents of conquest—Travelling and the merrymen of the woods—Wolves—Industries and woodland: salt manufacture, ship-building, iron-smelting and Scottish "bloomeries"—Agriculture and the forest—"Acts of God": the raging fire and the whirlwind—Final results.

III. EFFECTS OF THE DESTRUCTION OF THE FOREST UPON ANIMAL LIFE 327
Physical changes—Immediate results—The Scottish fauna originally a forest fauna—General effects on fauna—Some individual examples—Distribution of the Roe Deer, past and present—Decline of the Red Deer in range and in physique—Restriction and extermination of Reindeer—Scottish Reindeer probably a woodland race—History of the Elk or Moose in Scotland—Conclusions regarding the race of Deer—Some other forest dwellers: The Urus; Beasts of prey—The disappearance of the Common Squirrel, of the Capercaillie, and of the Great Spotted Woodpecker—Conclusions regarding the effects of forest destruction.

CHAPTER VII

INFLUENCES OF CULTIVATION AND CIVILIZATION . 363

I. DECREASE OF ANIMAL LIFE 365
Breaking in of waste land: Extermination of the Great Bustard; Reduction of numbers of Quail and other creatures; Disappearance of Butterflies—Reclamation of swamps: Former abundance of marshes; Disappearance of their frequenters and inhabitants; Extermination of Bittern and Crane—Interference of civilization: Toll of lighthouses and other lights; Railways and telegraph wires; River barriers and fisheries; Pollution of rivers and estuaries.

II. INCREASE OF ANIMAL LIFE 387
Vegetable food and feeders: Man creates his own agricultural pests; some illustrations—Animal food and insectivorous and carnivorous animals: Pests bring their own retribution—The refuse of civilization: Influence of garbage on animal life.

III. DISPERSAL OF ANIMALS 397
Effects upon the spread of animals of canals, of roads and bridges, and of railways.

IV. CHANGE OF HABITS 400
The habit of selecting a domicile—Influence of houses and of towns—Towns and song—Towns and nesting—The food habit—Faunas of civilization: Animals of waterworks and coal-pits.

CHAPTER VIII

CAMP-FOLLOWERS OF COMMERCE, OR ANIMALS INTRODUCED UNAWARES 417

I. HANGERS-ON OF MAN AND HIS DOMESTIC ANIMALS . 420
Undesirable aliens: Fleas and Bugs; Internal parasites—Parasites transported with live stock.

CONTENTS

xiii

PAGE

II. STOWAWAYS ON SHIPS 425

Introduction, prosperity and decline of the Black Rat—Alexandrine Rat—Arrival of the Common or Brown Rat; colonization of interior, abundance and destructiveness—Shipworms, Barnacles, and a Sea-Fir.

III. SKULKERS IN DRY FOOD MATERIALS 434

The Cricket and Cockroaches—Imports with wheat—Imports with flour—Aliens in biscuits, sugar, tobacco, peas and beans.

IV. FOUNDLINGS AMONGST FRUIT 447

The living freight of bananas—The apple as a smuggler—A stowaway in the seeds of the Douglas Fir.

V. CREATURES CONVEYED BY PLANTS AND VEGETABLES 457

Types transported by vegetables: Cabbages and faunas—Imports with nursery stocks and living trees—With plants of the flower garden—The tropical fauna of Scottish greenhouses—Naturalized Earthworms, Snails, etc.

VI. TIMBER TRANSPORTEES 467

Wood-wasps—Timber carries boring creatures—A few exceptions; Lizards and the Zebra Mussel—Long-horned invaders—Other beetle immigrants.

Final Remarks: The efficiency of commerce as an importer; A month's arrivals from abroad.

CONCLUSION

CHAPTER IX

CHAINS OF CIRCUMSTANCE 479

I. A RETROSPECT 480

The influence of man a developing factor—A contrast, the ways of Nature and Man—Main trends of man's influence: Influences tending to increase animal life; Influences tending to reduce animal life; Influences tending to modify structures and habits.

II. SOME FINAL CONCLUSIONS 492

An increase in numbers—An increase in variety—The great change.

III. SOME INDIRECT RESULTS 498

Enumeration not the whole story—The complex of life: Rabbits and vegetation; Influence of sheep, goats and rabbits; Gulls and the moorland.

IV. THE RECOIL UPON MAN OF HIS INFLUENCE UPON ANIMALS 506

Recoils on health: Flies and disease; Rats and disease; Ague in Scotland—Recoils on prosperity: Marshes and Liver Rot; Earthworms; Hive Bees—The recoil upon man's character and civilization.

INDEX 519

LIST OF ILLUSTRATIONS

FIG.   PAGE

Frontispiece—Contrast between Red Deer of prehistoric times and modern Red Deer

1. Bone harpoons of Azilian type from Oronsay (*Proc. Soc. Antiq. Scot.*, vol. XLVIII, 1914, p. 97) 7
2. Elk—a former inhabitant of Scotland 15
3. Old World Lynx 16
4. Ptarmigan—once more common than Grouse in Scotland 18
5. Mouflon—a wild ancestor of domestic sheep (*Trans. High. Agr. Soc.*, 1913, p. 12) 36
6. Soay Sheep—a primitive domesticated breed (*Trans. High. Agr. Soc.*, 1913, p. 20) 38
7. Turbary or Peat Sheep—a primitive domesticated race (*Trans. High. Agr. Soc.*, 1913, p. 17) 42
8. Cheviot Sheep—a modern result of selective breeding (*Trans. High. Agr. Soc.*, 1915, p. 402) 44
9. Black-faced Sheep—an illustration of development in length of wool (*Trans. High. Agr. Soc.*, 1915, p. 402) 48
10. Front view of skull of Urus (*Proc. Soc. Antiq. Scot.*, vol. IX, 1873, p. 650) . 50
10a. Side view of skull of Urus (*Proc. Soc. Antiq. Scot.*, vol. IX, 1873, p. 651) . 50
11. Urus—the native wild ox of Scotland (Griffith's Edn. Cuvier's *Animal Kingdom*, 1827, vol. IV, p. 411) 53
12. Horn-sheaths of Celtic Shorthorn, and upper part of skull, with fragments of skin and hair attached (*Proc. Soc. Antiq. Scot.*, vol. IX, 1873, p. 622) . 57
13. Highland Kyloes (*Trans. High. Agr. Soc.*, 1900, p. 167) 61
14. Aberdeen-Angus bull—a highly developed result of domestication (*Trans. High. Agr. Soc.*, 1911, p. 287) 63
15. Celtic Pony, probably resembles native pony of Scotland (*Trans. High. Agr. Soc.*, 1904, p. 249) 71
16. Hebridean Pony—Uist race—a primitive breed (*Trans. High. Agr. Soc.*, 1904, p. 258) 75
17. Shetland Pony (*Trans. High. Agr. Soc.*, 1915, p. 400) 77
18. Highland Garron—probably moulded by Norse influence (*Trans. High. Agr. Soc.*, 1909, p. 379) 81
19. Clydesdale mare—a modern product of selective breeding (*Trans. High. Agr. Soc.*, 1915, p. 397) 84
20. Wild Boar and primitive Scottish type of pig 91
21. Large White Boar—finished product of domestication (*Trans. High. Agr. Soc.*, 1913, p. 385) 94
22. Wild Rock Dove 96
23. Ancient Pigeon-House or "Doo-cot" near Leadburn (*Trans. Edinburgh Field Nat. and Micr. Soc.*, vol. VI, 1909, pl. VI, photo, B. Campbell) . . 99
24. Scottish breeds of poultry—Scots Greys and Scottish Dumpies (*The Feathered World*, London) 101
25. Grey Lag Geese—a wild species domesticated in Scotland . . . 103
26. Skull of Brown Bear from peat-moss in Dumfriesshire (*Proc. Soc. Antiq. Scot.*, vol. XIII, 1879, p. 362) 112

LIST OF ILLUSTRATIONS

FIG.		PAGE
27.	Brown Bear, a former native of Scotland	113
28.	Wolf—a Scottish scourge exterminated about two centuries ago	117
29.	Foxes (Vixen and cubs)	123
30.	Wild Cat	125
31.	Golden Eagle	129
32.	White-tailed Eagle or Erne, once common, now practically exterminated in Scotland	131
33.	Kite or Gled, once common, now exterminated in Scotland	135
34.	Garefowl or Great Auk (once a native of Scotland, now extinct) with its solitary egg	143
35.	Bones of extinct Garefowl from kitchen-midden at Keiss, Caithness (*Proc. Soc. Antiq. Scot.*, vol. XIII, 1879, pp. 78, 79)	145
36.	The decline of Tweed fisheries during fifty years (diagram)	152
37.	European Beaver—exterminated in Scotland	157
38.	Pine Marten—approaching extinction in Scotland	159
39.	Polecat and young—approaching extinction in Scotland	163
40.	Decline of Polecat, as shown by skins and prices at Dumfries Fur Fair (diagram)	165
41.	Destruction of Rabbits and Hares, as shown by skins on sale at Dumfries Fur Fair (diagram)	167
42.	Decadence of "vermin"—Wild Cat, Marten and Polecat—through twelve years' work of one gamekeeper (diagram)	176
43.	"Catching the Badger"—from a coloured plate after Alken	185
44.	Artaxerxes Butterfly—exterminated on Arthur's Seat	191
45.	Osprey or Fish Hawk—practically exterminated in Scotland	192
46.	Peregrine Falcon—formerly protected for Hawking	201
47.	Scottish Crested Tit—increasing in numbers under protection	231
48.	Little Owl—an introduction to Britain which has become a nuisance	255
49.	Cottony or White Scale attacked by Cardinal Ladybird	260
50.	Capercaillie—reinstated in Scotland after extermination	269
51.	Blue or Mountain Hares—established in many new areas in Scotland	281
52.	American Grey Squirrel—acclimatized in many parts of Britain	289
53.	Common Red Squirrel—a former native of Scotland, reintroduced	291
54.	Some deliberate introductions to Great Britain (diagram)	300
55.	Section of peat-moss showing relationship of prehistoric forests (after F. J. Lewis)	306
56.	Remains of Upper Forest of the Peat—roots of Fir Trees laid bare by wastage of peat, Findhorn Valley (photo by Geological Survey)	307
57.	Antlers and portion of skull of Red Deer, unearthed in the Meadows, Edinburgh (*Proc. Soc. Antiq. Scot.*, vol. XV, 1881, p. 41)	335
58.	Reindeer—formerly natives of Scotland—in Scottish Zoological Park (*Guide to Scottish Zoological Park*, 1917, photo, F. C. Inglis)	339
59.	Fragmentary Antlers of Scottish Reindeer of Glacial Period, found at Kilmaurs, Ayrshire	343
60.	Antlers and portion of skull of Elk, found at Airleywight, Perthshire (*Proc. Soc. Antiq. Scot.*, vol. IX, 1873, p. 319)	347
61.	Great Spotted Woodpecker—at one period exterminated as a nesting species in Scotland	359
62.	Great Bustard—formerly a native of Scotland	366
63.	Quails—once common in Scotland, now scarce	367
64.	Bittern—banished from Scotland with the marshes	374
65.	Crane—a former inhabitant of Scottish marsh-lands	375
66.	Colorado Potato-Beetles	387

LIST OF ILLUSTRATIONS

FIG.		PAGE
67.	Rabbits and Cultivation, correspondence between increase of agricultural activity and of number of skins of Rabbits on sale at Dumfries Fur Fair (diagram)	390
68.	Zebra Mussels—from a mass of some 90 tons removed from a water-main at Hampton-on-Thames (*Brit. Mus. Econ. Pamph.*, *The Biology of Waterworks*, 1917, p. 25)	415
69.	Some alien Cockroaches introduced into Scotland	437
70.	Hessian Fly and seed-like puparia, in which form it was probably introduced to Britain	439
71.	Granary Weevil and destroyed wheat	441
72.	Piece of Dog Biscuit perforated by Biscuit Beetles (*Trans. High. Agr. Soc.*, 1917, p. 151)	444
73.	Beans and Peas damaged by larvae of Bean and Pea Beetles (*Trans. High. Agr. Soc.*, 1914, p. 180)	446
74.	Bean Beetle (*Trans. High. Agr. Soc.*, 1914, p. 180)	446
75.	Snowy Tree Cockroach—frequently imported with bananas	448
76.	Branch of Apple Tree covered with Apple Mussel Scale (*Trans. High. Agr. Soc.*, 1917, p. 138)	449
77.	Section of damaged apple showing larva of Codlin Moth	452
78.	Codlin Moth	452
79.	Douglas Fir Seed Chalcid	454
80.	Douglas Fir Seeds from Peeblesshire—showing escape holes of the parasitic Chalcid	455
81.	Steel-Blue Wood-Wasp (*Sirex noctilio*), male (*Trans. High. Agr. Soc.*, 1917, p. 134)	468
82.	Steel-Blue Wood-Wasp (*Sirex noctilio*), female (*Trans. High. Agr. Soc.*, 1917, p. 134)	468
83.	Steel-Blue Wood-Wasp eating its way out of a pine stem after emerging from pupa stage (*Trans. High. Agr. Soc.*, 1917, p. 134)	468
84.	Timberman Beetles, imported to Aberdeenshire in Norwegian pine logs	471
85.	Some chance introductions to Great Britain (diagram)	473
86.	Comparison of surface features of Scotland, before the arrival of man and at the present day (diagram)	484
87.	Comparison between the livestock carried by 1000 acres of cultivated and uncultivated land (diagram)	495
88.	Alteration of moorland by Gulls, West Linton	503
89.	Near view of the same moorland	503
90.	Representation of Recoils on man's health, due to different types of his interference with animal life (diagram)	513

MAPS

			PAGE
I.	Introduction and spread of the Capercaillie in Scotland	to face p.	272
II.	Distribution of the Mountain Hare in Scotland	,,	282
III.	The Spread of the Common Squirrel in Scotland from various centres	,,	292
IV.	Distribution of Scottish "bloomeries" and slag furnaces	,,	320
V.	Decline of Red Deer—distribution at different periods	,,	334
VI.	Influence of man-made obstructions on Scottish Salmon Fisheries	,,	380
VII.	River pollution and fisheries	,,	384
VIII.	Distribution of Ague or Malaria in Scotland in the eighteenth century	,,	508

CHAPTER I

INTRODUCTORY

I. 1

MAN AND NATURE

There be many strange things, but the strangest of them all is MAN... Earth, Mother Earth, is from everlasting to everlasting...but Man fretteth and wearieth her; for he putteth his horse to harness, and his ploughs go to and fro in the furrow, even as the seasons come round. He spreadeth his snares for the silly birds; he gathereth the fishes of the sea in the meshes of his nets. Man surpassing in wisdom. By craft he over-reacheth the wild beast upon the mountain, and putteth to his yoke the long-maned steed, and the strength of the great bison.

THOMPSON'S *Sales Attici* (Sophocles).

SINCE Man came to his own upon the earth, he has exercised with little restraint the power of his new wisdom over all created things. So widely and deeply has his influence spread during the hundreds of thousands of years of his wanderings, that it is wellnigh impossible to gauge its effects or to distinguish them amidst the workings of Nature as a whole. Change is apparent in the interrelationships of the plants and of the animals of a country with the passing of years; but who can say that here the heavy touch of Man alone has fallen, and that there only are subtle traces of wild Nature, wrought out through cyclic changes, alternations of climate, and through the processes of natural evolution in living things? The complications due to the action of contemporaneous natural agencies, together with the difficulties of obtaining evidence regarding the earlier periods of Man's existence make the ultimate analysis of Man's influence on Nature no simple task.

INTRODUCTORY

SCOTLAND PARTICULARLY FITTED FOR OUR STUDY

In some respects Scotland is particularly well fitted for our study, mainly owing to its geographical situation and geological history. In the first place man arrived at a comparatively late date within its borders. There is no evidence that the country was inhabited by the human race until long after the period of rude stone implements, the Old Stone Age, when man was already established in South Britain and in the majority of the European countries in the same latitude. His influence in Scotland, therefore, is limited to the New or Polished Stone Period and succeeding ages, distant enough though the first may seem to our modern historical view.

In the second place, Scotland has undergone, and in comparatively recent geological times, an experience unlike that of neighbouring countries. During the Great Ice Age, it was completely buried beneath a continuous ice-sheet, some 3000 feet thick, which effectually blotted out its earlier plants and animals. The Scottish flora and fauna are therefore recent acquisitions due to the immigration of living things when the ice-sheets were dwindling or after they had entirely disappeared. Further, owing to the fact that Scotland has for long been bounded on three sides by a broad sea, the fauna with which Nature stocked her at the close of the Ice Age has remained isolated, suffering, it is true, fluctuations which Nature has ordained or man has induced, but unaffected by that constant immigration and emigration—except in a few cases of the more mobile creatures, such as birds—to which continental countries are constantly liable.

The original post-glacial fauna of Scotland may be likened to a limited capital upon which man has traded. So far as he has been satisfied with the natural interest of the capital, the capital has remained as it was in the beginning[1], but this has seldom been the case. Often he has trenched upon it, and at times so deep have been his overdrafts that some items of the account have been seriously diminished or exhausted. At other times he has added afresh to the old capital, but in a new currency of his own introduction. Could we but assess the original animal capital

---

[1] We are here ignoring natural fluctuations.

which the Neolithic invaders of Scotland had at their disposal, a great step would be made towards gaining a basis from which to compute the influence of man upon the animal life.

In the third place, from its small size Scotland gains advantages in such a study; and this partly because the fauna of a small country is more compact, and its changes, as a rule, are more readily marked; and partly because Scotland's few degrees of latitude eliminate the possibility of temperature barriers, one of the most important and far-reaching of the climatic influences which complicate the fluctuations of animal life in continental areas.

And lastly, since the study of Nature gained a firm foothold, Scotland has possessed a succession of observers and recorders such as few countries of similar size and population can claim, naturalists whose labours form a solid foundation for the accurate estimation of the later changes in animal life.

METHODS OF ENQUIRY

To enquire into the doings of man is to investigate History, and the historical method enters largely into this natural history study. The foundation of our enquiry must be such records as the past has left us. The chronicled history of Scotland begins with the advent of the Romans on their northward progress through these islands in the first century of our era, but since, at that time and for many centuries thereafter, the records of even the great political events, of the doings of man with man, are vague and unsatisfactory, it need hardly be said that the dealings of man with animals seldom encumber the written page.

Even in the "historic period" therefore, the beaten tracks of historical knowledge have to be forsaken, and appeal has to be made to the relics man has left in his long-forsaken homes, to the casual pictures he has carved, often with hand and eye of wonderful skill, to the tales of travellers, many from foreign lands, who described the features of Scottish animal life which struck their fancy as differing from those familiar to them, and to the records of unusually outstanding natural phenomena which, on occasion, our own political historians of former days condescended to notice.

But even the sparse and slender guide-posts of early chronicled history fail us in the ages (seven thousand years or more) which intervened between the coming of man to Scotland and the Christian era. Glimpses of this long-forgotten past can be gained only by piecing together the evidences left by animals and man himself, from bones and relics discovered by systematic excavation or by lucky chance in beds of marl, in the layers of peat-bogs, in the deposits of caves, in the kitchen-middens or refuse food-heaps of the early inhabitants, and in the structures built by man for defence, or for interment of his hallowed dead.

Pictures of Scottish animal life in successive ages having been gleaned from these varied sources, simple comparison of one with another and with the fauna as it is known to-day will reveal the vast changes which have taken place. Yet still a problem lies before us—that of sifting from the totality of change the effects due to the influence of man as distinct from the inevitable changes wrought by time in all Nature, animate and inanimate. In working out this problem reference will be made on occasion in the following pages to outstanding cases in other lands which help to illustrate man's influence and to explain the effects of his dominance in Scotland.

MAIN DIRECTIONS OF MAN'S INFLUENCE

Man has been described from one point of view as an instrument of destruction and from another as a creative agent. The truth of the matter as regards his relations with Nature is that he is neither all in all a destroyer nor a creator, but exercises his powers mainly as a transformer and a supplanter. Wherever he places his foot, wild vegetation withers and dies out, and he replaces it by new growths to his own liking, sometimes transformed by his genius for his own use. Where he pitches his tents and builds his cities, wild animals disappear, and woodlands and valleys where they sported are wrested from their prior owners and given over to the art of agriculture and to animals of man's own choosing, as well as to a host of camp-followers, which attach themselves to his domestication whether he will or no. Intentionally and unintentionally, directly and indirectly, man transforms and supplants both animal and vegetable life. Some animals he deliberately destroys, some he deliberately introduces, and the characters of some he deliberately transforms by careful selection and judicious interbreeding. Other animals find his presence uncongenial and gradually dwindle in numbers or disappear, while others are encouraged by his activities to increase in numbers, sometimes even to his own confounding.

I. 2

SCOTLAND AS MAN FOUND IT

> Heir agane sall ye se braid planes, thair wattirrie dales: heir a dry knowe, or a thin forrest, thair a thick wodd, all meruellouse delectable to the eye throuch the varietie baith of thair situatione, and of the thing selfe that thair growis.
>
> Historie of Scotland by Jhone Leslie, 1578.
> (Dalrymple's Translation.)

As a preliminary to the detailed consideration of man's influence upon Scottish animal life, let us try to picture the condition of the country as primitive man found it, when in his northward wanderings his communities ventured beyond the natural boundary of the Cheviot Hills. Only with such a picture at the back of our minds can we hope to realize the changes which man has wrought in the passing of time. Before trying to gauge the extent of man's trading, we must endeavour to assess the capital which Nature placed in his hands to begin with.

THE ARRIVAL OF MAN IN SCOTLAND

Notwithstanding that even in the more distant stages of the Early Stone Age, man had travelled dry-shod from the land that is now France, across the grassy valley that separated the main mass of Europe from its western prolongation which is now the British Isles, there is no sure sign that his wanderings in Palaeolithic times ever brought him to the southern limit of Scotland. For tens of thousands of years he dwelt on the plains of England, leaving his handiwork—rudely dressed stone implements of various types which fall into a long range of stages from the early Chellean to the late Magdalenian or Reindeer period—scattered over those southern portions which lay clear of the heaviest and most persistent ice-fields of the Great Ice Age. But the northern portions of these islands, still shrouded in their glaciers, offered no attractions to the hunters of the Early Stone Age, and the period of the great glaciation seems to have long passed away, with its mammoths, woolly rhinoceros, musk-oxen, cave-bears and lions, before man ventured to follow the retreating glaciers northwards beyond the Cheviots.

The earliest relics of man's handiwork in Scotland consist mainly of implements of bone or horn, flattened harpoon-heads, with long and well-shaped barbs on both

Fig. 1. Bone harpoons of Azilian type from prehistoric settlement in Oronsay. ⅜ nat. size.

sides, and rough pick-axes carved from the antlers of red-deer. Implements of stone and flint-chips, rudely dressed, have also been found, but their numbers are too few and their characters too indefinite to point clearly to their place in the recognized cultural stages of European man. The characters of the bone implements (Fig. 1) indicate in a general way the *Azilian* period, a stage regarding which little is known, although it may be placed at the opening of the Polished Stone or Neolithic Age, or at any rate between that and the definite later stages of the Palaeolithic epoch. These earliest inhabitants of Scotland were hunters, fowlers and fishers. They possessed no domestic stock and there is no evidence that they tilled the ground or cultivated corn. Yet their craft furnished a well-stocked larder, as their kitchen-middens and other relics in Scotland show[1]. They gathered from the sea-shore in great variety edible shell-fish—crabs, and mollusca such as limpets, whelks or buckies, periwinkles, mussels, oysters, cockles, scallops and razor-shells. The foreshore and sea-cliffs supplied them with many kinds of birds—wild-ducks, geese, shags and cormorants, great auks, razorbills, guillemots and gannets. From coastal waters they obtained dog-fish and rays, sea-breams, wrasses and the conger-eel; and by the river-banks and woodland glades they trapped and slew the otter, red deer and wild boar. Nor did they disdain the blubber and the flesh of whales and dolphins which fortune stranded on their coasts, or the seals which basked and bred there in abundance.

Of the personal appearance of the early settlers we can form a just estimate from examination of the skulls and other bones which have been preserved underground in the neighbourhood of their habitations at the MacArthur Cave near Oban, or of the relics in the horned and chambered cairns which the successors of the Oban fishermen, the men of late Polished Stone or Neolithic Age, built in scattered localities from Galloway to Caithness and the Orkneys, to protect and commemorate their dead. They were a short people, their thigh-bones suggesting that the men stood about 5 ft. $4\frac{1}{2}$ in. high, and the women about 5 ft. 1 in.—some 2 or 3 inches below the standard of a modern Briton. Their lower limbs differed from ours and resembled those of some of the more primitive races of man at the present day, with thigh-bones flattened, shins compressed from side to side, and the bones of foot and ankle more compact and stouter—all adaptations for strong and rapid movement, indicating that the people lived an active strenuous life. In facial expression, they differed only in small degree from ourselves. Their heads and faces were long and narrow, their foreheads

[1] The following account contains references to such animals and plants only as have been identified in Scottish deposits of the periods mentioned.

fuller and rounder than ours, the bridge of the nose and nostrils moderately narrow, and their eyes rather narrow and elongated. Their jaws were square and their front teeth, instead of overlapping as do ours, met firmly edge to edge. So regular and healthy were their teeth, a necessity for a primitive life, that they show only a wearing down due to constant use, and seldom or never any signs of the decay or caries which has given rise in our generation to armies of dentists and the science of dentistry.

The earliest traces of these primitive peoples in Scotland are associated with the so-called "Fifty-Foot Beach." Their canoes, simple dug-outs of pine, have been found at Perth in the Carse clays of this period, and frequently in similar deposits in the Forth and Clyde valleys. Some of their implements were left beside the remains of a whale, stranded in these far-off days on a shore which is now part of the fertile Carse of Stirling, and, as Dr B. N. Peach has pointed out, their kitchen-middens lie along the ridge of the Fifty-Foot Beach in the upper reaches of the Forth, never occurring in the lower seaward ground—a clear indication that at the time the refuse heaps accumulated, this Beach was the limit of high water whither the kitchen-middeners retired to feast upon the shell-fish collected at low tide.

THE PHYSICAL CONDITION OF SCOTLAND

What, we must ask, was the Scotland in which the Azilian or Early Neolithic peoples settled after their wanderings through Britain from the continent of Europe? As compared with its condition at the present day, the land was depressed relative to sea-level, all the shore area that lies beneath a contour-line varying in different areas from 35 to 65 feet being submerged by the ocean. Where the coast is bounded by high cliffs, this depression would have had little effect on the outlines of the country, but where the land shelves gradually to the sea, as in many parts of the Moray Firth and in the great valleys of the Tay, Forth and Clyde, the sea made considerable encroachments upon the land. So it is that while the Fifty-Foot terrace is generally represented on the west coast, as in the islands of Jura and Mull, by a comparatively narrow ledge, cut in the cliffs by wave action during what must have been an extended period of time, the old shore is represented in the midlands by the Carse lands of the Forth valley, a fertile plain more than three miles wide above Stirling, and extending as far as Gartmore, some 17 miles beyond that town, and on the Clyde by level terraces which can be traced beyond Dalmuir on the north and Paisley on the south.

To the first-comers the inland landmarks of Scotland must have appeared almost as they are to-day. The hills and valleys had already been carved into their present aspect. True, the rivers were swollen in volume and many of the lakes, ponded back by the debris of the old glaciers, were greater than now, while many low-lying areas, now fertile plains, were bogs and marshy flats; but the ice-fields of the Great Glaciation had disappeared, although a recrudescence of colder conditions had again clad the mountain-tops in snow and filled the higher valleys with moving sheets of ice.

CLIMATE AND VEGETATION

The period during which the Fifty-Foot terrace was being carved out or levelled up by the sea was a prolonged one, and in it Scotland underwent drastic changes of climate. At exactly what stage in the formation of the Fifty-Foot Beach man appeared upon our shores, it is impossible to say, but it is certain that the conditions which immediately followed the retreat of the great ice-sheet had long passed away. The Arctic climate had gradually been replaced by one at least as temperate as that of to-day; distinctive Arctic plants, such as Arctic Willows (*Salix repens* and *polaris*), Arctic Birch (*Betula nana*), Crowberry (*Empetrum nigrum*), Creeping Azalea (*Loiseleuria recumbens*), and Mountain Avens (*Dryas octopetala*), all of which occurred in late glacial times on low ground at Corstorphine in the neighbourhood of Edinburgh, had deserted the lowland valleys and followed the line of the dwindling snowfields to the hill-tops. With the rising temperature, forests of Silver Birch, Hazel and Alder clad the lowlands and spread up the mountain-sides, at least to an elevation of over 1500 feet above sea-level.

But these conditions too, had undergone modification before the arrival of man in Scotland, for the Fifty-Foot Raised Beach, which contains the earliest records of his settlements, appears to be contemporaneous with a lower succession of layers in our peat-mosses, the plants in which indicate a period of decreasing temperature and of increasing rainfall. Bogs of sphagnum moss, over which cotton-grass waved in abundance, gradually swamped the birch woods and buried the decaying trees under thick layers of peat. These inhospitable conditions culminated in a return of snowfields to the mountain-tops and glaciers to the high valleys, and the Arctic plants again crept down the mountain-sides. It need not be assumed, however, that the temperature which welcomed man to Scotland was very much lower than that of the present day, for the fact that even now snow-wreaths which are little short of permanent, lie year in, year out, in the corries of the higher Grampians and of Ben Nevis, indicates that a small fall in temperature would be enough to clothe the hill-tops once more with a permanent cap of snow and to fill the upper valleys with glaciers. Conditions so forbidding, however, did not extend to the lowlands or the coastal areas where man made his first homes. So far as we can judge from the seeds or leaves of plants found in lowland deposits contemporaneous with the men of the Later Stone Age (thanks to the researches of James Bennie and Clement Reid), a rich flora, very similar to that of the present day overspread the valleys. The meadows were covered with lush grass, chequered by the blossoms of buttercups, of spear-thistles, dandelions and sow-thistles, of the yellow ox-eye, scentless mayweed, the hemp-nettle and St John's wort; on dry banks, coltsfoot, tormentil and chickweed flourished; in, or near by running water grew the water buttercup and water blinks, mare's tail and water milfoil; by riversides blossomed valerian, meadow-sweet and the red campion; the thickets were starred by the wood-sorrel; the bugle, dog's mercury and dock flourished in the shade; the raspberry and bramble vied for a place with the wild rose; and the marshy places were enlivened with a wealth of flower and foliage, from the showy ragged robin, lousewort, buck bean and marsh marigold to the lesser spearwort and marsh violet, pondweeds, rushes and sedges.

Before man had spent many centuries in our northern land, the cold disappeared as unaccountably as it had come, the snowfields melted, the land rose from the waves, and, as if to make amends for its former rigour, the climate so ameliorated that forests of Scots fir spread over the drying peat-bogs and extended to an altitude of close on 3500 feet up the mountain-sides, where, in our own day, 2000 feet marks the upward limit of forest growth.

ANIMAL LIFE

What were the creatures which the first inhabitants of Scotland found installed in its glades and forests? For an answer we have to enquire of the relics which have been cherished for thousands of years in the depths of the marine clays, of the fresh-water marls and peat-bogs which were contemporaneous with or preceded by a short space the coming of man, or even of the food refuse which man himself cast carelessly aside all unthinking that it would reveal to his distant successors the animals he encountered and overcame. The dregs of these deposits reveal a strange mixture of animal types.

The native inhabitants of Scotland, with the exception of the mobile forms, such as birds, had reached Britain from the continental areas while the English Channel and a part of the North Sea were dry valleys in the western extension of Europe. The rigours of the Ice Age and its inhospitable glaciers had cleared the land of the preglacial fauna, but as the ice disappeared the cold climate of the later phases of the Great Ice Age attracted an Arctic fauna, which retreated northwards as vegetation sprang up on the heels of the shrinking glaciers. The grassy plains which superseded the Arctic vegetation of the great flats of clay and hillocks of gravel deposited by the ice-fields, led another series of animals, creatures of the plains, to make the northward pilgrimage. While the forest conditions, which, as we have seen, preceded the first peat period and covered the land even up to a height of 1500 feet above sea-level, enticed still another series, of woodland forms, there to make their homes.

Under the conditions prevailing in a continental area where the change of climate is gradual, these three types of fauna form no permanent stations; for they pass over such an area in waves, forsaking it on a northwards trek as the cold retreats, as easily as they entered it from the southward with the Arctic conditions. But Britain stands in a different case from such continental areas. Scarcely had the last of the immigrant contingents crossed from the main mass of their fellows on the continent, than the gradual sinking of the land relative to the sea led to the submergence of the valleys of the North Sea and of the English Channel, so that, as a twelfth century troubadour quaintly sings of the latter,

> That famous stretch of fertile land
> Is hidden now by sea and sand,
> No more will its venison grace the dish,
> The ancient forest yields nought but fish.

So the immigrants to Britain were cut off from access to the continent of Europe. The result of this isolation is plainly to be seen in the strange assemblage of animals which greeted man on his arrival in Scotland. The Arctic creatures, the beasts of the plains, and the forest lovers, each ranging northwards as the conditions which had attracted them ebbed towards the north, were checked in their migration by the sea-walls of northern Britain, and as a consequence were compelled to make the best of the changes of climate and vegetation which overtook them. Some, unable to adapt themselves to unusual climes, had died out before the coming of man, but representatives of each group remained to indicate the successive changes of Scottish conditions in the days before man.

It is not to be expected that the peat-bogs and other deposits should furnish a complete synopsis of the fauna, partly because the bones of the smaller animals are more liable to disappear through the ordinary processes of decay, and partly because the deposits have only in a very few cases been systematically examined with the investigation of their animal content in view; so that we have to be satisfied with identifications of the remains which have appealed most to the utilitarian excavators of peat and marl, usually the bones of the larger animals. Even so we can furnish a fair view of the general aspect of the animal life.

Imagine that from our fourteen million acres of cultivated land and mountain grazings the domestic stock had disappeared, that this acreage was given over to forest and wastes, and that over this wild area the present-day denizens of our mountain heaths and tiny woods spread in full possession. A picture arises of an old Scotland in which the wild creatures, freed from a hopeless competition with man's methods and advances, assorted themselves on mountain-side and plain, in the meadows and in the forests, as their natures determined and not as man decreed. But even such a picture is far from complete; for the present-day fauna is not the old fauna. Many additions have to be made and some subtractions, if our picture of the old times is to approach accuracy.

The fauna that greeted man was a rich one. In the meadows browsed the Reindeer, from southern Dumfriesshire to Caithness and even in Rousay in the Orkneys. It has long since vanished from our fauna, and is now confined to the northern portions of the Old and New Worlds. In its company the Giant Fallow Deer or so-called "Irish Elk" (*Megaceros giganteus*) may still have lingered in the southern districts, though its latest Scottish remains appear to have been contemporaneous with the marl deposits which preceded the formation of the great peat-mosses in our lakes. The Wild Horse probably still scampered over the plains of the southern lowlands, disturbing there the herds of the great Wild Ox (*Bos taurus primigenius*), which spread throughout the length and breadth of the land. Under tussocks of grass, Northern Voles (*Arvicola* or *Microtus ratticeps*), now extinct in Britain, made their nests in the company of the Common Hare and perhaps also of the last Scottish representatives of the Arctic Lemming (*Myodes torquatus*), whose bones have been found in an Arctic deposit near Edinburgh, and in the Bone Cave of Inchnadamph in West Sutherland. To-day the first is found from northern Europe and Asia southwards to Hungary; while the last is confined to the central mountains of Scandinavia, whence its migrations have been a source of wonder to naturalists for many a century. In the wilds, the Mountain or Variable Hare made its home, and in its spring and autumn colour-change it still betrays its former association with Arctic conditions.

Great variety of life lurked in the shade of the forests. The largest of existing deer, the Elk (*Alces alces*) (Fig. 2)

Fig. 2. Elk—a former inhabitant of Scotland. $\frac{1}{36}$ natural size.

was common in the Lowlands, especially in the Tweed valley, where its remains have been found on many occasions. Once it ranged from Wigtownshire to Strath Halladale in northern Sutherlandshire, but its scanty numbers are now confined to the woodland regions of northern Europe. Of other forest deer, Red Deer of large size (see Frontispiece), with magnificent antlers, sometimes bearing twenty-two points, were abundant throughout the country from Wigtownshire to Caithness and even in the distant Orkney and

Fig. 3. Old World Lynx. ⅛ natural size.

Shetland Islands, whence they have long since disappeared. Their degenerate descendants on the mainland are now confined to the waste Highlands north of the valley of the Forth and Clyde. Roe Deer, on the other hand, judging from the few remains which have been unearthed, were scarce; nevertheless, they must have been widely distributed, for their bones have been found in a peat-bog at Shaws, Dumfriesshire, as well as in a kitchen-midden, probably of Neolithic date, on the mainland of Shetland. In the thickets, the Wild Boar was plentiful, and occasional bones tell of the presence of the nocturnal Badger. By the river-banks, Otters were to be found, and the European Beaver, whose last surviving colonies linger on in a few localities in Scandinavia, Russia, Germany and Austria, ranged from Dumfriesshire at least as far north as the parish of Kinloch in Perthshire.

Many beasts of prey followed upon the trail of the creatures of the forests and plains. A few survivors of the European Lynx (*Lynx lynx*), now all but exterminated except in Scandinavia, still awaited in Scotland the coming of man, for the remains of this species have been found by Dr Peach and Dr Horne in a Bone Cave at Inchnadamph in Sutherlandshire[1] associated with relics of human habitation. In the same cave, a tooth of a Brown Bear (*Ursus arctos*) was discovered, but the distribution of the Bear in Scottish woods must have been a wide one, for a well-preserved skull was unearthed many years ago in a peat-bog at Shaws in Dumfriesshire, and an eye-tooth was once found in a Caithness broch. Of the larger beasts of prey, Wolves were the most common, and because of their abundance the most dangerous, and amongst the lesser carnivores the Common Fox, Stoat and Weasel shared in the smaller prey of the woodlands and meadows.

It goes without saying that great variety of birds inhabited the country, but their remains, owing to their small size, have seldom been recovered from the early deposits, and have still less often been identified with certainty. We know that on the moorlands the Ptarmigan (Fig. 4), of Arctic origin, was abundant, for the bones of hundreds of individuals occurred in the Inchnadamph cave, where they far surpassed in number the relics of their fellows, the Red Grouse. The Raven occupied lowland areas which it has long since deserted, and the Great Auk, now extinct, tenanted the Northern and Western Isles at least as far south as Oronsay, where its remains have been found in an early Neolithic or Azilian kitchen-midden. The same refuse-heap yielded remains of many sea and shore birds which occur in Scotland at the present day, such as the Cormorant and Shag, Gannet, Guillemot and Razorbill; Gulls, Terns, Wild Ducks, Wild Geese, the Redbreasted Merganser and the Water

[1] The identification of the animal remains from this interesting cave was made by Mr E. T. Newton, F.R.S., formerly of the Geological Survey.

Rail. But the peat-bogs of Scotland have not yet revealed that wealth of bird-life which has been recovered from the peat-bogs and superficial deposits of England, and which includes amongst others, the Crow, Snowy Owl, Golden Eagle, Buzzard, Pelican, Wild Swan, Eider Duck, Great

Fig. 4. Ptarmigan—once more common than Grouse in the valleys of Sutherland. ¼ natural size.

Crested Grebe, Coot, Bittern, Crane and Capercaillie. Yet we know from later prehistoric and early historic evidences that many of these certainly occurred at a later date in Scotland also, and had probably already been established.

As for the smaller fauna, insects must have been plentiful, but their remains, apart from multitudes of wing-cases of beetles, have not been recognized in the early deposits.

In the lakes, and possibly still in the rivers was abundance of the red-bellied Char whose present-day descendants have been imprisoned by succeeding changes of climate and conditions in a few of the deeper lakes; and the fresh waters swarmed with hosts of smaller fry, Crustaceans, Wheel-animalcules or Rotifers, Water-bears or Tardigrades, and such like, on which Char and Trout and migrant Salmon made comfortable diet.

Of the denizens of the seas which bounded the Scotland of early man, we have little direct evidence, but such as it is, it indicates a temperate marine fauna very similar to that of the present day. Apart from the contents of the estuarine deposits contemporaneous with the Fifty-Foot Beach, we have to depend on the refuse which Neolithic man accumulated in his kitchen-middens in different parts of the country. These, however, afford proof of a wonderfully varied seafare. The earliest inhabitants of Britain, the hunters and fishermen of the Oronsay shell-mound, varied their diet, as Mr Henderson Bishop's recent researches show, with many species of marine mammals, fishes, crustacea and molluscan shell-fish. In the refuse-heaps have been found remains of a species of Dolphin, of the Common Seal (*Phoca vitulina*), and of the Grey Seal (*Halichœrus grypus*) which still occurs on the rocky shores of the Outer Hebrides, but very rarely on the coasts of the mainland and on the Inner Islands. Of fishes, eight species have been preserved, such as the Conger Eel (*Conger conger*), the Black Sea Bream (*Cantharus lineatus*), the Sea Bream (*Pagellus centrodontus*), the Spotted Wrasse (*Labrus maculatus*), the Angel Fish (*Squatina squatina*), the Tope (*Galeus canis*), the Spiny Dog-fish (*Squalus acanthias*), and the Thornback Ray (*Raia clavata*). Only two species of Crabs were found—the Common Edible Crab (*Cancer pagurus*) and the Swimming Crab or Fiddler (*Portunus puber*), but enormous heaps, containing thousands upon thousands of molluscan shells, yielded a long list of 25 marine species, all of which are common on the West Coast at the present day. The molluscan fauna of Scottish seas, however, was not absolutely identical with that of to-day. For example, in a shell-mound composed of refuse from human meals, near Ardrossan on the Ayrshire coast, there have been discovered many specimens of a top shell, *Trochus lineatus*, which is now extinct in the Firth of Clyde. The contents of other kitchen-middens show that other species, such as Oysters, were common in localities where they no longer occur, and that, on the whole, the forms used by Neolithic man for food were larger than their present-day representatives in the same neighbourhood.

Probably too, the Neolithic seas of Scotland swarmed with herds of the larger mammals, such as can scarcely be imagined near our coasts, now that man has persecuted and slaughtered for centuries. At any rate, remains of Finner Whales (*Balænoptera*) have been found in the Carse clays to the west of Stirling, in some cases associated with implements of man's creation, and in situations many miles west of any point accessible to whales, even were they likely to venture nowadays towards the head of the Firth of Forth. Nor can it be doubted that Seals of various species bred on many islands and rocky portions of the coast-line which they have long deserted, and that the Walrus, half a dozen individuals of which were seen at different times on the coasts of Scotland and its isles even so late as the first half of the nineteenth century, was in Neolithic times a frequent visitor, and may even have bred on the northern coasts.

SUMMARY

Partial and incomplete as our survey of early Scotland must be, it yet affords a reasonably accurate picture of the country when Neolithic man—the long-headed, square-jawed, short but agile-limbed hunter and fisherman—founded his most northern settlements in the British Isles 9000 or more years ago. It was a country of swamps, low forests of birch, alder and willow, fertile meadows and snow-capped mountains. Its estuaries penetrated further inland than they now do, and the sea stood at the level of the Fifty-Foot Beach. On its plains and in its forests roamed many creatures which are strange to the fauna of to-day—the Elk and the Reindeer, Wild Cattle, the Wild Boar and perhaps Wild Horses, a fauna of large animals which paid toll to the European Lynx, the Brown Bear and the

Wolf. In all likelihood, the marshes resounded to the boom of the Bittern, and the plains to the breeding calls of the Crane and the Great Bustard. Yet the naturalist of the present day, could he be transported back to these far times, would notice strange blanks in the fauna. Many of the pests of our modern crops and woodlands were absent, for civilization and the easy communication of later ages have brought multitudes of noxious insects and other plagues in their train. We can scarcely imagine the golden days when the crops of the husbandman were free from the ravages of the Rabbit, and his stores secure from the depredations of Rats. Yet so these were, for, with many another nuisance, man introduced these pests, as he did also his domestic Oxen and Horses, Sheep and Goats, Dogs and Cats, domestic Fowls and Pheasants, in times that followed on the discovery of the new land of Caledonia.

PART I

MAN'S DELIBERATE INTERFERENCE WITH ANIMAL LIFE

PART I

MAN'S DELIBERATE INTERFERENCE WITH ANIMAL LIFE

THE deliberate interference of man with the wild creatures which possessed the land before him has become more and more marked with the passing of time and the development of civilization. In the primitive days, of the simple hunter and fisherman, when the population of the country was limited by the numbers of wild animals and plants available for food and clothing, the effect of this interference was at its lowest. The discovery of even a primitive cultivation of the soil resulted in a regular increase of food supplies and a consequent increase of population. This was followed by the discovery that certain wild animals could be brought under the yoke to become man's coadjutors in the task of tilling the soil, or could be reared as a dependable source of food; and this discovery led again to a new and great increase in numbers of the human race. To the needs of this vast and still multiplying population, far outnumbering the stock which a wild country could support, can be traced most of the influences which have played directly upon animal life in Scotland. Domestic stock had perforce to be increased in numbers, its useful qualities improved; the undefended flocks and herds had to be protected from beasts and birds of prey and smaller vermin, against which a war of destruction and often of extermination had to be waged. Other creatures of the wilds were slaughtered for the value of their carcases as food, or their pelts as clothing; others again were introduced or protected because of the services they rendered man as a grower of crops, or of the pleasures they afforded him as a sportsman. The following chapters endeavour to trace the main effects of man's direct interference with animal life in Scotland, though it must be kept in mind that there can be no direct interference without indirect results, for, since by Nature's laws the animal world is no loose aggregate of living things, but a closely woven fabric of interdependent lives, man's crude meddling many a time brings in its train changes he little thought of, his snapping of a thread in the fabric deranges more than he could have dreamed, the pattern of the whole.

CHAPTER II

THE DOMESTICATION OF ANIMALS

GENERAL EFFECTS

THROUGH long ages man wandered upon the plains of the Old World, content if he could meet the needs of the day by the primitive arts of the chase and of fishing. Much experience had been gathered by the race of men before the irresponsible hunter and fisherman settled down to the responsibilities of the herdsman. Nevertheless so many years have gone since our common domestic animals were led by the hand of man from the wilds to the fold, that, as the learned Dr Campbell admitted long ago, "it is no easy Matter to penetrate so far through the Gloom of Antiquity as to discern any Thing distinctly on this Head." Yet it is easy to imagine that the breaking in of animals for his own use raised civilized man from the slough of barbarism, in which still flounder those races to whom the art is even now unknown. By the labour of his Oxen, land beyond his own powers of cultivation was brought under crops; his Horses relieved him of the tedium of the march and transported his goods to fresh fields; his Sheep supplied him with clothing; and the constant presence of his flocks and herds banished the distractions of the morrow's food supply. So his mind, freed from immediate anxieties, turned to new pursuits, and the care of his living possessions stimulated a sense of responsibility and a sympathy with and forethought for their welfare and improvement.

No influence has been more potent in changing the surface features of Scotland and in altering the relationships of the wild life of the country than this forethought bred of the care of domesticated animals. To supply his stock with pastures, man has levelled forests, drained swamps, and turned the wildernesses of mountain and moor into fertile grazings. To gain for them scope and security, he has restricted the wild fauna and has destroyed beasts and birds of prey, which once found easy victims throughout the length and breadth of the land. But these results followed indirectly upon the attainment of domestication and the gradual increase of domesticated animals, so that they may be more properly dealt with in a future chapter. The direct influence of man, with which we are more immediately concerned here, is limited to the effects he has wrought in the animals he has brought under control, to the changes of temperament and habit, of structure and of function which by long ages of careful breeding and selection he has induced in the creatures which he chose to share the land with him.

LINES OF ARGUMENT

It is generally agreed that the greater share in the original domestication of animals was accomplished by the Aryan races of the Old World, though the American Indians brought under subjection the alpaca and the llama, the guinea-pig and muscovy duck, and possessed a host of cultivated plants, including some of such importance as maize, potatoes, tomatoes, kidney beans, pineapples and strawberries, tobacco, quinine, cascara sagrada and cocaine, cotton and rubber. It is also generally stated that our familiar domestic animals were first broken in in that convenient home of mysteries—the East—and that thence they were carried by the Neolithic peoples in their wanderings across western Europe. It seems probable, however, that there may have been many centres of domestication in countries where wild Oxen and Horses, wild Boars and Sheep, wild Geese, Ducks and Pigeons were common. At any rate, in the case of Scotland, there is evidence that the early domesticated animals were half-wild creatures, roaming at large in the woods and on the hills, exposed to peculiarities of soil and climate, and in some cases to admixture of blood with the wild representatives of their races. So that, even if our domestic herds and flocks sprang from stock transported by the men of the New Stone Age from the continent of Europe before the English Channel yet existed, they must soon have assumed distinctive territorial characteristics. That in these later days of careful and selective breeding, Scottish domestic animals possess features of their own is shown by breeds of such world-wide repute as Clydesdales amongst horses, Aberdeen-Angus, Ayrshires and Galloways amongst cattle, Cheviots and Black-faces amongst sheep. These distinctive features, it need scarcely be said, are due to the deliberate influence of man; and so great changes has he wrought by his forethought and experimenting that in the realm of domestic animals he may almost be looked upon as playing the rôle of creator.

In endeavouring to illustrate the influence of man on domestic animals from the Scottish point of view, it is desirable that we limit ourselves to the discussion of such creatures as inhabited the country on his arrival, or such as show characters peculiarly Scottish. A few of the wild creatures he almost certainly domesticated within our own borders, while of the remaining domestics he most likely influenced some from the time of their arrival, even if they did not form a native nucleus of his domestic stock.

The changes in the habits and temperaments, functions and structures of the creatures subjugated by man, will be most easily appreciated by a comparison of these characteristics, so far as they can be estimated, in the wild prototypes and in their later domestic modifications.

THE BEGINNINGS OF DOMESTICATION

The domestication of wild animals must have been a slow, and in its early stages, to a great extent, an involuntary process. Probably it began along one line with the congregating of certain kinds of animals in the vicinity of inhabited sites, where they found an abundant and constant supply of food in the refuse cast by primitive man on his kitchen-middens. In the case of the less dangerous animals, these encroachments would gradually be tolerated by the human inhabitants, partly because familiarity breeds contempt, and partly because of the value of the raiders in removing evil-smelling garbage. We can easily imagine steps whereby some form of Wolf or Jackal, through association and growing familiarity with man, became a sort of half-tolerated, half-domesticated Pariah Dog, driven aside half-heartedly by man, but constantly returning to feed on the refuse of the streets; and from this a further step which led to its definite recognition as a constant and valuable companion which might be trained to subserve the purposes of its guardian, and on account of its usefulness became worthy of some care and protection.

Along another line and with another kind of creature, domestication may have arisen from the chance capture of young animals, which, treated in the first instance as pets, became closely associated with their captors' families, feeding and even breeding in semi-captivity, so that they too in the course of a few generations fell into the ways of domesticated things.

Although it is generally stated that the domestication of animals began with the men of the Neolithic or Polished Stone Age, recent investigations indicate that even in the later stages of the Palaeolithic cultures, a few animals may have been trained to definite uses. At Saint Marcel, Indre, in deposits containing relics of Magdalenian culture, was found a stone pendant bearing on one side the representation of a deer, probably a Reindeer, at a gallop, and on the other side lines which represent very fairly the runners and cross-stays of a sledge. It seems no unwarrantable assumption to regard the figures on the two sides as related to each other, for Palaeolithic engravers frequently carried their artistic motives around the surfaces of the bones or antlers on which they worked. So it has been surmised that the Magdalenians had led the Reindeer captive, and harnessed it to a sledge. Other Palaeolithic sculptures have been discovered in France, whereon representations of wild Horses bear lines round nostrils and neck and other markings, which have been interpreted by M. Piette as halters and rude harness. It seems possible, therefore, that the secrets of the domestication of animals had been tapped before Neolithic man made his great conquest of the animal world.

The earliest inhabitants of Scotland seem to have possessed no domestic animals; for the shell-mounds of the Azilian or early Neolithic settlers in Oronsay yielded traces of none to the careful examinations of Mr Henderson Bishop and Mr Ludovic Mann, notwithstanding that abundance of bones of wild creatures, including those of Red Deer, Boar, Otter and several species of marine mammals, were discovered. But Oronsay is an island lying in the western seas, and it may be said that its isolation nullifies any conclusions regarding domestic stock drawn from its relics, since it would be impossible to transport animals of any size across the straits that separate the island from the mainland. While it must be admitted that the ultimate decision as to the first appearance of domestic animals in Scotland must rest upon detailed investigations on the mainland, it seems to me that the evidence afforded by Oronsay cannot be ignored, for excavations by Mr Symington Grieve in the neighbouring and contiguous island of Colonsay have shown that at an early period Sheep and Oxen were familiar to the inhabitants. Yet remains of Oxen are absent from the *oldest* deposit in Colonsay. (See Table, p. 33.)

The mainland yields scanty evidences of the status of the animals of early Neolithic times, but here also the indications are against the presence of domestic stock. Antlers of Red Deer, dressed by man, have been found associated with stranded Whales in the Carse of Stirling and near Kincardine-on-Forth, but no implement manufactured from a bone of any domestic animal has been recovered. In the upper bone layer of the Bone Cave of Allt nan Uamh, near Inchnadamph, Sutherlandshire, Dr Peach and Dr Horne found traces of man's presence, in burnt stones, hearths, burnt and split bones, and sawn antlers of Reindeer. In the same deposit with these were remains of many wild animals including Red Deer and Reindeer, but no trace of domestic animals. I mention Deer in particular because, if early Neolithic man could catch these swift and wary animals for food, there is little likelihood that domestic animals, had they been known, would have escaped. Similar testimony may be gathered from some of the earliest kitchen-middens. In the Bone Cave at Duntroon, Argyllshire, in an extensive shell-mound buried 12 feet below the surface on the promontory of Stannergate near Dundee, and in a kitchen-midden found at a depth of 4 feet near the North Sutor, Cromarty, remains of Red Deer have been found, associated in the last case with antlers of Roe Deer, but in no case with traces of domestic animals. The food refuse in the places mentioned, which have been chosen in widely separated areas, represent the accumulations of long ages of human occupation, and it seems fair to suppose that had domestic animals been known, some relics of their existence could scarcely have avoided a last resting-place in the kitchen-midden. The absence of such evidences is the more striking when it is recalled that there are not wanting in the same neighbourhoods similar but later deposits yielding abundant proof of the presence of many domestic creatures.

That the people of the later Neolithic period were well

acquainted with the majority of our present-day types of domestic animals is abundantly clear from the bones which have been found in the chambered or horned cairns; but at what precise stage of Neolithic culture these were introduced or subjugated is difficult to decide, owing to the impossibility of placing such heterogeneous deposits as kitchen-middens in a connected chronological sequence. One point seems tolerably clear, however, that all our domestic animals did not appear in Scotland at one and the same time. I have gathered a definite impression from examination of the records of kitchen-middens of Neolithic date that while the remains of Oxen, Sheep and Pigs are common and are often found together, those of the Horse and Dog are either rare or absent. But this may be partly accounted for by the different frequencies with which the animals were used as food. The evidence most satisfactory in dealing with such a question is that derived from long occupied sites of human habitation, where excavations have been carefully planned and carried out with the object of interpreting the separate periods of occupation indicated by distinct layers of debris. Such excavations afford most valuable chronological information, but unfortunately few Neolithic sites in Scotland have been investigated with the necessary precision. One series of excavations may be cited—Mr Symington Grieve's explorations in the shell-mound of Caisteal-nan-Gillean on Oronsay and in the Crystal Spring Cavern of Colonsay—on account of the actual information it yields, and as an illustration of the method which will decide the sequence of the introduction of domestic animals and of prehistoric culture in general. Mr Grieve's researches show that in both the shell-mound and the neighbouring cave, distinct layers or periods of occupation can be traced in the deposits, and that the formation of the shell-mound was begun and completed some time before the cave settlement was formed. The two together, therefore, supply a series of successive strata covering a long period which begins in early Neolithic times. Moreover, since passage dryshod can now be made between Colonsay and Oronsay at low tide, and since even in Neolithic times their separation could only have been slight, the two islands may be regarded as a geographical unit from a faunistic point of view.

What information do the strata yield? Mr Grieve distinguishes three series of layers—lower, middle and upper—in both accumulations. The lower layer in the mound is the oldest, and obviously the superimposed layers in the shell-mound, and following upon that, in the cave, belong to more and more recent periods of occupation.

When man first arrived in Colonsay, domestic animals were absent, but Red Deer were abundant, although they became scarcer with the continued presence of man. Their bones are plentiful in the lowest layer of the shell-mound, but decrease in number in the middle and upper layers, while only one fragment of an antler was discovered in the cave, and that in the lowest stratum. Along with the Red Deer occurred the Wild Boar, which also became extinct, so far as we can gather, before the later strata were formed. It is represented by several bones throughout the mound but occurs only in the lowest stratum of the cave. The Sheep seems to have been the first domestic animal introduced, for it is represented by a single bone in the highest layer of the shell-mound, and its remains become common in the layers of the cave. Then followed the Ox, in the middle and upper layers of the cave, where some bones were found embedded in the stalagmite which encrusted the cave floor. Lastly came the Horse, represented by bones solely in the upper stratum of the cave and on the surface of the floor. No traces of the Dog were found.

The results of this very interesting excavation may be scanned at a glance in a simple tabular statement:

Succession of Animal Life in Oronsay and Colonsay as Revealed by Kitchen-middens

Strata	Red-deer	Wild Boar	Sheep	Cattle	Horse
Crystal Spring Cavern, Colonsay — Upper			××	×××	×
Crystal Spring Cavern, Colonsay — Middle			××	××	
Crystal Spring Cavern, Colonsay — Lower	×	×	×		
Caisteal-nan-Gillean, Oronsay — Upper	×	×	×		
Caisteal-nan-Gillean, Oronsay — Middle	××	×			
Caisteal-nan-Gillean, Oronsay — Lower	×××	×			

× present.   ×× common.   ××× abundant.   blanks indicate absence.

No excavation on the mainland has afforded so complete a series of successive occupations of early date. In the three layers of the MacArthur Cave discovered near Oban in 1894, a site of very early Neolithic culture, the bones of Oxen were found at all levels, those of the Pig in the lowest and highest strata and those of the Dog in the highest layer. In a rock-shelter near West Kilbride in Ayrshire, described in 1879, three floors of occupation were also distinguished; bones of Sheep, Oxen and Red Deer were found at all levels, the remains of a Dog were discovered between the oldest and middle floors, and those of a Horse between the middle and most recent floors. Goat remains were also found, but their definite location in the series of deposits was not noted by the excavators. Professor Clelland of Glasgow University, who examined the animal remains, came to the conclusion that the earliest inhabitants of this rock-shelter used the Sheep as food.

There are many difficulties in the way of drawing definite conclusions regarding the appearance of domestic animals, from such observations as have been made. The sequence on an island such as Colonsay may not apply to the mainland; for various reasons, the remains of some animals may find their way more readily into the kitchen-midden than those of others; and there is difficulty in distinguishing between the bones of the early domesticated animals and their wild representatives. But my own opinion, founded on such evidence as I have been able to gather, is that the domestic animals which first appeared in Scotland were Sheep and Oxen, and of these the Sheep was probably the earlier. Then followed in doubtful sequence the Dog, the Pig, the Goat, and perhaps last of the larger domestic animals, the Horse. The domestication of the Pigeon, Duck and Goose, and the introduction of poultry and other foreign fowls belong to a period much later than the Neolithic Age when all the larger and more important domestic animals made their appearance. In the succeeding pages, I propose to indicate the changes which man has wrought in the domestic animals which are most interesting from the Scottish standpoint.

II. 1

SHEEP IN SCOTLAND

ALTHOUH a large-horned wild sheep (*Ovis savini*) was a native of Eastern England in the early days of the Great Ice Age, no remains of sheep have been found in the interglacial deposits of Scotland, nor in the post-glacial beds of clay, marl mosses and peat-hags which were formed ere Neolithic man reached the northern confines of Britain. This absence of remains may in part be accounted for by the fact that sheep in a wild state prefer rocky fastnesses, and are little likely to have been entrapped in the bogs which yield so many skeletons of deer and wild oxen; but so far as fossil evidence goes, it seems probable that sheep were absent from the host of animals which invaded Scotland when the ice-fields of the Pleistocene Age disappeared, and were unknown in the country until they were introduced by herdsmen of Neolithic culture. Nevertheless the sheep of Scotland present so many interesting and unique features, and have been, even since the Middle Ages, so famous for their wool, that, as Bishop Leslie said of them in the sixteenth century, they may "nocht be slipit over with silence."

The uniqueness of Scottish sheep lies in that in the small compass of our northern kingdom there survive two forms of outstanding interest—types of the earliest known domesticated stocks. In other countries these early races have disappeared owing to improvement and continued crossbreeding, so that the modern breeds give only the vaguest hints as to their wild ancestors. But in the fastnesses of Scotland, in the uninhabited isle of Soay in the North Atlantic, and in the isolated Shetlands, there still exist two breeds which are living links with the past—that in the former being a close relative of the Wild Mouflon of Corsica and Sardinia, that in the latter representing the domesticated "peat sheep" which the Neolithic peoples made familiar over the greater part of Europe.

THE WILD ANCESTORS OF DOMESTIC SHEEP

The wild blood which has gone to form our Scottish breeds seems to be mainly that of the Mouflon (*Ovis musimon*) (Fig. 5), the last wild remnants of which are confined to Corsica, Sardinia and Sicily, and of the Asiatic Urial (*Ovis vignei*). These wild species more resemble goats than sheep as we are accustomed to think of them. They are extremely active and agile, moving from one ledge to another when pur-

Fig. 5. Mouflon—a wild ancestor of domestic sheep. $\frac{1}{12}$ nat. size.

sued, with such sureness and rapidity that they are almost unapproachable. In both species the rams carry heavy, wrinkled horns which curve backwards and then down and forwards in a fine regular sweep, but while the ewes of the Mouflon are hornless, those of the Urial bear short upright goat-like horns. Differences in size and colour also distinguish the two species: the Mouflon race stands $27\frac{1}{2}$ inches at the shoulder, the Urial 32 inches; while in both the hair on the body is short and close with a thick underwool, the general colouring of the Mouflon is foxey-red and that of the Urial reddish-grey or fawn in summer, and in winter greyish-brown. In both, the under parts of the body, the sides of the short tail, the rump and the lower parts of the legs are white. In both species also, the rams bear a short mane on the neck and a goat-like beard, but the beard of the Mouflon is confined to the lower part of the throat and chest, while that of the Urial extends from the chin to the chest and in old rams is white in front and black behind.

In what respects has man influenced these creatures of nature in adapting them for his own use?

THE SHEEP OF SOAY

The sheep of Soay (Fig. 6, p. 38) may be regarded as an early stage in the domestication of the Mouflon, though in the characters of some of the ewes and in the offspring which he has raised by cross-breeding, Professor J. Cossar Ewart detects an admixture of Urial blood. Indeed the sheep of Soay may be taken as an illustration of how little the habits and characteristics of domesticated animals alter from the wild state, where they are freed from close association with man, and are not subject to his constant interference. In the uninhabited isle of Soay, one of that group of rocky islands which lies out in the Atlantic 40 miles west of the Outer Hebrides, and of which St Kilda is the greatest, the remnant of a once widely distributed race of sheep finds a congenial home. There they live a wild life, seeing man once or twice a year at most, when some of the St Kildans endeavour to gather their sparse crop of wool by hunting them down with dogs. When they were established on this island no man knows, but they belong to a large-horned race which was widely spread in Europe in the Bronze Age, was represented in the Swiss lake-dwellings, in the settlements of the Romans in Britain, and was identified by Professor Ewart from the Roman Camp at Newstead near Melrose. To their inaccessible habitation we owe the survival of these last representatives of a great race.

The name "Soay" itself is said to be a Norse word signifying "Sheep Island," and for many centuries the peculiarities of the Soay sheep, as compared with the more

38 THE DOMESTICATION OF ANIMALS

familiar domestic breeds, have been recognized by Scottish writers. In the early half of the sixteenth century (1527) Hector Boece, the Bishop of Aberdeen, drew attention to this curious breed.

> Beyond thir Ilis [i.e. beyond Hirta or St Kilda] is yit ane uther Ile, bot it is not inhabit with ony pepill. In it ar certaine beistis, nocht far different fra the figure of scheip, sa wild that they can nocht be tane but girnis [except with snares]: the hair of them is lang and tallie, nothir like the woll of scheip nor gait. (Bellenden's Translation, 1536.)

Fig. 6. Soay Sheep—a primitive domesticated breed preserved only in Scotland. ⅛ nat. size.

The unnamed isle is evidently Soay. I have no doubt that at this period the island of St Kilda was inhabited by the same race of sheep, for of it Boece says,

> This last Ile is namit Hirta, quhilk in Irsche is callit ane scheip; for in this Ile is gret nommer of scheip, ilk gretar than ony gait buk [goat buck], with hornis lang and thikkar than ony horne of ane bewgill[1], and his lang talis hingand down to the erd.

[1] The original reads "cornua bubulis crassitudine æqua, sed longitudine aliquanto etiam superantia"—with horns equal in thickness to oxen horns, but exceeding them even considerably in length.

Another description of the Soay sheep of the sixteenth century is very interesting and rather amusing on account of the perplexity into which these strange goat-sheep threw the writer, Bishop Jhone Leslie of Rosse. His account, published in Rome in 1578, is here given in the translation made by Father Dalrymple in 1596:

> Neist this [the island of "Hirth" or St Kilda] lyis another Ile, bot nocht inhabited, quhair nae kynd of cattail is fund, excepte sum verie wylde, quhilkes to cal scheip or gait, or rathir nouthir scheip nor gait, we knawe not, nor wat we weil: forby thair wylde nature, nathir haue thay wol lyke a scheip; nathir beir thay hair lyke a gait, bot for nane of the twa [literal translation: but they have something between the two], I can nocht tel quhat.

These accounts lay hold of the main features of the strange Soay sheep as they still exist—the wild nature, the goat-like carriage and movement, the "tallie" or drab colour, the hair overlaid by wool, the long, curved and massive horns. Boece's reference to "long tails" is most likely an error of description, but may indicate that a character induced by earlier domestication has been lost during the intervening four centuries of wild life, for to-day the tails of Soay sheep are as short as those of their wild ancestors. To these characteristics it may be added that the Soay sheep are less than they once were and are gradually becoming smaller, so that, instead of being "gretar than ony gait buk," if we are to believe Boece, they are now regarded by Mr H. J. Elwes as the "smallest aboriginal sheep now known to exist as a pure breed."

How then has the influence of man told on ancestral characters in this early stage in the domestication of the sheep? The characters of the domesticated race are still essentially those of the wild Mouflon, whence it mainly derived its inheritance, for Professor Ewart finds no marked difference between Soay sheep and the Mouflon in skeleton, horns or throat fringe. The size is somewhat less, due no doubt to scanty fare, and to long confinement on a small island and consequent close interbreeding. The predominance of hair in the Mouflon has been replaced by a predominance of wool in the Soay race, whose coat was recently declared by Mr H. Sanderson, Galashiels, a manufacturer of tweeds, to be "finer in the staple than any other wool grown in Scotland at the present time," though its shortness makes it difficult to spin alone. The fleece is a uniform pale brown or fawn instead of the patchy foxey-red of the Mouflon, though Soay lambs still retain the ancestral tint. Thus slight are the changes wrought on the characters of the wild stock by an early stage in domestication, as shown by Scotland's unique inheritance on the island of Soay.

THE PEAT OR TURBARY SHEEP OF SHETLAND

As Soay sheep represent, in the main, the influence of Mouflon ancestry, so the rare examples of Peat Sheep, which till recently occurred in Shetland, and which rank with them as survivors of one of the two earliest domesticated races of sheep, show the predominance of Urial blood.

Of the two races, the Turbary or Peat Sheep (Fig. 7, p. 42), with long slender limbs and erect goat-like horns in the ewes, is the older, at least as regards central and north-western Europe, for Professor Ewart has stated that the Neolithic peoples of these parts seem to have had no other domestic breed.

In Scotland sheep bones occur in kitchen-middens associated with dressed flints, and these bones, in the few cases where they have been carefully examined, have been found to belong to a "slender-legged variety." Bones of sheep have also been found in the Neolithic chambered cairns of Orkney; and in Caithness in similar structures there have occurred remains attributed to "sheep or goat"—a natural alternative on the part of the excavator, if, as I suspect, he had observed skulls bearing the erect goat-like horn-cores which are typical of the "Turbary" race of sheep. There is ground for believing, therefore, that even in Neolithic times the Turbary or Peat Sheep (*Ovis aries palustris*) was widely distributed in Scotland.

In most of the excavated sites of later ages in Scotland, these slender-limbed, goat-horned sheep have been represented. They occur in lake-dwellings or crannogs of Bronze and subsequent periods, in underground "Eird" or "Picts'" houses, in Roman camps, in hill-forts and in the brochs inhabited well into the early centuries of our era. Probably they contributed largely to the great flocks of small dun-coloured sheep which spread over Scotland in the Middle Ages and later, and were the wonder of the early travellers in our country. Witness the exclamation of Don Pedro de Ayala, Ambassador from Ferdinand and Isabella of Spain to James IV of Scotland in the fifteenth century, regarding the "immense flocks of sheep" which he found "especially in the savage portions of Scotland" or the description by the Scottish chronicler, Robert Lindsay of Pitscottie, who in the sixteenth century states that:

there was great Peace and Rest a long Time; wherethrough the King [James V] had great Profit, for he had ten thousand sheep going in Ettrick Forest.

Already however there must have been a considerable amount of crossing in the Scottish flocks, for numerous individuals appeared possessing supernumerary horns, the rams carrying four and sometimes six, and even the ewes up to three and four. This peculiarity especially struck the Italian doctor, Cardan, who, induced to come to Scotland to offer medical advice to Hamilton, Archbishop of St Andrews, took the opportunity of making a tour of the country. Sheep with four horns, he wrote in 1552, were frequent, but not like those he had seen at Milan, for the Scottish sheep had one pair curved and the other pair straight.

More detailed is the account of Bishop Leslie of Rosse, published in Rome in 1578 (Dalrymple's translation, 1596):

Tuedale [Tweeddale] nochtwithstanding because of the gude Wol in quhilke it abundes by all vthiris sulde nocht be slipit ouer with silence. In this cuntrie ar fund, evin as with thair nychtbouris, that sum of thame are knawen to haue four or fyue hundir, vthiris agane aucht or nyne hundir, and sum tyme thay ar knawen to haue a thousand scheip[1]. The scheip indeed ar litle, and hornes thay beir lyke rames; bot the yewis twa, thrie or four, and the Ramis at sum tymes sax: Thay beir verie schorte tailis, as schorte as the tail of ane hine [hind]. In tendirnes of thair flesche thay ar lyke the cattel that ar fed in the rest of the south cuntreyes of the Realme, bot farr excelis thame that feid in the pastoure of the nerrest cuntreyes. The cause is thocht to be this, that the knowis of thir cuntries abundes in a certane schort and bare grase, quharin scheip properlie delytes.

---

[1] This sentence is a serious mistranslation of the original, which reads "quorum alii quatuor aut quinque, alii octo aut decem nonunquam millia ovium habere noscantur"—some of whom are known to have sheep four or five others eight, nay even on occasion ten *thousand* in number.

It shall be shown when we come to discuss the influence of domestic stock on the wild fauna of Scotland that the intensive cultivation of sheep which resulted in these enormous flocks in Tweeddale and the neighbouring parts was one of the chief factors in banishing Red Deer from the Lowland hills.

What were the characteristics of the ancient "Turbary" race which for thousands of years, from Neolithic times to the Middle Ages, formed the main part of the domestic sheep of the Scottish peoples, and how had domestication and the

Fig. 7 Turbary or Peat Sheep (ewe)—a primitive domesticated race preserved only in Scotland. $\frac{1}{7}$ nat. size.

rude selective breeding of the older periods affected the characters of the wild Urial whence they seem to have obtained the greater part of their inheritance?

We can scarcely do better, in supplementing the characters derived from the skeletal remains of the Neolithic kitchen-middens and of the later prehistoric and early historic deposits, than make appeal to the last representatives of the Turbary race which till recently lingered in the isolated islands of Shetland.

The chief difference in general appearance was one of size, for while the domestic Turbaries still retained the slender limbs and light agile build of the Urial, they stood only some 22 inches high at the shoulder, as against the 32 inches of the wild Urial of the present day. The Scottish Lowland breeds in the sixteenth century still retained the short tails characteristic of the wild species, as Leslie noted, and even to-day the Shetland survivors possess short tails with only thirteen vertebral bones in contrast to the twenty or more tail-vertebrae found in modern improved British breeds. In another and significant respect the Turbary sheep retained the Urial characters, for, in contrast to the hornless ewes of the Mouflon, their ewes possessed a pair of light erect goat-like horns, very different from the heavy, curved but not spiral horns of the rams. But in one respect, other than size, the domesticated race differed materially from its wild ancestors, and that in regard to its coat. For whereas the wild species possesses an undercoat of wool concealed beneath a longer coat of fawn hair, in the Turbary sheep, the wool, though still short and fine, predominated over the hair, if we can judge from the Shetland survivors, and was of pale brown colour, known as " moorit."

CHARACTERISTIC IMPROVED SCOTTISH BREEDS

The characteristics of the races of domestic sheep which first appeared in Scotland having been summarized, it remains to indicate the changes which selection of suitable stock and careful in-breeding and crossing have wrought in typical improved breeds of the present day in Scotland, such as the Cheviot Sheep of the south country uplands and the Highland Black-faced Sheep (Figs. 8 and 9, pp. 44 and 48). It says much for the reality of man's influence in altering the characters of his domestic stock, that in little more than three-quarters of a century, since Youatt described the " Black-faced Sheep," in 1837, and Low the "Black-faced heath Sheep," in 1842, local conditions, different modes of treatment and to some extent crossing with other breeds have, according to Professor Wallace, split the breed into "at least seven very distinct sections which might rank as breeds."

The Highland Black-faces and the Cheviots of to-day differ in several important respects from the primitive domestic breeds. As a rule, neither Cheviot ewes nor rams have horns, though the latter may possess a smooth pair with a simple curve. On the other hand, both sexes of Black-faces almost invariably carry horns which spring horizontally from the skull and are curved; but those of the ram are especially luxuriant, the horns being rough, strongly ridged, and forming corkscrew spirals, the forward directed points of which have frequently to be cut in old rams to allow them to feed comfortably. Professor Ewart regards these peculiar horns in the Black-face as evidence of the presence in the breed of the blood of the Argali (*Ovis ammon*)—the magnificent

Fig. 8. Cheviot Sheep—a modern result of selective breeding.
(Champion, Highland Show, 1914.)

wild sheep of the Pamirs, Tian-Shan and Altai Mountains of Asia.

In both Cheviots and Black-faces, the tail is much longer than that of wild sheep, but while in Black-faces it reaches not lower than the hocks, in Cheviots it is so abnormally lengthened that it is found advisable to dock it.

Both breeds have gained in size of body as compared with the primitive domestic races of Scotland. But though the Cheviot is the larger and heavier of the two (a fat tup weighing at least 200 lbs. live weight), the Black-face is the hardier, and makes a living more successfully on the heathery moors of the Highlands. Modern breeds also show the influence of man's selection in a subtle quality—that of fattening rapidly when placed on suitable pasture—a quality absent in primitive Scottish breeds.

Comparison with the ancestral forms and early breeds shows that in a very important respect a great advance has been made, for the coat of hair, furnished by Nature, has been almost altogether subordinated to the development of the original undercoat of wool. In both breeds the predominant colour is white in place of an ancestral shade of brown, but while in the Black-face the wool is long, loose, shaggy and rather coarse, in the Cheviot it is shorter, closer and finer. In both breeds, it is unnecessary to add, a fleece far exceeds in weight those of the primitive races of domesticated sheep.

IMPROVEMENT OF WOOL IN SCOTLAND

Perhaps no single character in the domesticated races of animals affords so clear a demonstration of man's continuous influence as the wool of sheep. Even in a limited area like Scotland, the results point convincingly to the power of selection. It must be remembered that in the earliest days of civilization the flesh of domestic sheep was seldom used for food though the milk was drunk. The main value of sheep lay in furnishing fleeces which, prepared as skins, formed the clothing of barbarian tribes. Almost in our own era Caesar described the Briton as clothed in the skins of animals, of which no doubt the sheep was the chief; and Pliny the Younger says of his own time and country, as translated by Philemon Holland:

> Sheepe likewise are in great request, both in regard they serve as sacrifices to appease the Gods, and also by reason of their fleece yielding so profitable a use: for even as men are beholden to the boeuf for their principal food and nourishment which they labour for, so they must acknowledge that they have their clothing and coverture for their bodies from the poore sheepe.

Even from the outset of the domestication of the sheep, therefore, there was strong inducement for the herdsman to improve the quality of the fleece. In the wild species from which the domesticated breeds of sheep have sprung, the undercoat of fine wool is invariably shorter than even the close outer crop of coarser hair, and the general colour varies from foxey-red to fawn. Since it is clearly an advantage for clothing that the softness and heat-retaining properties of the fleece should be increased, early selection tended to the lengthening of the woolly undercoat; so that even in the primitive domesticated breed represented by the Soay sheep, the natural proportions are reversed, and while the hair is rather under 2 inches long, the wool reaches a length of $2\frac{3}{4}$ inches. Nevertheless Soay fleeces average under a pound in weight, and at the present day are considered scarcely worth shearing.

Excavations in early human sites of occupation, though they yield evidence of the presence of sheep, give no indication of the nature of their coat, so that we have to content ourselves with a few quotations from historical records. Already in the twelfth century, a great wool industry had been developed in Scotland, for in the reign of David I, woollen cloth was manufactured on a large scale in many of the villages, and the enumeration among the burgher classes, of weavers, "litcters" or dyers, and pullers indicates manufacture of some skill and delicacy. In several directions the improvement of the fleece had progressed, in an increase in the length of the wool, in its fineness, and in its colour. Boece in the sixteenth century writes enthusiastically about the quality of the Scottish product:

> Quhat may be said of our wol? quhilk is sa quhit [white] and smal [fine], that the samin is desirit be all peple, and coft [bought] with gret price, speciallie with merchandis quhair it is best knawin [known]. Of this wol is maid the fine skarlettis with mony uthir granit and deligat [grand and delicate] clothis. (Bellenden's Translation.)

And in the following century, William Lithgow, who travelled over the southern parts of Scotland in 1628, says of Galloway wool that it was better than any he had seen in Spain. "Nay," he writes, "the Calabrian silk had never a better lustre or a softer gripe than I have touched in Galloway on the sheep's back." Truly in these days, it was "Galloway for woo'." It need hardly be said, however, that all wool was not of this high standard, witness the estimate of Aberdeenshire wool by Schir Robert Egew, Chaiplan to My Lord Sinclair, who in his account of his stewardship in 1511 writes:

Item thar wilbe of tendit woll this yeir of your scheep Fyve stane. It will gif ilk stane, vij schillings and that is ane gud price for Buchane woll considering the ter[1] that is in it.

The demand for Scottish wool in the following century in countries beyond the borders is some index to the improvement that had taken place, and to the estimation in which the Scottish product was held. From a charter found in the Charter Chest of the Earl of Mar and Kelly, Professor Hume Brown quotes the average annual exports for the years 1611 to 1614 "Of Woll, 10,374 staneis wechte at £5 the stane, £51,870."

While in most countries and on the mainland of Scotland throughout the centuries, selection was constantly made with the object of attaining a high standard of white wool, for no other reason it would seem than that fashion favoured whiteness, a curious and reverse tendency is to be noted in the island flocks where fashion favoured coats of many colours. In 1794, Dr James Anderson wrote:

> In all the remote parts of Scotland and the isles where sheep have been in a great measure neglected, and allowed to breed promiscuously, without any selection, there is to be found a prodigious diversity of colours; and, among others, dun sheep, or those of brownish colour tending to an obscure yellow, are not infrequent....It is for this reason, and to save the trouble of dyeing, that the poor people in the Highlands propagate black, and russet and brown, and other coloured sheep, more than in any country where the wool is regularly brought to market.

The tendency of fashion to guide the influence of man and to regulate the colour of the fleece is still dominant in Shetland and amongst the breeders of Shetland sheep. Three types of Shetland shawls are in demand—a brown, a white and a grey, the last colour being also used as an edging to shawls of one of the other colours. The result has been that owing to deliberate selection for the purpose of meeting this demand three colours of fleece have come to predominate in the Shetland breed—a "black" or brown variety (known as "moorit," said to be from a Norse word signifying "moor-red"), which still retains the colour of the primitive domesticated breed, a fine snowy white variety, and a bluish grey variety known as "Sheila," having longer and coarser wool.

---

[1] *ter*, probably the Aberdeenshire dialect word for *turf*; and in this connection, therefore, signifying *grass*, *earth* or, generally, *dirt*.

In modern times, with increased attention to the care of flocks and to breeding, the rate of improvement in the qualities of the fleece has been even more striking. In the wild species, the wool is hidden beneath short hair; in modern improved breeds, so successfully has fine wool been encouraged at the expense of the rougher hair that it is almost or quite impossible to distinguish any hair amongst the wool. The wool of the primitive Soay breed is shorter

Fig. 9. Black-faced Sheep—an illustration of development in length of wool. (Champion, Highland Show, 1914.)

than 3 inches and the fleece averages under one pound in weight; often enough the shaggy wool of a Black-face trails upon the ground and the average fleece of a good hill flock weighs from $4\frac{1}{2}$ to 5 pounds, while a Cheviot ram may bear a fleece of 10 to 12 pounds weight.

The parti-colours of the wild species have been replaced by a coat of uniform colour, and in most breeds the original shades of brown have been eliminated in favour of white wool.

II. 2

CATTLE IN SCOTLAND

THE farmer's family, according to Hesiod, one of the earliest writers on agriculture, consisted of the Husband, the Wife and the Ox, the Minister of Ceres. The ox was the "constant Companion of Man in the Labours of the Field," as well as the mainstay of the food supply of the early communities. On account of these particular uses to which cattle were put, the influence of man has had a less striking effect on the outer aspect of oxen than on that of the wool-bearing sheep. Yet for us the ox gains an additional interest in that at an early stage the domestication of oxen was probably more intimately connected with Scotland than that of sheep. For there can be no doubt that when Neolithic man reached these lands the forests still sheltered herds of wild cattle which here or elsewhere formed a nucleus of our domesticated breeds.

THE NATIVE WILD CATTLE OF SCOTLAND

In the time of the Ice Age, perhaps even of the earlier Forest Bed of Norfolk, there appeared an Ox of large size, the great Urus (*Bos taurus primigenius*) (Figs. 10 and 11, pp. 50 and 53), which before the close of the Ice Age had spread from the north of Scandinavia to Sicily and from the Siberian Steppes to the west of Scotland. For many centuries, under climatic conditions of great diversity, it inhabited the Scottish plains. When the snows of the Ice Age disappeared from the Lowlands in one of the mild interludes which broke the severity of an Arctic climate, and a coat of verdure spread over the plains, the Urus made one of the small band of animals which ventured into the southern counties of Scotland in quest of new pastures. Its earliest Scottish remains have been found in interglacial deposits near

Crofthead in Renfrewshire, where they were associated with bones of the Giant Fallow Deer or "Irish Elk" and of the Horse, as well as in the valley of Cowden Burn, in the same

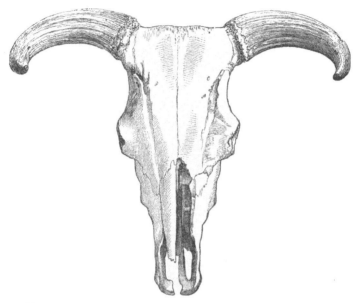

Fig. 10. Front view of skull of Urus, the native wild ox of Scotland, from Fifeshire. ¼ nat. size.

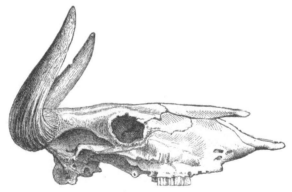

Fig. 10 a. Side view of above skull of Urus. ¼ nat. size.

county. In the marl deposits formed on the floor of the lakes which succeeded the Glacial Period, skulls and other bones of the Urus have been commonly found from Wigtownshire to Caithness, although the headquarters of the race, as indicated by the frequency of its occurrence, appears to have been in the plains of the eastern coast, especially in the valleys of the Tweed in Roxburghshire, of the Tay in Perthshire, and in the flat lands of Caithness. Even in later ages, the Urus was still common throughout Scotland, for its remains are abundant in the peat-bogs which accumulated under conditions of great humidity about the time of and subsequent to the arrival of the first Neolithic settlers in Scotland. At this period the headquarters of the race appears to have been in the Lowlands, for although isolated records occur as far north as Belhelvie Moss in Aberdeenshire, the majority of the remains have been recovered from Ayrshire, Berwickshire, and particularly from the higher grounds drained by the Tweed and its tributaries.

The subsequent history of the Giant Ox in Scotland is one of gradual decline, and it is reasonable to assume that the dwindling of the great herds was connected with the appearance of man in the country. Nevertheless, the Urus lingered on in association with man in Scotland for many centuries, being gradually driven northwards into the wilds, until within the confines of the northern counties it finally disappeared. Its remains have been found in juxtaposition with relics of Neolithic man from the Clyde Valley to Caithness: in the former case near the mouth of the Kelvin, in laminated beds of silt where many dug-out canoes, hollowed from solid trunks of oak, have been found, and in the latter, in "horned cairns" belonging to the later period of the Polished Stone Age, at Camster, Ulbster and Clythe. During the succeeding two thousand years, till approximately 1000 B.C., it may have lingered in the Lowlands of Scotland, for in 1781 six skulls were found in a "merle moss" at Whitmuir Hill, Selkirkshire, in the neighbourhood of many "brass axes." The writer of the Statistical Account of the parish of Selkirk actually states that along with the skulls was found "a Roman spear with which these animals were destroyed," but this statement is more than doubtful, for Professor Ewart has found no trace of the Urus amongst the abundant animal remains of the Roman Camp at Newstead near Melrose. It is almost safe to assume, therefore, that long before the Romans invaded southern Scotland, in the early centuries of our era, the pressure of civilization had driven the survivors of the Giant Ox beyond the bounds of the Scottish Lowlands.

Nevertheless it appears still to have survived in the northern parts of the kingdom, for remains which seem to be identical with those of the marls and peat-bogs, have been found in underground buildings or "Eird" houses apparently belonging to the period of the brochs or "Pictish Towers," at Skara in Orkney and in an ancient mound at Keiss in Caithness, as well as in a broch itself, at Kintrawell beyond Brora in Sutherlandshire. There is some reason to believe, therefore, that the Broch Period, lasting towards the ninth or tenth century, saw the last wild British survivors of this great race of cattle, long before they had disappeared from the dense forests of Central Europe, where they were believed by Professor Nilsson to have existed in a wild or half-wild state even to the beginning and middle of the sixteenth century, many years after the exploits of the redoubtable Siegfried in the woods of Worms, sung in the *Niebelungen lied*:

> Then slowe the dowghtie Sigfried a Wisent[1] and an Elk,
> He smote four stoute Uroxen and a grim and sturdie Schelk[2].

Characteristics of the Urus

There is no Scottish written evidence to guide us in determining the characters of this Scottish wild ox, for the only definite relics of its existence are the bones of the prehistoric and early historic deposits. Their evidence, however, is clear as regards the general character of the animal. Its bones in every respect, in their proportions, contours and even in the details of their ridges and muscle-impressions, agree so closely with those of recent oxen as to show that the Urus was in reality no more than a variety of *Bos taurus*. In one striking respect the skeletal remains differ—in size. From a skeleton which he compared with that of a recent ox, Principal Sir William Turner estimated that the Urus must have easily stood six feet high at the shoulder, a size sufficiently great, though not to be compared with Caesar's exaggerated description of

---

[1] Bison.
[2] Red Deer, or perhaps the now extinct Giant Fallow Deer.

CATTLE IN SCOTLAND

the Uri in the Hercynian Forests—"magnitudine paullo infra elefantos"—'in size little less than elephants.' The horn-cores, borne on the massive, flat-fronted skull (Figs. 10 and 10a), a third larger than the skulls of domestic cattle, indicate that the horns were of great length, even larger, it is said, than those of the long-horned breed of cattle found in the Campagna of Rome.

As regards colour, there is little trustworthy evidence, and for want of better we must appeal to an oil painting, supposed to represent the Urus, which was discovered in Germany about a century ago by Major Hamilton Smith:

> We found an old painting on pannel of indifferent merit in the hands of a dealer in Augsburg, which represents the animal, and judging from the style of drawing, etc., may date from the first quarter of the sixteenth century. It is a profile representation of a bull without mane, but rather rugged, with a large head, thick neck, small dewlap entirely sooty black, the chin alone white, and the horns turning forward and then upward like the bull of Romania; pale in colour with black tips. In the corner were remains of armorial bearings, and the word *Thur* in golden German characters. We made a sketch of the figure.

Fig. 11. Urus—the native wild ox of Scotland. $\frac{1}{60}$ nat. size.

The sketch formed the basis of a plate in colour in Griffith's edition of Cuvier's *Animal Kingdom*, and this is here reproduced (Fig. 11). In one point fossil evidence testifies to the accuracy of the painting, for a horn, found in peat in Pomerania, was pale horn-coloured with a black tip. Probably the Urus was of a dark reddish-brown colour verging on black, with long, black-tipped horns, and with hair short and comparatively smooth, except on the forehead where it was long and curly.

One other character of the Urus is worth recording in view of its modification in the domestic breeds derived from this wild race—namely, its temperament. Caesar said of the Uri inhabiting the Hercynian forest of Central Europe " Great is their strength and great their speed; they spare nor man nor wild beast on whom they may cast their eyes."

THE URUS AND DOMESTICATION

Notwithstanding that man lived for many centuries in Scotland in company with Wild Oxen, he seems to have made little or no progress in domesticating them. This may have been due in part to an exceptionally wild strain in the Scottish race, and in part to the fact that there was little inducement for him to break in a new race of cattle, since he had brought with him to Scotland a smaller and more amenable breed, the Short-horned Celtic Ox (*Bos taurus longifrons*), which had already been domesticated on the continent. It is not surprising, therefore, that kitchen-middens of Neolithic and later ages yield many bones of the smaller domesticated breed, but afford little evidence of the presence or domestication of the Urus, a creature of the wilds and remote fastnesses. Unfortunately, however, in the majority of the early excavations, examination of the animal remains was of a more or less cursory nature, bovine bones being simply recorded as "oxen," without attempt to arrive at a critical estimation of their further significance.

It seems very unlikely that two closely related races of cattle could exist in the limited area of Scotland without a certain infusion of wild blood into the domesticated stock; and it is just possible that some of the larger ox remains found along with the bones of the Short-horned Celtic Ox in the Roman Camp at Newstead as well as in other Roman settlements, and the few Urus-like skulls of the brochs, may represent more or less remote descendants of crosses between the Urus and the Celtic Shorthorn. It is possible also that the sixteenth century "wild" White Cattle of the Caledonian

Forest may have been direct descendants of the Urus, though the weight of evidence seems to show that their relationship is more distant, and that they are rather the offspring, which have run wild, of a breed domesticated from the Urus. Whether or not the blood of the indigenous cattle of Scotland may still linger in direct lineage in our modern stock through cross-breeding with the Celtic Shorthorn, it is generally allowed that the Urus, domesticated perhaps on the plains of Europe, is the ancestral form of the larger breeds of cattle in Britain at the present day.

THE EARLIEST DOMESTICATED CATTLE OF SCOTLAND—THE CELTIC SHORTHORN

Not until the Ice Age had passed away did there appear on the plains of Europe a small race of cattle which formed the nucleus of the earliest domesticated breed. There is no evidence that this small race, the Longfronted Ox, or, as it is commonly called, the Celtic Shorthorn (*Bos taurus longifrons*), existed in Scotland at any period before the time of man's arrival. Its remains have been found in river gravel of doubtful age, near Currie at the northern end of the Pentland Hills, and in 1870 there were discovered bones of several of these oxen, which, before cities and villages were dreamt of, had been entrapped and engulfed in the shell-marl of the Nor' Loch which formerly lay in the hollow now occupied by the Princes Street Gardens, Edinburgh. Even in the later peat deposits remains are exceedingly scanty, though they are distributed in bogs from Roxburghshire to Ross. Indeed, in Scotland the history of the Celtic Shorthorn traces a course exactly the reverse of that of the Great Wild Ox, for while the latter is abundantly represented in the early deposits and decreases with the coming of man until it disappears, the former is absent from the early deposits and increases with the spread of Neolithic man until it occurs throughout the length and breadth of the land.

The Celtic Shorthorn does not appear to have been known to the earliest settlers in Scotland, who, as we have seen, left traces of no domestic animals, yet it appears in the deposits of very early Neolithic times. Horns characteristic of the Celtic Shorthorn were discovered in 1816 near the surface of the clay of the Fifty Foot Beach in the Tay Valley at Blair Drummond in Perthshire, and bones have been found in the deposits which have yielded Neolithic dug-out canoes in Rutherglen near Glasgow. Actually associated with the handiwork of men of the Polished Stone Age, skulls and bones have been found in caves and shell-mounds on the Ayrshire coast, in kitchen-middens, as in Oronsay and in the MacArthur Cave at Oban, as well as in the chambered or horned cairns of late Neolithic Age at Camster, Ormiegill near Ulbster, Garrywhin near Clythe and Hill of Bruan, all in Caithness, and in a chambered cairn at Loch Stennis in Orkney.

In Scotland, then, the Celtic Shorthorn appeared in the Neolithic period, shortly after the first settlement of man in North Britain, and this, together with the fact that its remains are most frequently found in deposits accumulated on the sites of human habitation, point to its presence as a domesticated animal which had followed in the train of the men of the Polished Stone Age from the regions of the south.

There can be no doubt of its close relationship to man in subsequent ages, for its remains have occurred in almost every prehistoric site of occupation which has been excavated in Scotland. In the Bronze and Iron Ages, which together extended to the beginning of the Christian era, it was common, and the numerous herds of cattle, "pecoris, magnus numerus," observed by Caesar on his arrival in Britain just before the first century of our era, and remarked upon in his *Commentaries*, belonged to this race. According to Professor Boyd Dawkins, it was the only breed in existence in Britain at the time of the Roman Conquest. In spite of the fact that the Romans almost certainly brought with them new races of cattle from the Continent, the Celtic Shorthorn still predominated in Roman and Romano-British settlements in Scotland, as at Newstead near Melrose, Traprain in Haddingtonshire and Inveresk near Musselburgh, and remained for long the only domestic cattle of the native population. Its importance as food to the inhabitants of Scotland during or shortly after the period of Roman occupation is indicated by the contents of the refuse-heap of a cave at Borness in Kirkcudbrightshire, excavated in 1875, for there the recognizable ox-bones numbered 1112, as contrasted with 630 bones of Sheep, 26 of Red Deer, and three of the Horse.

There is no need to detail its history in later times,— it is sufficient to say that the bones of the Celtic Shorthorn have been found in plenty in the Scottish crannogs or lake-dwellings, as at Loch of Dowalton in Wigtownshire, Lochlee and Lochspouts in Ayrshire, and Isle of Eriska in Argyllshire; in cave deposits, on the south coast at St Medan's and St Ninian's Caves in Wigtownshire and Borness Cave in Kirkcudbrightshire, on the east coast

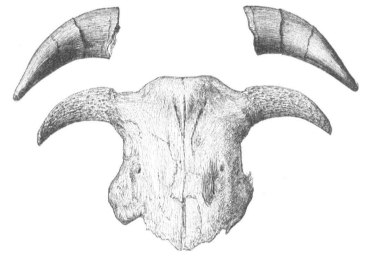

Fig. 12. Horn-sheaths of Celtic Shorthorn and upper part of skull with horn-cores and with fragments of skin and hair attached, found deep in Irish bog—relics of a primitive domesticated race of cattle. ¼ nat. size.

at Wemyss in Fifeshire, and on the west in caves on the Ayrshire coast; in duns or hill-forts from Tiree to Burghead; in underground "Eird" or "Picts'" houses even in the outer islands, on the mainland of Orkney near Kirkwall and at Skara, and in the outer Hebrides in Harris; in kitchen-middens or shell-mounds and in the "Pictish Towers" or brochs throughout the length of the land. The Celtic Shorthorn was therefore the one well-defined race of domestic cattle familiar to the Scottish peoples till the close of the broch period towards the ninth or tenth century of the Christian era, and probably it remained dominant to a much later day.

Characters of the Celtic Shorthorn

What was the nature of this domestic race, which, for some 6000 years, was outstanding in the history and development of the early peoples of Scotland? Two sets of characters infallibly single out its bones from the remains of other cattle—those of its skull and those of its limbs. The skull was long and narrow, more like that of a deer than of a modern ox, and the forehead had a median ridge, a prominent bony crest, and carried two short tapering horns, about nine inches long, curved gently forwards and downwards. The limbs also were deer-like in character, the bones being slender in proportion to their length, as compared with those of modern oxen. From the skeleton, Professor Nilsson estimated that the Celtic Shorthorn "was 5 feet 4 inches long from the nape to the end of the rump-bone, the head about 1 foot 4 inches, so that the whole length must have been about 6 feet 8 inches." The skeleton indicates that the Celtic Shorthorn was a long-bodied but light and agile ox, well-fitted to protect itself by speed of limb from the many beasts of prey of the Scottish forests.

A few fortunate finds give a clue to the nature and colour of its coat. On a skull found in an Irish bog and having the characters of the Celtic Shorthorn, Dr John Alexander Smith found part of the skin and hair still attached (Fig. 12). The hair was of rough shaggy nature like that of Highland Kyloes (Fig. 13, p. 61), and was of dark red or brownish tint. Confirmatory evidence is furnished by the contents of the strange masses of "bog butter" which, stored in wooden kegs or wrapped in skin or birch bark, have been found often at considerable depths below the surface in peat-mosses in the counties of Argyll, Inverness, Banff, Moray, Sutherland and in the islands of Skye and North Yell. The apparent age of these butter masses suggests the probability that the butter was made from the milk of the Celtic Shorthorn. The butter invariably contains abundance of cow-hairs, and these are always red in colour. Again, Dr Joseph Anderson in 1878 discovered on a dagger of the Bronze Age, found in a large sepulchral cairn at Collessie in Fifeshire, a mass of agglutinated hairs, remains of the hide which covered the wooden sheath of the dagger; and these under the microscope showed "the same appearance and structure as the dark hairs of a Shetland cow taken from one of the rivlins or Shetland shoes of untanned hide in the Museum [of Antiquities in Edinburgh]."

Although nothing is known from written records regarding the nature of the early domestication of cattle in Scotland, the bone remains seem to point to the wild or half-wild nature of the herds. In the first place, the limb bones are those of an active mobile creature, and this activity could only mean that the cattle were not restricted in any serious degree, but ranged over large areas and depended on their movements for escape from the Brown Bears and Wolves which shared the forest with them. In the second place, in the Scottish bone deposits which I have examined, and which range from Neolithic times to a period when Christianity was already firmly established on the east coast, the large majority of the remains of the Celtic Shorthorn are those of young animals, as the presence of milk-teeth and of bones not completely ossified clearly shows. This I take to mean that the inhabitants found it easier to slay the young than the old animals—that indeed adult animals were hard to slay, for other things being equal, their greater food value should have made them preferred, and their bones to preponderate in the refuse-heaps. The indications are that the cattle were captured by a sort of hunting which found the young animals ready victims, and therefore that the herds of the Celtic Shorthorn were little better than wild.

It may indeed be said that part of the influence of man upon domestic cattle has been expended in gradually narrowing their range of freedom, and with this, their activity, so that from the lean muscular oxen of the wilds, the fat ox of the stall has been developed. The process of enclosing has been a gradual one, as old Scots Laws demonstrate, for the "Leges Forestarum" generally ascribed to William the Lion (A.D. 1165–1214) invoke penalties upon cattle found straying in the King's forest—a superfluous provision had the cattle been enclosed or even carefully herded. That they were not so herded even in the seventeenth century is shown by the fact that it was found necessary in 1686 to pass a law that cattle should be herded in winter as well as in summer to prevent destruction of plantations and enclosures. And as everyone familiar with the history of Scottish agriculture during the past two centuries knows, it is only within comparatively recent times that open "out fields" and "in fields" gave place to the enclosed fields of the present day.

MODERN SCOTTISH CATTLE AS EXEMPLIFYING THE INFLUENCE OF MAN

The domesticated breeds of cattle existing in Scotland at the present day fall into two distinct groups: a small type, of which the Highland Kyloes may be taken as examples, and a large and heavy type such as the Aberdeen-Angus and Shorthorn breeds. Now the two races from which all the modern Scottish breeds have sprung are the Great Ox or Urus—*Bos taurus primigenius*—and the Celtic Shorthorn; but so distant is the ancestry, so mixed has been the breeding in an effort to obtain new and better stocks, and so potent has been the influence of man in perpetuating the characters of his choice, that it is impossible with certainty to attribute any modern breed to its originator. At the best, we can only say that probability lies in the suggestion that the Celtic Shorthorn, the only domestic race of the early Celts, was, with its owners, ultimately driven into the refuges of the mountains and islands by the influx of Romans and Saxons, and there probably gave rise to the characteristic mountain and island breeds—the Highland Kyloes and Shetlanders; while the lowland races became more and more permeated by the blood of the larger cattle derived from the great Urus, which the invading peoples brought with them from the Continent. Thus at one pole of the Scottish breeds the cattle of the West Highlands approach most closely the type of the Celtic Shorthorn, while the Shorthorn and Aberdeen-Angus show the largest proportion of the blood of races which owe their ancestry to the Urus. Of other well-known Scottish breeds, the old established Galloways seem to share in great part the same origin as the Highland Kyloes, and the Celtic Shorthorn had also much influence in the moulding of the extinct Orkney and original Shetland breed, as well

CATTLE IN SCOTLAND 61

as of the Ayrshires, though in the last case, even in comparatively recent times, there has been a great admixture of Urus blood through crossing with modern Shorthorn and other races.

In spite of these complexities of descent, there yet stand out clearly several main lines along which man has influenced the characters of the original races.

Of all the modern breeds, the Highland cattle, in their build, in the nature and colour of their coat, and in their

Fig. 13. Highland Kyloes (cow and calf)—a primitive domesticated breed ("Mhaldag," First Prize, Highland Show, 1886).

habits, approach most closely to their wild prototypes. They still retain the hardiness which one would expect in the descendants of a race inured to the climate of Scotland for many thousands of years, and in the very picturesqueness of their long shaggy coats and bushy forelocks they suggest the unimproved creatures of the wild. A description by Bishop Leslie, published in 1578, of the "fed" or domesticated "ky, nocht tame," which in his day ranged the mountains of Argyll, the very area from which the most characteristic of modern Highlanders are derived, gives point to the comparison, for there can be no doubt that he refers to the Kyloes of the sixteenth century:

> In the mountanis of Aargyl, in Rosse lykwyse, and sindrie vthiris places, ar fed ky, nocht tame, as in vthiris partes, bot lyke wylde hartes, wandiring out of ordour, and quhilkes, throuch a certane wyldnes of nature, flie the cumpanie, or syght of men: as may be seine in winter, how deip saeuir be the snawe, how lang saevir the frost ly, how scharpe or calde how evir it be thay nevir thair heid sett vnder the ruffe of ony hous. Thair fleshe of a meruellous sueitnes, of a woundirful tendirnes, and excellent diligatnes of taste, far deceiues the opiniounis of men, that nevir tasted thame.

But how great have been the changes in most other breeds. The Giant Urus, six feet high at the shoulder, has been reduced to the much smaller proportions of the modern Shorthorn and Aberdeen-Angus (Fig. 14, p. 63). His long graceful limbs, which gave him a speed surpassing that of most of the animals of the prehistoric forests, have, at the demand of the market, become reduced and embedded in an over-developed body in which have been lost the supple lines of the wild ancestor.

In colour no less than in form, man's selection has worked great changes. In place of the uniform dark reds, browns or blacks of the primitive races, modern breeds range through white and shades of yellow, red, brown, red and black, black and white, to unbroken black. Even such a strange freak of colouring as a broad white band like a white sheet tied round the animal's black body, has become perpetuated in the definite race of "belted" or "sheeted" Galloways. It is surprising in how short a period such colour changes may take place under man's guidance. At the beginning of last century, the Aberdeen-Angus breed contained individuals of the most diverse colours, brindled, red and black, black and white, red, brown and yellow. Yet to-day the only recognized colour of pedigreed stock is black.

Horns, too, have been modified through man's selection. Not only are their sizes and shapes more varied, but in Aberdeen-Angus and Galloways, they have actually been bred out of existence. That the polled or hornless condition did not originate in recent times is shown by a polled skull identified by Professor Ewart from the Roman Camp at Newstead; and it is said that the first historical reference to polled cattle was made in the ninth century when King Kenneth MacAlpine (A.D. 844–860) in promulgating the laws at Scone in Perthshire, specifically mentions "black homyl," in modern Scots "humle" or "humlie," that is to say, hornless cattle. Yet the creation of a polled race under man's influence seems to one familiar with the slowness of nature's processes, a thing of marvellous celerity; for even in the middle of the eighteenth century, a large proportion of purebred Galloway cattle had horns of considerable length, and the complete disappearance of horns since that time is simply due to the efforts of breeders to meet the demands of English

Fig. 14. Aberdeen-Angus bull—a highly developed result of domestication. ("Metaphor," Champion, Highland Show, 1910.)

graziers, who, compelled to drive their purchased herds a considerable distance across the borders, found that horns merely contributed to accident and damage, and accordingly, when possible, selected hornless individuals.

Other characteristics, less apparent than the external features just described, have suffered change under domestication, for man's influence extends even to traits of character and to the deeper physiological activities. How can we compare the untameable ferocity of the Urus of Julius Caesar's day with the docility of the large modern breeds, or the wildness of the Celtic Shorthorn, adults of which were seldom captured by our prehistoric predecessors, with the mildness of Ayrshires and Galloways? This loss in character may be connected with the increased weight of body and general lassitude of mind bred by a sheltered existence, and these have been fostered by the gradual reduction of the free-way of the herds until their exercise is confined to the narrow limits of an enclosed field. The breed of the mountains—the Highlanders—still retains more than any other the spirit of the wild.

A curious development, to which much attention has been paid since the improvement of cattle became a science, is that of early fattening, and the extraordinary tendency of some breeds to gain flesh rapidly and at an early age, speaks wonderfully of the power of human selection. Of all breeds the Shorthorn seems most to have developed the tendency to early maturity. The official weights of the prize-winners at the National Hereford Shorthorn Show at Kansas in 1900 afford an excellent illustration of this trait. The average of eight prize-winners is given in each case. While heifers under six months old averaged 571 pounds, the weight at a year was 810 pounds, at two years 1270 pounds, and cows at three years or over weighed 1806 pounds. At six months, a bull averaged 588, at a year 966, at two years 1467, while at three years or over the weight averaged 2298 pounds. It is a striking fact, emphasizing the significance of artificial improvement, that the breed which comes most slowly to maturity for the market is that which remains nearest in type to the original stock—the West Highland—for in reaching its maximum of weight the Kyloe lags about a year behind most of the other British breeds.

In a last subtle respect domestication has had an astonishing effect—in the development of the milk-supply of cows. So utterly has nature been circumvented in this respect that records, gathered in 1905 by the Fenwick Society from 18 dairies in the south of Scotland, comprising 443 cows, show that the lactation period ranged from thirty-eight to nearly *forty-six* weeks, and that the annual yield per cow averaged 875 gallons of milk.

THE WILD WHITE CATTLE OF SCOTLAND

An account of the cattle of Scotland cannot be concluded without some reference to the ancient White Cattle of the parks, though from the point of view of domestication, they are neither flesh nor fish, neither the wild progenitors of our domestic stock nor a direct link in the chain between indigenous oxen and modern breeds. Yet they have a romance of their own. There can be no doubt that this fine race, with all the characteristics of its modern descendants, existed in the woods of Caledonia at a very early date:

> Mightiest of all the beasts of chase,
> That roam in woody Caledon,
> Crashing the forest in his race,
> The Mountain Bull comes thundering on.
>
> Fierce, on the hunter's quivered band,
> He rolls his eyes of swarthy glow,
> Spurns with black hoof and horn the sand,
> And tosses high his mane of snow.
> SIR WALTER SCOTT.

To-day the remnants of the White Cattle are preserved in a few parks of which the chief are in Scotland, and in these they have developed into more or less distinct races; but Harting in his account of British Wild White Cattle refers to twenty-two herds enclosed at one time or another by Royal assent, and Rev. John Storer, in his *Wild White Cattle of Great Britain*, records as many as "forty localities where wild white cattle or their domesticated descendants are proved to have existed."

Our earliest historians regarded these White Cattle—"spotless bulls," as Ossian calls them, with their strange black muzzles, ears and hoofs—as truly wild, though it is curious that all their early accounts seem to refer to animals kept more or less under protection in woods and parks. "The great wood" of Chillingham in Northumberland is referred to as early as 1220, and in records of the year 1292, wild cattle are distinctly mentioned as inhabiting it, though their distinctive features are not specified. The earliest description with which I am acquainted is that of Hector Boece, published in 1527, which in Bellenden's translation runs:

At this toun [Stirling] began the gret wod of Calidon. This wod of Calidon ran fra Striveling throw Menteith and Stratherne to Atholl and

Lochquhabir as Ptoleme writtis in his first table. In this wod wes sum time quhit bullis, with crisp and curland mane, like fiers lionis; and thoucht thay semit meik and tame in the remanent figure of thair bodyis, thay wer mair wild than ony uthir beistis, and had sic hatrent aganis the societi and cumpany of men, that thay come nevir in the woddis na lesuris quhair thay fand ony feit or haund thairof: and, mony dayis efter, they eit nocht of the herbis that wer twichit or handillitt be men. Thir bullis wer sa wild, that thay wer nevir tane but slicht and crafty laubaur: and sa impacient, that, eftir thair taking, thay deit for importable doloure. Als sone as ony man invadit thir bullis, thay ruschit with so terrible preis on him, that thay dang him to the eird, takand na feir of houndis, scharp lancis, nor uthir maist penetrive wappinis.

King Robert the Bruce is said to have endangered his life in hunting the white bulls of Torwood, for one being "sair wundit be the hunteris, it ruschit feircelie on the king, howbeit he had na wappinis in his hand to debait himself fra the dint theirof." Yet by the service of a retainer he escaped.

Even in the sixteenth century, however, the White Cattle were exceedingly few in number, for Lindsay of Pitscottie concludes his account written in years preceding 1565, by stating that "because the flesch was pleasant and daintie to the mouth, the haill race of them almost is extinguished," and Bishop Leslie, writing only a few years later (1578) is still more doubtful of their survival, which was vouched for only by hearsay: "In quhilke ["Tor Wod," the Caledonian forest], onlie, eftir the commoune speiking, war the quhyte kye fund, of quhilkes now restes verie few or nane." (Dalrymple's Translation.) Leslie's account of the "kye, oussin [oxen] and wilde bullis" is worth quoting from the same translation, on account of the description and quaint Scots expressions:

In this Wod war nocht onlie kye bot oxne and Bules snawquhyte with a mane thick and syde, quhilke thay beir lyke the mane of a lyone. Thay mairouer war sa cruel and wylde that frome mankynde thay abhored in sik a sorte that quhateuir thing the handis of men had twechet, or the air of thair mouthis had blawne vpon or endet as we speik, frome al sik thay absteined mony dayes thaireftir. Farther, this oxx or Bull was sa baulde, that nocht only in his yre or quhen he was prouoked walde he ouircum horsmen, bot euin feiret he nathing nathir tyred he, commonlie al men to invade baith with hornes and feit ye the dogis, quhilkes with vs ar maist violent, he regardet nocht bot walde clate him with his cluifes or kaithe him on his hornes. His flesche was all girsslie bot of a trim taist. He was afortymes a frequent beist in this Torr Wod, but now consumed throuch the gluttonie of men only in thrie places is left, in the Park of Striuiling, the Wod of Cummirnalde and of Kinkairne.

It is quite clear from these and other contemporaneous descriptions, that however numerous and truly wild the White Cattle may once have been in Scotland, and there is no evidence on these points except the hearsay of the "commoune speiking," they were scarce enough in the sixteenth century, when they were already the guarded inmates of parks. The latter point is emphasized in a document of 1570 recording a charge brought against retainers of the Regent Lennox of having

slain and destroyed the deer in John Fleming's forest of Cumbernauld, and the white kye and bulls of the said forest, to the great destruction of police and hinder of the commonweal; for that kind of kye and bulls has been kept there many years in the said forest, and the like was not maintained in any other part of the isle of Albion, as is well known.

During the intervening four centuries the numbers of White Cattle have been increased through careful protection and breeding, until there are now several well-known races, but the early records have some bearing on the controversy as to the origin of the race as a whole.

It is impossible and would be profitless to enter here into the details of this controversy. It is sufficient to say that it is held on the one hand, that the White Cattle are the direct descendants and representatives of the Wild Urus (*Bos taurus primigenius*) which inhabited the wilds of Scotland from prehistoric times till perhaps even the days of the brochs or "Pictish Towers"; and that, on the other hand, such high genealogy is denied them, and they are said to be descendants of a domesticated race of the Urus, which were preserved on account of their peculiarities of colour, and were allowed the run of the Scottish forests. There is little reliable evidence, as the quotations given above show, that the White Cattle had ever a wide distribution in Scotland, and the early bone deposits are silent as to their presence. Indeed, the evidence, in strong contrast to that regarding the Urus and the Celtic Shorthorn, rather points to their presence only as animals more or less under protection in limited reserves; on the whole, the latter of the two contrasted views seems to be favoured by the weight of opinion at the present day.

II. 3

THE HORSE IN SCOTLAND

> With flying forelock and dishevelled mane,
> They caught the wild steed prancing o'er the plain,
> For war or pastime reined his fiery force.

"THE Horse in his Nature is as gentle and docile, as in Appearance he is a noble, majestic and well-proportioned Animal, but his peculiar Excellencies are determined by the Service for which he is designed," wrote Dr John Campbell in his *Political Survey* (1774), and therein he epitomizes the influence of man, for he goes on to say, "We require Horses for various Purposes, and to suit these they must have various Properties, indeed so various, that what are regarded as Excellencies, in some, would be Defects in others." Our present problem lies in endeavouring to discover in what ways in Scotland man has played upon the structure and nature of Horses for his own "various Purposes."

THE HORSE NATIVE IN SCOTLAND

It has often been taken for granted that the Horse, like the Celtic Shorthorn and the Sheep, first made its appearance in Scotland in the train of the herdsmen immigrants of the New Stone Age; but long before man placed his foot in these northern regions, their plains were the sporting grounds of herds of wild native horses. Whether the primitive horses whose remains have been found in the Forest Bed of Norfolk, ever made their way to Scotland, it is impossible to say, for the glaciers of the Ice Age have long since scoured away possible evidences of that highly probable invasion; but ten bones of small horses were discovered during the cutting of a line of railway near Crofthead in Renfrewshire, in a series of deposits five feet below a layer in which were found remains of the Giant Fallow Deer or "Irish Elk" (*Megaceros giganteus*) and of the Urus (*Bos taurus primigenius*). There can be little doubt that the horses represented in this deep deposit entered Scotland during one of the milder inter-glacial periods which broke the continuity of the Age of Ice.

Deposits of later date also afford evidence of the presence of horses at an early stage in the development of modern Scotland. Bones of two horses, along with those of the Wolf, the Reindeer and the Fox, were discovered in a rock fissure of post-glacial age in the Pentland Hills; the marl mosses of Forfarshire, according to Lyell, entomb many of their remains; and the bones of horses were found in 1868 embedded at the depth of between six and eight feet in a peat-moss at Balgone near North Berwick. I have seen in the Hunterian Museum in Glasgow a lower jaw and cannon-bone found, probably in the river-gravels of the Clyde, during excavations in Stockwell Street, Glasgow. These later relics, however, have scarcely the same significance as those of the Crofthead deposits, for it is possible that some of them may represent the domesticated herds of the Neolithic inhabitants.

What was the nature of the horse which inhabited Scotland before the coming of man? The Crofthead bones were said to represent an animal with the essential characteristics of the horse (*Equus caballus*), but a third smaller than ordinary horses of to-day. The solitary cannon-bone of the Stockwell Street horse shows that it belonged to a slender-limbed race, for its length is seven times its breadth (the length being 238 mm. and the width at the middle 34 mm.), a characteristic feature of a type of small horse, designated by Professor Cossar Ewart the desert or plateau type. This race frequented the valleys of Italy in the preglacial Pliocene Period, and in Glacial times wandered over the plains of France, and thence into the still-connected land of Britain. From the remains found in various parts of Western Europe, a fair estimate can be made of the appearance of the Scottish predecessor of the domesticated races. It was a small animal, as horses go, from about 12·2 to 14 hands, that is 50 to 56 inches, high at the shoulder, with long slender limbs. Its head was small, its face fine and narrow, with a straight profile. Of the colour of its coat and the nature of its mane and tail, there is naturally no direct evidence from the prehistoric remains, but Professor Ewart is of opinion that it was closely related to a modern type which he has designated the Celtic pony (*Equus agilis celticus*) and which attains its most characteristic development in the small yellowish dun, long and strong maned, short docked, shaggy ponies of northern Iceland (see Fig. 15). These lack any trace of "chestnuts" or callosities on the inner surface of the hind legs, of "ergots" or fetlock callosities on all legs, and possess highly specialized tails with a thick bunchy tail-lock which develops in winter and shields the thighs from rain and storm.

Did this slender-limbed graceful horse which predominated on the plains of the Glacial Age—the "plateau" type of Ewart (*Equus agilis*)—survive in Scotland at the advent of man? If man did find it when he first settled on the coasts of North Britain, did he succeed in taming it and turning it to his own uses, or did the Neolithic wanderers bring with them from the continent, horses already domesticated, which were to form the foundation of our modern breeds?

To these questions it is impossible to give dogmatic answers, mainly owing to the scant examination given to the treasures of many an early Scottish excavation yielding animal remains. But the presumptive evidence is strong that the "plateau" horse was domesticated in this or another country, and formed part of the domestic stock of the people of the New Stone Age, for small slender-limbed horses have been characteristic of Scotland from time immemorial. That the horse still existed in Scotland just prior to or contemporaneous with the arrival of the Neolithic peoples, is hinted at by the discovery of bones, of what particular race of horse has not been determined, in the Fifty-Foot Raised Beach at Shewalton near Irvine in Ayrshire. This Fifty-Foot Beach, it will be remembered, is that which has yielded the earliest evidence of man's appearance in Scotland, and though it contains, in its development on the east coast, skeletons of Whales and implements manufactured from Deer antlers and used by man, it has hitherto afforded no indication of the presence of domestic animals.

THE HORSE IN PREHISTORIC AND EARLY HISTORIC SCOTLAND

Neolithic man in Scotland, however, was acquainted with the horse, as is clearly shown by the occurrence of characteristic bones in the shell-mound settlement at Ardrossan, and in the late Neolithic horned or chambered cairns of Caithness at Ormiegill near Ulbster, Garrywhin near Clythe and Hill of Bruan, and in the chambered cairn at Loch Stennis in Orkney.

Fig. 15. Celtic Pony (in rough winter coat)—probably bears close resemblance to the native pony of Scotland. $\frac{1}{15}$ nat. size.

In later ages, the horse became more and more familiar to the Scottish races, as is revealed by the frequency with which its remains appear in contemporaneous deposits. In Scottish crannogs or lake-dwellings it is rarely found, though where its remains occur, as at Lochlee near Tarbolton in Ayrshire, they indicate a "small variety." In the few clearly defined deposits of the Bronze and Iron Ages in Scotland, the horse is represented, but by no means universally nor so frequently as oxen or sheep. Passages have often been quoted from Julius Caesar and other Roman historians, to show that on the arrival of the Romans, the Britons possessed active horses of great speed and dexterity. Furthermore, that they had domesticated them in considerable numbers is shown by the statement that "Cassivelaunus, when he dismissed his Army, retained four thousand of his Chariots to harass the Romans when they attempted to forage." The men of Caledonia were equally well supplied with small active horses, for Tacitus in his *Life of Agricola* recounts that at the Battle of Mons Graupius, somewhere to the north of the Tay,

the intermediate space between both armies was filled with the charioteers and cavalry of the Britons, rushing to and fro in wild career, and traversing the plain with noise and tumult.

That the Romans themselves added races of their own to the ponies already established in Caledonia, is clearly demonstrated by Mr James Curle's and Professor Cossar Ewart's researches at the Roman station of Newstead near Melrose; for

in addition to well-bred ponies under 13 hands at the withers, the auxiliaries who held the Border fort during the first century had 14 hands horses as fine in head and limbs as modern high-caste Arabs. In all probability, the better bred horses, measuring about 14 hands, belonged to the cavalry and mounted men (about one in four) attached to the infantry regiments, while the coarse-headed animals of more powerful build (measuring nearly 15 hands) were as a rule used for transport.

In the centuries following the Roman occupation, the native settlements all tell of the presence of a small domesticated race of horses, the bones of which, with those of other domestic animals, occur in Romano-British settlements, as at Borness Cave in Kirkcudbrightshire; in underground "Picts'" or "Eird" houses, at Kildrummy in Aberdeenshire and even in the outlying islands as at Nisibost in Harris; in hill-forts, as at Dunsinane in Perthshire; occasionally in the "Pictish Towers" or brochs, as at Keiss in Caithness, and Burray and Sandwick in Orkney; and in shell-mounds and kitchen-middens of various periods down to the sixteenth century pre-Reformation accumulations of the monastery and nunnery of Iona.

Examination of the remains and records of many refuse-heaps of man shows clearly that the horse never became so common or so widely distributed as oxen or sheep, and that amongst the tribes of cave-dwellers who frequented caverns by the sea, and subsisted mainly upon the shell-fish of the beach, it was almost unknown. In one of the few caves in which it has been found—that at Borness in Kirkcudbrightshire—three horse bones were identified as against 630 bones of sheep and 1112 of cattle. In interpreting such evidence as the deposits yield, however, it is well to keep in mind that the bone heaps as a rule represent the remains of human food, and that although in early times horse-flesh was commonly used as food, in later times it was proscribed by Christians and ceased to be used about the year 1000 of our era.

INFLUENCES WHICH HAVE MODIFIED THE NATIVE RACE

It is unfortunate that little is known of the finer racial characteristics which the relics of the bone deposits might have yielded to expert scrutiny. As regards the majority of the prehistoric and early historic periods the most we can say is that the horses were invariably small in size, and were yet strong, hardy and active. Much can be learned, however, by reading backwards, as it were, from the representatives of modern breeds, and from their present-day developments tracing the influences which have played upon them.

In the endeavour to eliminate the results of the later influence of man upon animals by discovering the modern races which most nearly approach in character to the original indigenous forms, it is a safe rule to begin the search in those areas most remote from the centres of cultivation and civilization. In discussing the sheep and oxen of Scotland we have seen that the primitive breeds, like the earlier races of man himself have been gradually ousted from the highly cultivated lowlands by newcomers, and have been driven to the strongholds of the mountains and outlying islands. This rule which governs the distribution of the primitive Highland cattle and the old Hebridean, Orkney and Shetland sheep, holds also for horses—the horses which survive at the present day in the Hebrides and in Shetland offer the nearest approaches to that native race or races which peopled Scotland in days before the arrival of man. To these island races we appeal in order to discover the minimum effects of man's influence on modern Scottish breeds.

THE PONIES OF THE HEBRIDES

A characteristic race of small horses, now approaching nearer and nearer to extinction, is still to be found in the outer islands on the west coast. As its characters show, the race is one of great antiquity; and since the isolation of the islands is sufficient almost to eliminate chances of any considerable cross-breeding, we can safely assume that the bones of a very small horse found in an ancient underground "Picts'" or "Eird" house at Nisibost in Harris are those of the earliest known representative of the Hebridean race. Perhaps from the same ancient line were descended the horses whose remains were found by Mr Symington Grieve in the upper strata of the Crystal Spring Cavern in Colonsay, where they had evidently been used as food, for the bones of young animals were more plentiful than those of adults.

Early historic references to the horses of the outer isles are few in number. In 1549, Monro noted that amongst the western isles there

> layes ane little ile, half ane myle lang, callit be the Eriche, Ellannaneache, that is in English, the Horse isle, guid for horse and uther store, pertaining to the Bishope of the iles.

Martin Martin, in 1703, said of Lewis that

> the Horses are considerably less here than in the opposite continent, yet they plow and harrow as well as bigger Horses, tho' in the Spring time they have nothing to feed upon but sea-ware.

Of the St Kilda individuals the same writer says merely that they are "of a lower size than in the adjacent Isles," a diminutiveness no doubt due to the severe conditions of their life on this isolated and storm-swept island. In 1876 the horses of St Kilda had become extinct, though at that time middle-aged men could still recall their former presence. Even in 1764, when Macaulay described them, they were verging on extinction for, said he, "all the horses of St Kilda are only ten, including foals and colts, they are of a very diminutive size, but are extremely well cast, full of fire and very hardy."

The Hebridean ponies are commonly of a brownish-black or a foxey-red colour, though there are occasional duns and greys, and one race, that of Uist, like the ponies of the Faroe Islands, is distinguished by a striking silvery mane and tail

(Fig. 16). They bear a coat of thick rough hair, are of small size, only 50 to 54 inches (12·2 to 13·2 hands) high at the shoulder, and in the old, almost extinct race of Barra, have fine limbs, and rather large heads with straight profile and flat nose. In this Hebridean race Professor Ewart considers that we have "a remnant of a very old and once widely distributed variety, the origin of which is never likely to be revealed."

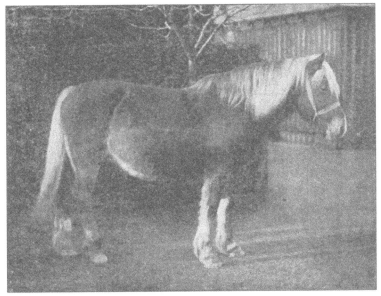

Fig. 16. Hebridean pony—Uist race—a primitive breed approaching extinction. $\frac{1}{20}$ nat. size.

Yet the Hebridean race, primitive as are many of its features, has not escaped the later influence of man, for Mr J. H. Munro Mackenzie, a close student of Highland ponies, considers that some of its members on Mull, Tiree, Skye and Uist, and parts of the western mainland,

> show a very strong cross of Arab blood, which is believed to have come by Spanish Armada horses, or Arab chargers brought home by Highland officers. The ponies have beautiful heads and good shoulders: are good all over; famous for staying through long journeys under heavy weights and on poor keep.

Individuals of dun or grey colour generally show most

trace of Arab blood, and their height at the shoulder ranges from 54 to 58 inches (13·2 to 14·2 hands); so the results of the cross are still apparent in size as well as in build.

SHETLAND PONIES

No race of horses stands higher in the affection of the Scottish people than the "Shelties" or "Shulties" of the Shetlands (Fig. 17). Their neatness and grace no less than their gentleness and docility have contributed to this popularity, and to these they add the distinction of long lineage. Strangely enough, though remains of horses are common in prehistoric and early historic dwellings in Orkney, relics of the early horses of Shetland have come to light in only one excavation—in the kitchen-midden of the "Pictish Tower" or broch at Jarlshof, Sumburgh, explored in 1911; but these Professor Ewart regards as sufficient to indicate a pony of ancient type not more than twelve hands high—the earliest known representative of the "sheltie." The characteristics of the early race are indicated in a figure of a small horse sculptured in relief on the early Christian monument of Bressay, which probably belonged to a period between the ninth and eleventh centuries. The evidence of these discoveries makes it tolerably certain that Shetland had its native race of ponies before the invasions of the Norsemen in the ninth and succeeding centuries. However closely the prehistoric pony of the Shetland Isles may have approached in character the slender-limbed Celtic pony of the Ice Age, it is clear that the incursions and settlement of the Scandinavians must have influenced its later history. For with them they brought their own Norse ponies, small, hardy, stout-limbed, yellow-dun animals, which, reared under similar stern conditions of food and climate, would be able to survive in islands which have proved fatal to most other introduced breeds. So it came to be that before written history took up the tale of the shelties, the original race of the "plateau" type already contained an admixture of Norse blood of the heavier "forest" type.

The earliest description with which I am acquainted is that of Jerome Cardan, an Italian doctor, who spent a few weeks in Scotland in the summer of 1552, and was struck by the very characters which Shetland ponies still preeminently exhibit—diminutiveness and docility. The islands of Shetland, he writes, possess " equi exiles et asinis quasi similer tam patentia quem magnitudine." One other quotation may be made, because, although written more than 200 years ago, it exactly describes the ponies of the present day, and shows incidentally how persistently these natives of the stern kingdom of Shetland have retained the characters bequeathed them by their " plateau " and "forest" ancestors.

Fig. 17. Shetland Pony. ("Sovereign," Champion, Highland Show, 1914.) $\frac{1}{18}$ nat. size.

In his *Brief Description of Orkney, Zetland, Pightland-Firth and Caithness*, published in 1701, the Rev. John Brand writes in his breathless style of the people of Shetland:

> They have a sort of little Horses called Shelties, then which no other are to be had, if not brought thither, from other places, they are of a less Size than the Orkney Horses, for some will be but 9 others 10 Nevis or Handbreadths high, and they will be thought big Horses there if eleven, and although so small yet are they full of vigour and life, and some not so high as others often prove to be the strongest, yea there are some, whom an able man can lift up in his arms, yet will they carry him and a Woman behind him 8 miles forward and as many back; Summer or Winter they

never come into an House but run upon the Mountains in some places in flocks, and if at any time in Winter the storm be so great, that they are straitened for food, they will come down from the Hills when the Ebb is in the sea, and eat the Sea-Ware (as likewise do the Sheep)....They will live till a considerable Age as 26, 28, or 30 Years, and they will be good riding Horses in 24 especially they'll be the more vigorous and live the longer if they be 4 years old before they be put to Work. These of a black Colour are Judged to be the most durable, and the Pyeds often prove not so good; they have been more numerous then now they are, the best of them are to be had in *Sanston* and *Eston* also they are good in *Waes* and *Yell*, these of the least size are in the Northern Isles of *Yell* and *Unst*.

The Coldness of the Air, the Barrenness of the Mountains on which they feed and their hard usage may occasion them to keep so little, for if bigger Horses be brought into the Countrey, their kind within a little time will degenerate; And indeed in the present case, we may see the Wisdome of Providence, for their way being deep and Mossie in Many places, these lighter Horses come through when the greater and heavier would sink down: and they leap over ditches very nimbly, yea up and down rugged Mossy braes or hillocks with heavy riders upon them, which I could not look upon but with Admiration, yea I have seen them climb up braes upon their knees, when otherwise they could not get the height overcome, so that our Horses would be but little if at all serviceable there.

This Lilliputian breed, which still retains the wild habit of moving in droves, and of which the smallest recorded specimen was only 26 inches high at the shoulder, while the average is only 40 inches, is the nearest modern representative of the small race which in prehistoric days and for long ages inhabited the mainland of Scotland. Yet the effects of man's interference are evident in the diversity of its individuals, for while some are still slender-limbed riding ponies, others have assumed the thick-set characters of diminutive draught horses.

Examination of the horses of the mainland will show how much further the influence of man has gone in modifying the characters of the original race.

THE HORSES OF THE SCOTTISH MAINLAND

In the opening portion of this account, the nature and distribution of the horses which inhabited Scotland in prehistoric and early historic days has already been indicated; the present section shall therefore be devoted mainly to an account of the methods and results of man's influence on the primitive types. It may be stated in a sentence that the modes of influence are the ordinary means of the breeder—

the careful selection of the best of his stock, and the intermixture of new blood derived from races possessing qualities more desirable for the immediate purpose in view; and that the general result of man's influence has been an increase in size, and in specialization for particular forms of labour. It would simplify the intricacies of the history of the mainland horses if the reader were to keep in mind that the foundation upon which subsequent developments were built was that race of small, active, strong and hardy horses which the Romans found yoked to the chariots of the Caledonians at the battle of Mons Graupius, in the sixth year of Agricola's conquests in North Britain—the 85th year of the Christian era.

We have already seen that the researches at the Roman station at Newstead, conducted by Mr James Curle, led to the discovery that the Roman legionaries brought with them horses larger and heavier in build than the native Scottish race. There is little likelihood, however, that this importation of what was probably the "bad and ugly" native German breed known to Caesar, would exert a deep influence upon Scottish horses, for the Roman occupation was limited in space and in time.

Norse Influence

It was a different matter with the invasions of the Scandinavian peoples in the ninth and succeeding centuries, for they formed permanent colonies and amalgamated with the native races in a way which the Romans never did. One result of this close contact is apparent in the influence the Norse peoples exercised upon the Scottish breed of ponies through their own characteristic horses. The Norse horses (*Equus caballus typicus*) were greater in stature than the Scottish native race, they were heavier in build, were of a yellow-dun colour and were noted for their intelligence and tractability. Their heads were short and broad with prominent eyes, their neck and shoulders heavy, their quarters rounded, their tails low-set, their limbs short, sturdy and large-jointed, and their hoofs broad. "In neck and shoulders, trunk and limbs," says Professor Ewart, "the Norse variety may be said to be intermediate between a true pony and a small cart-horse of the Suffolk type." To the influence of this "dœlehest" or valley horse, as Mr Leonhard Stejneger calls it, from the interior of Norway, we probably owe the characteristic horses of the Scottish uplands—the Highland "Garrons" (Fig. 18), which average between 14 and 15 hands high, and are, according to Professor Robert Wallace,

> unequalled hill ponies for staying power at slow speed, sure-footedness, and for carrying heavy loads of deer or smaller game on rough hillsides and mountainous places, and bearing the sportsman to the shooting ground.

"Wild Horses" in Historic Times

There are several references in early Scottish literature to the presence of "wild horses" which existed in the Highlands even to the early part of the seventeenth century. Boece (1527) says of them (as translated by Bellenden) "Beside Lochnes, quhilk is xxiv milis of lenth, and xii of breid, ar mony wild hors," and in another place:

> In all the boundis of Scotland, except quhair continewall habitation of peple makis impediment thairto, is gret plente of...wild hors...Thir wild hors ar not tane but [except by] crafty slicht; for, in time of winter, the landwart peple puttis certane tame cursouris and maris amang thir wild hors; and, be thair commixtioun and frequent cumpanie makis thaim so tame, that thay may be handileit.

The Forest of Birse in Aberdeenshire, also sheltered a herd of wild horses in 1507 as the *Records of the Sheriffdom of Aberdeen* (vol. I, pp. 106–7) show, and it is known that there inhabited the hills of Sutherland till after 1545, a herd of "wild meris, staigs and folis" which was established as belonging in right to Sutherland of Duffus, when the Bishop of Moray laid claim to the herd. Again as late as 1618, Taylor the Water Poet alleges that he saw "wild horses," along with "deere, wolves, and such like," in the "Brea of Marr."

Boece indicates that these "wild horses" were something different from the ordinary "cursouris" of his time, and it can be no rash step to assume that they were either native ponies, or crosses between these and the Norse horses, which in the unenclosed state of Scotland had, like representatives of the Park or Fallow Deer (see p. 285), run free and become established in the wilder districts.

May it not also be that the many Highland traditions of mythical yellow horses with bristling manes and flowing tails, and a touch of Satanic elusiveness, are simply vague memories harking back to the last survivors of the yellow-dun native and Norse horses of the wilds? Certainly in some cases the traditions are localized in areas, such as the Forest of Birse, or Loch Lundie in Sutherlandshire, where wild ponies are known to have roamed at will in centuries not long gone.

Fig. 18. Highland Garron—probably moulded by Norse influence. "Braulin," Champion, Highland Show, 1908. $\frac{1}{28}$ nat. size.

BREEDING AND INTERBREEDING

While a natural admixture of blood was taking place between the horses of the Norse invaders and those of the invaded land, primitive attempts were also being made to improve the races at hand, by a rough and ready method of selective breeding. This early system offers one easy explanation of the origin of "wild" herds. The oldest reference to Scottish horse-breeding of which I am aware, is contained in the Charter of Kelso whereby, before 1200 A.D., Gilbert de Imfraville granted the monks of Kelso a tenth of the foals bred in his forest and studs. The horses were bred by being let loose in the forest, where the foals ran wild for three years, until they were broken in. The best horses were selected and kept apart in the parks about the Baron's castle. Can there be any doubt that this system of forest-run horses—the *silvestres equi* possessed by the Kings as well as by their Barons—gave rise to the "wild horses" of later centuries?

In the thirteenth and fourteenth centuries, the selective breeding of horses in Scotland became more general, but with one main end in view, the building up of a race fitted for carrying a horseman and the accoutrements and impediments of war. There is yet no trace of the development of a heavy agricultural breed, for still "the slow team of steers reluctant" pressed the yoke and the horse was reserved for the lighter toils of carrying the fruits of the harvest to mill or market. There was nevertheless considerable store of horses in the country, for every burgess had to keep in stable, ready for the public service, a horse worth 20 shillings; and the export to England so increased that in 1396, King David Bruce felt compelled to put a heavy tax—one sixth of the value—upon each horse exported, lest the country should become impoverished of an essential adjunct of war. Similar and more stern restrictions on export were passed periodically by Parliament during the succeeding centuries.

Already in the fourteenth century, the persistent selection for war-horses had resulted in the differentiation of two breeds, for at the battle of Halidon Hill in the reign of King Robert Bruce, Froissart, who was present and describes the whole Scottish army of 3000 men as mounted on horseback, mentions that the knights and squires rode coursers but the peasants small horses. Yet, more than a hundred years later, Aeneas Silvius, the Pope's Nuncio, describes our horses as mostly small-sized pacers, that were never dressed by brush nor comb.

In the beginning of the fifteenth century, there was inaugurated a new policy—of improving the old breed by the infusion of new blood of good quality from abroad. From this time on, two distinct and divergent tendencies are observable in the policies of the Kings: one towards improving the quality of horses as regards speed, the other towards increasing their size. Probably with the former end in view, James II (1437–1460) brought mares from Hungary, as Bellenden puts it, to "mend the breed," while his successor, James III (1460–1488), aiming at size, imported the "Great" or "War" Horses of England, which Sir Walter Gilbey regarded as the forerunners of the modern Shire. James IV (1488–1513), favouring the swift and lightly built horses of the hunt and race-course, brought to Scotland the best of the horses of Spain and France, as well as heavier horses from Poland; whereas James V (1513–1542) "seeing the Realm standing in much Peace and Tranquillity," as Lindsay of Pittscottie relates:

rejoiced at the same thinking daily that all Things should increase more and more: To that Effect gart send to *Denmark*, and bring more great Horse and Mares, and put them in Parks, that, of their offspring, might be gotten to sustain Wars in Time of Need.

James also applied to Gustavus of Sweden for heavy horses, and in the "Black Acts" of 1535, enjoined that "in order to raise the size of the native breed in Scotland all manner of persons should plenish their studs with stud mares and great stallions."

So, on the whim of a ruler, the pendulum swings from, one might almost say, quality to quantity and back again: at one time Henry VIII of England sends Scotland "small but well-proportioned" Spanish Jennets and African Barbs; at another time the *Magnus Equus* or *Dextrarius*—the English Great Horse or War Horse—becomes the type to mould the race.

So the work of selecting and interbreeding and improving has gone on from the sixteenth century, gradually raising the quality as well as the stature of the heavy breeds till it has reached its climax in the evolution of the Clydesdale.

A MODERN SCOTTISH BREED—THE CLYDESDALE

The Clydesdale of to-day stands as an example of the power of man's influence over the characteristics of the horse. Compare it with the prehistoric ponies of Scotland which probably formed the basis of its evolution: they stood little over 48 inches high at the shoulder, the Clydesdale stands 64 inches and sometimes exceeds 68; their build was fine and light, the Clydesdale's is heavy and expresses the concentration of power; their strength in dragging the light chariots of the Caledonian warriors was accounted marvellous for their size, the Clydesdale can haul the heavily laden waggons of the farmer with an easy vigour; their nature was wild, the temper of the Clydesdale should be docile and mild.

From a minute examination of their external characteristics, as well as of the less patent features of the skeleton, modern investigators have come to the conclusion that the

Fig. 19. Clydesdale mare—a modern product of selective breeding. "Dunure Chosen," Champion, Highland Show, 1914. $\frac{1}{32}$ nat. size.

heavy breeds have inherited their qualities from several widely different original stocks; from the original "plateau" or moorland type, which inhabited Britain in interglacial times—short of stature and fine of limb, with small fine head, narrow and almost straight in profile; from a "forest" type, large and coarsely built, with short broad face—a prehistoric dweller in the forest-land of Central Europe; and from a long-faced, long-limbed, "steppe" type, which in preglacial times lived in the upland valleys of the East.

However this may be, the influence of man is singularly

manifest in the history of Scottish horse-breeding, selecting here and rejecting there, bringing now one race, then another from across the seas to "mend the breed," till inch after inch up to a cubit is added to the stature, and the first clear step made by the incoming of the Norse pony is capped by the importation of the great black Flemish stallion of Lochlyoch which about 1750 founded the race of Clydesdales.

II. 4

SOME LESSER DOMESTICATED ANIMALS

Of the lesser animals which have shared man's protection in return for services demanded of them, some have fallen under his influence to a much greater extent than others, but there are few which do not exhibit in some character or other, the effect of his control. It is possible to discuss here only those which have been brought into closest touch with the civilization of Scotland.

THE DOG

> Mastiff, greyhound, mongrel grim,
> Hound or spaniel, brach or lym,
> Or bobtail tyke or trundle-tail.

Whatever their characters be, dogs one and all owe their parentage to wild species of the races of Wolves and Jackals. Their origin is lost in the haze of ages; some 3000 to 4000 years B.C., the Egyptians had their distinct breeds, and the Lake-dwellers of the Swiss valleys possessed a domesticated form of the Indian wolf. The chances are, therefore, that the Neolithic immigrants to Scotland brought with them dogs already domesticated; but the difficulty of distinguishing between the bones of the early wolf-like dog and the wolf itself, tends to obscure such information as might be gleaned from the prehistoric deposits of Scotland. I have already shown that the first comers to Scotland possessed no domestic animals, but bones attributed to early dogs have been found in river gravel near Currie in Midlothian, and buried in peat on Morbhaich Moor near Tain in Ross-shire. It is highly probable, moreover, that the dog was familiar to the people of the late Neolithic period in Scotland—some 2000 years before our era—for in 1870 Dr Joseph Anderson recognized the bones of dogs, along with those of other domesticated animals, in the horned cairns of Ormiegill and Garrywhin in Caithness; and in 1885, a skilled anatomist, Dr J. G. Garson, of the Royal College of Surgeons of England, identified remains of the dog from the chambered cairn near Loch Stennis in Orkney.

In later days the dog became more common. It occurs in the Roman deposits at Newstead; the early Christians carved its figure upon their symbol stones; in a few underground "Picts'" or "Eird" houses its remains have been found—jaws of dogs at Kildrummy in Aberdeenshire, and bones at Nisibost in Harris; shell-mounds and kitchen-middens of various ages contain its bones, at Kirkoswald in Ayrshire and near Seacliff in East Lothian, where several dogs "of a large size" were represented, even to the sixteenth century refuse-heap in the cloisters of the Nunnery at Iona, which revealed the presence of a small dog, probably a pet of the inmates. The caves and rock shelters tell of the presence of the dog, not only by actual remains as at St Cyrus, at Oban, at St Ninian's Cave in Wigtownshire, and in the rock-shelter at West Kilbride, where Prof. Clelland identified bones as belonging to an individual as large as a shepherd's dog of to-day, but also, as at the Wemyss Caves of Fife, by the abundant traces of tooth-marks and chewed ends of bones of food animals, which leave no doubt that a carnivore was constantly present on the rubbish heap of the settlement. In the brochs, covering a period almost to the ninth century, the dog is by no means common, though it has been found in Sutherland (Cinn Trölla), Caithness (Kettleburn and Keiss) and Orkney, but an interesting feature of the broch remains revealed both at Keiss and Kettleburn, is that two breeds of dogs were in the possession of the inhabitants, some bones indicating "a large species, larger than a pointer, others being of smaller dogs."

Even if it be admitted that the dog reached Scotland a domesticated animal, it seems nevertheless true that the Scottish peoples paid much attention to and exercised considerable influence on its development, a fact to be correlated with another, that, as Boece tells, "the Scottis...set their ingine [ingenuity] to precell [excel] all uthir pepill in the craft of hunting." Laws were made regulating the breeding of hunting dogs, and watch-dogs were held in such esteem that an Act of David I (1124–1153 A.D.) demanded that a man who slew another man's house-dog should watch on his midden a year and a day, and should be responsible for any loss during that time.

The result of such attention was that at an early date, according to the ancient writers, Scotland came to possess dogs peculiarly her own. "Bloodhounds were known in England at least as early as 1570," wrote Mr R. Lydekker, but many years earlier the "sleuth" was a familiar Scottish dog. It is impossible to deal here with the many influences of selection and cross-breeding brought to bear in the creation of new races. It will be enough to indicate the main results of the earlier Scottish influence as recounted by Hector Boece in 1527.

In Scotland ar doggis of mervellus nature: for abone the commoun nature and conditioun of doggis, quhilkis ar sene in all partis, ar thre maner of doggis in Scotland, quhilk ar sene in na uthir parts of the warld. The first is ane hound, baith wicht, hardy and swift. Thir houndis ar nocht allanerlie feirs and cruell on all wild beistis, bot on thevis and ennimes to thair maister, on the same maner. The secound kind is ane rache, that sekis thair pray, baith of fowlis, beistis and fische, be sent and smell of thair neis. The thrid kind is na mair [larger] than ony rache; reid hewit, or ellis blak, with small spraingis of spottis: and ar callit be the peple sleuthoundis. Thir doggis hes sa mervellus wit, that thay serche thevis and followis on thaim allanerlie be sent of the guddis that ar tane away; and nocht allanerlie findis the theif, bot invadis him with gret crueltie, and thoucht the thevis oftimes cors the watter, quhair thay pas, to caus the hound to tine the sent of thaim and the guddis, yit he serchis heir and thair with sic deligence, that, be his fut, he findis baith the trace of the theif and the guddis. The mervellus nature of thir houndis wil have na faith with uncouth peple; howbeit the samin ar richt frequent and rife on the bordouris of Ingland and Scotland. Attour it is statute, be the lawis of the Bordouris, he that denyis entres to the sleuthound, in time of chace and serching of guddis, sal be haldin participant with the crime and thift committit.

Bishop Leslie (1578) adds further details regarding the first—clearly an early deerhound—that it is "gretter than ane tuelfmoneth [twelvemonth] alde calfe; and this sorte commonlie huntis the gretter beistes, as ye sall sie, athir the harte or the wolfe." Leslie also describes several other Scottish races of "senting dogs" and "slwthhundes," and concludes, "Of the varietie of Messen dogs, w^t quhilkes

SOME LESSER DOMESTICATED ANIMALS 89

gentle women vses to recreate thamselfes, althoch be mony and infinite, I will nocht heir make mentione."

These quotations are sufficient to show that Scotland took its share in that marvellous juggling with the endowments of nature which, from two or three wild species, of similar shape and character, equally ferocious, bloodthirsty and deceitful, has created the docility, gentleness and faithfulness of the deerhound and the "toy," dogs that bark and silent dogs, dogs that rely upon their sight and dogs that, as Boece quaintly says, rely upon the "sent and smell of their neis [nose]," dogs whose ears are their chief guide, dogs that point the game and dogs that retrieve it, dogs that depend on length of limb and dogs that depend on strength of jaw, dogs for the fray and dogs for the home.

THE WILD BOAR TURNED SWINE

" Swine, though never esteemed for their Beauty, in their Nature rather disagreeable," nevertheless possess a special interest, since there can be little doubt that they belong to the limited number of domesticated animals which were led from the forest to the fold within the bounds of Scotland. And Scotland can show, better than most countries, the steps of that sorry progress.

The Wild Boar (*Sus scrofa*) (Fig. 20, p. 91), common in the great forest areas of Europe, especially in the woods of the Vosges, as well as in Asia and North Africa, was at one time also a common denizen of the forests of Britain. Its remains have been found in peat at Balgone in Haddingtonshire, as well as in the marl mosses; and numerous tusks, found amongst the food refuse of the prehistoric peoples of Scotland, tell of the far-off times when

> The sad barbarian, roving, mixed
> With beasts of prey, or for his acorn meal
> Fought the fierce tusky boar.

At each period of early Scottish development, it is represented. The frequent occurrence of its bones in the Neolithic chambered and horned cairns of Caithness and Orkney; in cave deposits, as at Colonsay, where its gradual extinction can be traced; in kitchen-middens and shell-mounds; in crannogs or lake-dwellings, as at Lochlee in Ayrshire, Black Loch in Wigtownshire, and the Loch of Forfar; in the Roman station at Newstead; and in the brochs of Orkney and Caithness—these show that it was a common article of food even to about the ninth century of our era. There are many references in tradition and in place-names to the presence of the wild boar in Scotland: "Swinton" and the neighbouring "Swinwood" in Berwickshire, clearly hint at a time when the district was overrun by its herds. The town of St Andrews stands upon the "Cape of Boars"—Muckross; the village of Boarhills lies a few miles along the coast to the south-east, and in the Boar's Chase in the neighbourhood, a district eight miles wide by about four in breadth, the Kings of Scotland made sport of the ancestors of the domestic pig. History tells us too, how, in the twelfth century, King Alexander I "dotat [presented] the *Bairrink*, because ane bair [Scots for *boar*] that did gret injurie to the pepyle was slain in the said field," or as Wyntoun puts the story of Bellenden:

> That land thai oysyd all
> The *Barys rayk* all tyme to call,
> Wes gyvyn on that condytyowne
> To found there a relygyowne.

Not many years passed before the persecution of the boar by royal and less noble hunters led to its gradual disappearance. The accounts of the Sheriff of Forfar for the year 1263 bear witness to a charge made in that year, for 4½ chalders of corn for the support of wild boars, "*porci silvestres*," which are grouped with the King's horses and dogs—an indication that life was less easy than formerly for the wild boars in the forests of Strathmore.

From this fine creature of the primeval forest,

> That cruell boare, whose tusks turned up whole fields of graine
> And wrooting, raisèd hills upon the levell plaine,

man in his wisdom has bred the common swine. When the transformation was wrought in Scotland I have been unable to determine, for there is little to distinguish the odd bones of the early domestic pig from those of its wild ancestor, but the process must have been a gradual one. I believe, however, that this very indistinctness of the early remains, and the ease with which the wild boar can be domesticated—for if wild pigs be taken young from their mothers they become

SOME LESSER DOMESTICATED ANIMALS 91

Fig. 20. Wild Boar (left) and primitive Scottish type of Pig (right)—the latter an early stage in the domestication of the former.

nearly as domesticated as the hereditary domestics—together with the presence of a primitive breed of pigs in the Western Highlands and Islands, give good grounds for supposing that the prehistoric peoples of Scotland domesticated their own pigs.

Domesticated Pigs in Scotland

It is clear that at whatever early age the domestication of the pig took place in Scotland, for many centuries its herds were allowed to roam at large, picking up what food the forest yielded in the autumn, and encouraged to glean probably not too precarious a living by performing the work of scavengers in the towns and villages. Their enthusiasm in this latter labour made them in the end such a nuisance that the Assize of Haddington felt compelled, on the 12th of July 1530, to check their activities, for under that date

> The Sys [Assize] ordains that the hangman sall escheit to hymself all swyne doggs and catts at he fyndis one [on] the gait [street] fra this nycht furcht [forward].

And this statute not having the desired effect of clearing off the roving herds, the Council further ordained: "Penult of Octr. 1543—Item all muk to be put off the Gait and all swyne to be put off the Towne." The estimation in which the pig was held, and its domesticated quality about this time, may be judged from the then Food Controller's price, fixed in 1551, when young swine ("gryse") were valued at 18*d*. each, less than the price of a couple of capons (1*s*. each) and little more than the price of a goose (1*s*. 4*d*.).

The ancient "Leges Forestarum," generally attributed to the reign of William the Lion (1165–1214 A.D.), but perhaps of somewhat later date, give evidence of the roving nature of the early herds of swine; for while they provided that proclamation should be made in the parish kirk that swine as a rule were prohibited from entering the forest, they summoned burghers and husbandmen to bring their herds in the autumn to where oaks abounded, so that the king might benefit from the *pannagium*, a payment in kind due for pasturage on the feasts of acorns—the Crown claiming the best of ten swine and the forester a hog.

What became of these wild, roving, worthless Scottish

pigs? In the Highlands and Islands of Scotland there existed till after the middle of the nineteenth century a race of pigs of primitive character (Fig. 20, p. 91). They are generally spoken of as domesticated pigs which had run wild, and under natural conditions had reverted towards the type of the wild boar. But there is no definite evidence that they were ever anything better than they became, and I prefer to regard them as survivors of the early stages of domestication, which, like the primitive breeds of Sheep, Oxen and Horses, had been driven by the advances of civilization to the refuges of the outlying parts of our island. In many ways they resembled the wild hogs from which they sprang, and in this their interest lies, for they are a stepping-stone between the Wild Boar and the Common Pig.

Look at their characteristics as Prof. D. Low found them in 1845. Like their wild ancestors they had erect ears, an arched back with coarse bristles along neck and spine, and they were of dusky brown colour. They retained many wild habits, foraged for themselves on heath-clad hills or moors, grubbing up the roots of plants with their strong snouts, devouring, when they could find them, eggs and young of hill-birds, such as Plovers and Grouse, and even defenceless new-born lambs. Like the wild boars of north-eastern France to-day, they were the plague of the cultivated lands, now raiding potato-fields, now destroying corn crops. Lastly, they resembled wild pigs in their general build, having the small bodies and long legs of creatures whose food and safety depend upon their activity.

Similar primitive pigs existed also up to the latter half of last century in Shetland and in Orkney, where the small huts which used to afford them shelter on the moors may still be seen. It is interesting to recall that the hair of the Orkney pig was so long that the men of Hoy, as Rev. Geo. Low tells (1774), preferred it for making the ropes on which they risked their lives collecting birds' eggs on the cliffs; for the elasticity of the swines' hair rope hindered it from cutting on sharp rock edges, though this advantage was somewhat counterbalanced by a proneness on the part of the rope to untwist, to the destruction of its human burden.

Yet domestication had its influence even on this primitive breed: the simple pigs of the Highlands and Islands fed by day instead of keeping to the hours of twilight and night preferred by their wild ancestors; they were more sociable, too, for the male kept with the herd instead of retiring periodically to his solitary lair, as the wild boar does; and though the Highland pigs were wilder and more active than Lowland breeds, they never had the ferocity, the strength, nor the swiftness of the wild animals, which can keep pace for a time with a horse at speed.

Nevertheless it is a far cry from the primitive Highland race to the modern improved breeds of to-day (Fig. 21).

Fig. 21. Finished product of domestication—Large White Boar, "King of the Groves," Champion, Highland Show, 1912.

By reducing his domesticated pigs' activity, by supplying food far more abundant than the wilds yield, by selecting such as fell in with his views of what a pig ought to be, and by cross-breeding the descendants of the European Wild Hog with the Chinese breed of the Wild Indian Pig, *Sus indica*, man has profoundly modified their appearance and habits. Their bodies have increased in length as well as in depth, their legs have shortened and become more spread to bear the heavy bulk; their ears, no longer given to catching the first whisper of danger, have flapped uselessly across the ear-opening; their noses, no longer required for strenuous grubbing, have sunk backwards upon their faces as the face muscles have degenerated; the skull has changed profoundly; the tusks of the male, no longer tested in savage duels, have dwindled; and the food canal, exercised by overmuch food, has lengthened. Even habits have changed: the domestic pig feeds by day, the male no longer seeks solace in seclusion, the female bears more young at a birth and bears them more frequently, the descendants of the hardliving boar of the forest have become the types of gluttony.

THE PIGEON

................The dow
Heich in the lift full glaide he gan behald,
And with hir wingis sorand mony fald.
GAWAINE DOUGLAS.

Amongst domesticated birds none other has taken so distinctive a place in Scottish life as the pigeon. To Charles Darwin we owe a clear demonstration of the fact that the innumerable and extraordinary varieties of modern domesticated pigeons owe their origin one and all to the Rock Dove (*Columba livia*) (Fig. 22, p. 96). With this great variety of form due to man's influence I do not propose to deal, the curious reader will find it described and discussed in full in Darwin's *Plants and Animals under Domestication*. Here, confining myself to the Scottish aspect of the subject, I shall endeavour to show that in Scotland the wild Rock Dove was domesticated, and shall give some account of the significance of the early domesticated race.

The Rock Dove was at one time a common dweller on the sea-coast of Scotland, and although constant slaughter has reduced it in some areas almost to extinction, it is still to be found from St Abbs' Head to the Orkneys and Shetland, in the Hebrides and on the West coast, in places where exist caves suitable for its tenancy. In these deep caverns, the "doo-caves" familiar to many a district, the Rock Doves congregate in flocks, roosting on the ledges within, and there also building their slight nests during the long breeding season from March to September. From the caves they issue during the daytime to feed upon the farmer's grain or to render him good service by destroying grubs and the seeds of weeds.

Among pigeons, the Rock Dove is distinguished by the fine bluish-grey tints of the upper parts of its body, which shade into a pearl-grey spreading from the under surface to the tail feathers, tipped with leaden grey. The throat glistens with metallic purple and green; but the most striking and distinctive markings are the white rump and two black bars across each wing. The birds are rather less than

Fig. 22. Wild Rock Dove—ancestor of Scottish dove-cot "doos." ⅛ nat. size.

a foot in length, almost a third smaller than the familiar Wood Pigeon. Those who have no opportunity of seeing the true Rock Dove in its native haunts can form an excellent idea of its colouring and characteristics from many of the half-wild pigeons of our larger towns; for the interbreeding of domestic pigeons run wild, frequently results in a "throw-back" to the original type. Such blue-grey reversions to the Rock Dove type can be seen any day in the streets of Edinburgh.

There are many reasons for supposing that successful

efforts were made, and that at a very early date, to bring the Rock Dove under the influence of man in Scotland. These early efforts resulted in the formation of a primitive domesticated race—the pigeon of the "doo-cots"—which with little variation is found in almost all the maritime countries of the Old World. The steps by which the Rock Dove became the dovecot pigeon of Scotland were of the simplest nature and show how close was the relationship between the two.

It is probable that at first Rock Doves themselves were killed in quantity for food, but that indiscriminate slaughter threatened the birds with extinction, as in recent times in Fair Isle, and that wise men, observing the danger of rough and ready methods, resolved to encourage the multiplication of this useful bird. So in the very caves frequented by the Rock Doves, pigeon-holes were built and the original "doo-cave" became a primitive pigeon-house accommodating increased numbers of breeding birds. Of seven caves in the parish of Wemyss in Fife, hollowed by a former sea in rocks now far above high water mark, four "were long ago fitted up for pigeon houses." The Rock Doves which inhabited these had taken a first step towards domestication.

Man led them to a second step by the erection of independent pigeon-houses after his own design. It is impossible to trace the actual sequence of the erection of pigeon-houses, which, in all their differences of shape, and with all their conveniences of entrance and of innumerable nesting holes, still retain the hollow darkness of the ocean caverns. Nevertheless it is highly probable that the first pigeon-houses were erected along the coast near the "doo-caves," like those which crest the cliffs in the neighbourhood of the famous pigeon-caves of St Abbs or that which faces the "doo-caves" of Wemyss. As the pigeons became more and more accustomed to their artificial homes, became indeed distinctive "doo-cot" pigeons, the pigeon-houses would gradually extend along the coast to areas uninhabited by wild Rock Doves, and farther and farther inland. So many districts even far from the sea possessed one or more of these curious structures. Yet the coastal origin of the Scottish pigeon-house, and by implication, the coastal origin of its first tenants, are still indicated by the distribution of the "doo-cots": for while they are thickly scattered in coastwise parishes and counties—Fife alone possessed 360, with 36,000 pairs of breeding birds—they diminish in numbers with increasing distance from the sea, until in the Highlands they are exceedingly rare.

If evidence other than the history of the dovecots themselves were required to show that the Scottish dovecot-pigeon owes its ancestry to the Rock Dove, it is suggested by the fact that Rock Doves are easily tamed. Professor Macgillivray of Aberdeen has recorded that he completely tamed a Rock Dove in the Hebrides; Darwin relates that there are several records of those pigeons having bred in dovecots in the Shetland Islands, and that for more than twenty years Colonel King of Hythe kept in his dovecot the progeny of young wild birds taken in the Orkney Islands.

The dovecot pigeons themselves give good evidence of their ancestry, for they bear close resemblance to the wild species, and although they vary in the darkness of their plumage and in the size and thickness of their bill, they do so little more than Colonel King's Rock Doves did after twenty years of dovecot life. Further, although many of them have chequered wings due to the presence of large dark spots on the sides of each feather, in this respect they exactly resemble a chequered variety of the Rock Dove which occurs in Orkney and Islay.

In view of the interest of this primitive domesticated pigeon, I may be permitted to give a short account of its significance in Scotland, based mainly upon a paper by Mr Bruce Campbell in the *Transactions* of the Edinburgh Field Naturalists' Society.

When the Rock Dove first became a "Doo-cot" Pigeon it is impossible to say, but by the fifteenth century the pigeon-house was valuable enough to be reckoned along with "cunningares" or rabbit-warrens, as deserving the protection of the law. So it was ordained in 1424 that breakers of

mennes Orchardes, steallers of frute, destroyers of Cunningaires and Dowcattes...sall paie fourtie shillings to the King for the unlaw and assyith [indemnify] the partie skaithed [harmed].

But if the breaking of the doo-cot was one sin the taking of "ony foules of utheris Dowcattes" was another, and was "to be punished as thieft" (1474).

SOME LESSER DOMESTICATED ANIMALS 99

As doo-cots became more common and the "breakings of Dowcattes" increased, the fine was raised in 1503 so that

the unlaw thereof be ten pound, to-gidder with ane amendis to the partie, according to the skaith. And gif ony Children, within age, commit ony of thir things foresaid, [for children then were as they are now, the] Fathers or Maisters sall pay...or else deliver the said Child to the Judge, to be leisched, scourged, or dung, according to the fault.

Fig. 23. Ancient Pigeon-House or "Doo-cot" near Leadburn, Midlothian.

To the ordinary mind these penalties would seem sufficient to deter any potential "dowcatte breaker," but that they did not supply sufficient restraint subsequent legislation clearly shows. In 1567 shooting "dows" with gun or bow was forbidden under pain of forfeiture of moveables, or for the first fault 40 days imprisonment, and for the second loss of the right hand; and the climax of severity was reached, when in 1579, James VI ordained that, in addition to paying "the avail of the skaith done," the offender was to pay for the first fault "ten punds," for the second "twentie punds," for the third fault "fourtie punds" and

in case the offendours be not responsall in guddes...he sall for the first fault be put in the stokkes, prison, or irones auct [eight] dayes, on bread and water: And for the second fault, fifteene dayes...and for the third fault hanging to the death.

These Acts, some of which were repealed only in 1906, clearly imply that proprietary right in the dovecot pigeon was becoming more definitely recognized, and that the keeping of pigeons was becoming more and more common: that indeed the domesticated status of the pigeon of the dovecot was rising. So rapidly did dovecots increase in number, and so grievous was the destruction of grain caused by their inhabitants—the 36,000 pairs of breeders in the 360 dovecots of Fife are said to have consumed 3000 to 4000 bolls of wheat a year—that in 1617 the law put a check upon their erection, ordaining that only one dovecot should be allowed to each estate of the yearly rental of ten chalders, a chalder being equivalent to almost eight Imperial quarters.

The heydey of the dovecot is long gone by, for farmers found that their wheat was of more value than many pigeons (in Roxburgh in 1813 the birds could be bought for from 1s. 6d. to 2s. 6s. a dozen), and this together with the transforming of much arable land to pasture, and the decrease in the acreage of pease, emptied many a dovecot of its tenants. So it came to be that most of the pigeon-houses fell to ruins, and while in some counties a few remain to hint at a time when the primitive domesticated race of pigeons swarmed in the land from Dumfries to Caithness, in others even the place where they stood is forgotten—how many Aberdonians of to-day are aware that for more than two centuries the ground now occupied by Union Terrace was spoken of only as the "Dove Cott Brae?"

Although in England there have been evolved characteristic races of such divergent breeds as Pouters, Carriers, and Tumblers, I am not aware that Scotland has given rise to any breed other than her own simple "doos."

SOME LESSER DOMESTICATED ANIMALS

THE BARNYARD FOWL

Though clock,
To tell when night draws hence, I've none,
A cock
I have, to sing how day draws on;
A hen
I keep, which, creeking day by day,
Tells when
She goes her long white egg to lay.   HERRICK.

At what early period of their civilization the people of Scotland added the fowl to their domestic wealth is unknown,

Fig. 24. Scottish breeds of poultry—Scots Greys (to right), and Scottish Dumpies (to left). About $\frac{1}{10}$ nat. size.

although it was found in Britain by Julius Caesar in 55 B.C. In any case as the varieties of the barnyard fowl have arisen from the Red Jungle Fowl (*Gallus bankiva*) of India and south-eastern Asia, and as it was there domesticated more than a thousand years before our era, there would be little need to mention it in connection with Scotland, were it not that in our land the influence of man has created two peculiar local races.

The least distinctive of these—the Scots Grey (Fig. 24)—is a bird with upright and graceful carriage, moderately long legs, and plumage finely barred in black and white like that of a Barred Rock. The shape of the bird and its sprightly bearing suggest something of the style of English Game. The Scots Grey has been called the Scottish "Dorking," but its appearance suggests rather that it may correspond to an early stage in the evolution of such a breed as the Dorking. It may indeed be a developed representative of the early progenitors of the Dorking breed (itself perfected in our Islands), a surviving link with the undeveloped poultry of the good old days, which sold at 4*d.* each (as at Braemar in 1567). The carriage of the Scots Grey is that of the Jungle Fowl, but how great a change has transformed the reds and golden yellows of the wild bird into the sober chequer of the domesticated stock.

Even greater divergences from the wild stock mark the second distinctive Scots breed—the Scottish Dumpie (Fig. 24) —in which the plumage is also of a sombre grey due to fine alternating bars of darker and lighter colour, but in which the agile pose and graceful build of the Jungle Fowl have been entirely lost in a maximum of clumsiness and awkward bulk. For it is distinctive of Scottish Dumpies that their bodies are large, deep and remarkably long, and that this great bulk is carried on shanks of extreme shortness, rarely exceeding an inch and a half in length. Domestication has led to an extraordinary increase in productiveness, for the egg-laying of even the poorest breeds is a marvellous tribute to the selective influence of man, when contrasted with the limited clutches of seven to twelve eggs laid once a year by the ancestral Jungle Fowl. Comparison of the wild Jungle Fowl of eastern Asia with a typical Scottish Dumpie shows the surprising power which man has wielded over size and structure in perfecting the domesticated fowl. As great change he has wrought in colouring of plumage also. It must not be inferred, however, that this comparison implies that the changes were all wrought in Scotland. This is not so, since in Scotland the Dumpie and the Scots Grey as well, were evolved, not directly from their original wild ancestor, but through the intermediary of some domesticated race imported from the East.

SOME LESSER DOMESTICATED ANIMALS 103

THE DOMESTICATED GOOSE

Far abune the Angus straths I saw the wild geese flee,
A lang, lang skein o' beatin' wings, wi' their heids towards the sea,
And aye their cryin' voices trailed ahint them on the air.
                                            VIOLET JACOB.

From these very Wild Geese that in former ages frequented Scotland in "innumerable droves" the Scottish peoples have created that gabbling brood, the "grave, un-

Fig. 25. Grey Lag Geese—a wild species domesticated in Scotland. Illustration from Solway specimens. $\frac{1}{15}$ nat. size.

wieldy inmates of the village pond." And how little the change has meant in transformation of shape or habits; for of all his domesticated stock, geese have perhaps most resisted the selective influence of man.

Like the dovecot pigeon, the goose has been domesticated in many countries, but there is general agreement that our ordinary breeds of domesticated geese are descendants of the wild Grey Lag Goose (*Anser anser*) (Fig. 25) which is common in northern Europe and Asia, though in the former continent its numbers are now much reduced. In Europe the summer haunts of the Grey Lag are mainly confined to the north—from northern Scotland to Iceland, Scandinavia and Russia—but in winter it descends to warmer regions on the western and southern coasts of the continent. It is the largest of our wild geese, a fine bird with a general grey and brown plumage on the upper surfaces, shading backwards. into ashy grey and into creamy white on the under parts.

Several facts suggest that the Grey Lag was domesticated in Scotland. In the first place it is a frequent winter visitor to Scotland, generally arriving from its northern breeding haunts in September and November, and remaining with us over the colder months. More important, it still breeds here, in Ross-shire, Sutherland, Caithness and in the Outer Hebrides, and it is moreover the only kind of wild goose which nests in Scotland. Even so, it was at one time much more common both as a resident species and as a visitor than it is to-day. So we gather at any rate from its relative price as established in a controlled food list of 1551, where the value of a "wild guse of the great bind"—a Grey Lag—is fixed at 2s., while 18d. is the value set upon the smaller kinds, "claik, quink and rute"—the Barnacle Goose, the Golden Eye Duck and the Brent Goose. Besides, even so late as the latter half of the eighteenth century it was still breeding in the fens of England.

In the second place, the Grey Lag is easily tamed, so that its suggested domestication in Scotland need imply no special skill on the part of its domesticators. The Laplanders regularly tame the wild Grey Lag; and there are several instances of individuals captured young or hatched from eggs having become half-domesticated in various parts of Scotland, England, Wales and Ireland. In the last named country there is a numerous colony of semi-domesticated Grey Lag Geese which has been in existence since about 1700 at Castle Coole, near Enniskillen, and only in this condition does the species breed in Ireland. In Scotland well-known flocks were established in 1886 in South Harris, and in 1888 at Blair Drummond. The birds paired and nested and hatched their broods year in, year out, in perfect contentment. In fact our wild geese show a particular aptitude for losing their wild identity even with slight encouragement, and I have seen a Pink-footed Goose, which had been found injured

SOME LESSER DOMESTICATED ANIMALS 105

and had been cared for, moving and feeding contentedly with a flock of fărm ducks.

In the last place there is historic evidence that in the sixteenth century (and how much earlier no one can tell) the people of Scotland did actually tame and domesticate some kind of wild goose. I quote from Father Dalrymple's translation of Bishop Leslie's *Historie* (1578), this very interesting passage indicating the abundance of the wild birds and the method adopted for taming them.

In fauour of the reidar, I thocht gude, heir of the geis to speik a few wordes, for thair meruellous multitude in our cuntries, cheiflie in the west yles and lykwyse for the raritie and fewtie or scant of sum of thame in vthiris cuntries.

Excepte the Solande geise, of quhilkes afor we haue maid mentione, how thay ar bredd at hame; with vs vthir sax kyndes of geis ar funde, quhilkes only in thrie things ar fund to differ, to wit, in the sownde of thair voce, in gretnes of thair bodye, and taist of thair fleshe, quhilkes al ar seine in innumerable draues to flie to thir farthest Iles, in the spring of the yeir, eftir midday [Lat. *a meridie*—from the south] and thairfor, this opinione of thame is haldne, that athir iň thir westir Iles, or in Grundlande, quhen toward the South anothir land is no^t knawen, thay big thair nestis. Sum of thame no^t-withstandeng, throuch a certane craft, ar allured and prouoked to remane amang the lochis, and myrie places and amang the hathir and mures, amang ws, quhil thay haue laid thair eggis, and clekit thair burdes: for sum of thame, quhilkes w^t nettis ar takne, thair wingis ar clipit, and fed in the hous, quhil thay be tame: Thaireftir out and in frilie thay flie and swome, and ņocht only returnes hame agane, to thair accustumed and vsed fude, bot lykwyse thay bring vthiris with thame of thair awne kynde, as called to the banquet and commone feist with thame selfes, quhilkes quhen thay haue baytet, thay at last leir thame to sit, without al feir in the mid feild, and yardes, and plane places; and haldes thame stil besyde thame as neir nychtbouris, quhil al thair nestis be bigit, and thair young clekit.

Perhaps as a result of this easy domestication, geese seem to have been remarkably abundant in parts of Scotland in the seventeenth century, for Lowther during his tour in 1629, discovered that the Scots

have good meat, fish, flesh and fowl in great store, but dress it not well; in the South it is as dear as in the South of England, but in the north, about Dumbarton and thereabouts wondrous cheap, a goose for 4d, and so proportionably of other things.

It is possible, however, that these were wild geese.

Having given reasons for suggesting that Scotland was one of the countries in which the Grey Lag Goose was domesticated, let us glance at the influence which domestication has had upon the nature of the wild species. As we might expect, the colour of the plumage has changed, for the common goose seldom shows the varied browns and greys of the Grey Lag, although it ranges from dusky grey to pure white, the latter colour being almost invariably assumed by aged ganders. Selection for the table has led to an increase in the size and depth of the body, and this, added to compulsory inactivity, has brought with it increasing inability to fly and the exaggerated waddling walk of a creature whose weight has outgrown the intentions of its limbs. Selection for egg-laying has resulted in a great increase of productiveness, for the clutch of five to eight eggs of the wild goose is insignificant compared with the year's produce of the domesticated breeds.

SUNDRY OTHER DOMESTICS

The domestication of the Common Duck probably stands in similar relationship to Scotland as that of the Goose. It is agreed that the wild species whence sprang the "clam'rous duck" that "on the brink of her foul puddle quacks" is the common "wild duck" of our Islands—the Mallard (*Anas boscas*)[1]. The abundance of this species as a resident throughout Scotland, and the ease with which it can be tamed suggest that, like the goose, it may have been domesticated in our country at an early date. The influences that have played upon the duck in domestication have had effects similar to those referred to in the case of the goose, changes in the colour of plumage, increase of body, decreased power of flight, an exaggerated waddling walk and enormous increase in egg-production compared with the clutches of eight to twelve eggs of the wild bird.

The Turkey and the Guinea Fowl—the former descended from a wild Mexican species (*Meleagris mexicana*), the latter from East Africa (*Numida ptilorhynca*), were introduced to Scotland in a domesticated state, and have shown no distinctive changes under our care.

Domesticated Rabbits, descended from the common wild Rabbit (*Lepus (Oryctolagus) cuniculus*), which itself was introduced to Scotland ultimately from south-western

---

[1] So evident was the relationship in the early days of its domestication that the domestic duck is actually termed (in *Munimenta Gildhallae Londonensis*) a "dunghill Mallard."

Europe (see p. 247) owe none of their peculiarities to Scottish influence. Such is true also of our Cats, descended from the Egyptian Wild Cat (*Felis ocreata*), although in the Highland areas there has probably been some admixture of blood with our own wild species (*Felis sylvestris*), especially in former times when the Wild Cat was more common than it is to-day.

The Goat, offspring of the Wild Goat (*Capra aegagrus*) of the Mediterranean Isles, Asia Minor and Persia, notwithstanding its early introduction to Britain, and its former abundance in the Scottish Highlands, has suffered at our hands no change worthy of remark; and the same may be said of the Ferret, long regarded as a domesticated variety of the Common Polecat (*Mustela putorius*) but recently found to be related to the Polecat of Turkestan and Siberia (*Mustela eversmanni*).

Hive-Bees are domesticated in the sense that man houses and cares for them, that he feeds them in winter, and has exercised his limited powers of selection and interbreeding in perfecting the race. Ordinary hive-bees are little altered derivatives of the Wild Bee (*Apis mellifica*) of southern Europe. They were domesticated by the early Greeks, and were common throughout Europe at an early date. There is no record of their introduction to Scotland, but in the sixteenth and subsequent centuries many Acts were passed by the Scottish Parliament protecting bees from the "stealers of hives, and destroyers thereof." As the first reference to Bees which I have found in these Acts occurs in 1503, it is probable that they were becoming common in Scotland only towards the close of the fifteenth century. Their appearance in some of the islands belongs to a much later date. In Orkney they were so little known at the end of the seventeenth century that Mackaile can record the exploit of a youth who "stopt the skep (which a lady had taken thither from Angus) with a piece of a peat" on the plea that he found the Bees all flying away; and it was not till 1909 that Lewis was stocked with its present race of Hive-Bees, previous importations having become extinct. By frequent introductions of fresh blood from Italy and other parts of Europe man is constantly endeavouring to influence the nature and increase the productiveness of the Hive-Bee.

CHAPTER III

DELIBERATE DESTRUCTION OF ANIMAL LIFE

> The Hart, the Hynd, the Dae, the Rae,
>   The Fulmart and false Fox,
> The beardit Buck clam up the Brae,
>   With birssy Bairs and Brocks:
>
> Sum feiding, sum dreiding,
>   The Hunter's subtle snairs,
> With skipping and tripping
>   They playit them all in pairs.
>
> *The Cherrie and the Slae.*

NO other aspect of man's interference with the animal world bulks so largely in the imagination as his deliberate destruction of life. It is not that he has thus exterminated more creatures than have been banished by the felling of the woodland or the reclamation of moor and marsh; it is scarcely that the effects of gun and snare are more deadly than the removal of breeding places and the destruction of food supplies. Rather it is that the indirect influences are gradual in their working, that man's attention is fixed upon the fields which prosper under his care or the forests that fall under his axe, while the creatures which inhabited them wane and disappear unnoticed. But deliberate destruction is prompt, obviously merciless and final.

It has not always been so. In the earlier periods of man's development, in the days of the hunters of the Old Stone Age, the slaughter of animals was a necessity for protection as well as for food. But this was no uneconomic slaughter. It is true that wild animals were trapped in numbers in hidden pits, and it is true that in the aggregate many animals were killed, for more than two thousand molar teeth of the Mammoth have been found gathered together at the Palaeolithic settlement at Předmost in Moravia, and the broken bones of Horses, left over from many a feast, form a solid mass 100 yards in length and ten feet high at the Palaeolithic station of Solutré, in the Rhone Valley. Nevertheless primitive weapons almost limited the destruction to the absolute necessities of the sparse population.

The essence of the question of uneconomic slaughter is a simple one of capital and interest, where the breeding stock of any race of animals may be regarded as the capital and the year's young as the annual interest. So long as destruction is kept within the limits of the yearly interest and depreciation of capital is made up, all is well with the race, but so soon as the full interest is usurped and the capital stock begins to be entrenched upon, then the race is on the downgrade of reduced numbers, and, provided the destruction is kept up, of final extinction.

Several causes led to growing intensity in the slaughter of wild animals. The first in time was probably the domestication of wild creatures, and the consequent necessity for their protection. So the casual slaughter of prowling marauders developed into enmity and a blood feud against the larger beasts and birds of prey, an enmity which increased as feudal rule decayed and the people gained a new will and new powers to protect the crops and herds which their labours had created. In the second place, increased perfection of weapons and the invention of powder and the gun placed in man's hands new powers which he was not slow to use to the utmost. Many creatures have been banished from different areas of Britain since "weapons of precision" made their appearance, and nothing could witness more clearly to their influence than the fact that after the general disarming of the peasants of Poland by the Russian Government there was an enormous increase in the number of Wolves. In our own country similar effects have followed upon the absence of guns during the Great War, for never in the memory of man have the creatures of the wild, Deer and Rabbits, birds of prey, Stoats, Weasels and other "vermin," been so abundant as they are to-day.

The third and most fatal stage in the development of

destructiveness belongs to the greed of gain. Reckless destruction regardless of waste has followed upon the discovery of a profitable commodity. Garefowls or Great Auks were slaughtered by thousands on account of their oil and feathers and their bodies were burnt in great fires; the Buffaloes of North America and the Wild Cattle of South America were slain by millions, the former for their skins and tongue, the latter for their hides and horns, and their carcases were left to rot upon the plains. So too, and in even less worthy cause, the Egret has been slaughtered for its "osprey plumes," and precious birds all the world over for their brilliant plumage. And what is one to think of the wasteful slaughter proceeding at the present day in the islands of South Georgia, where, for the sake of their oil, Whales innumerable are being killed and their flensed carcases cast adrift, so that in the neighbourhood of the whaling stations masses of festering flesh spread solid for miles out to sea?

Many human motives have given rise to serious depletion of the animal world, and the chief of these I have endeavoured to illustrate from the Scottish point of view in the pages that follow. But the general warning must be added that a hard and fast classification must be looked upon only as a guide to clearness, that motives are seldom unmixed—that the creature slain in sport may also be used as food, just as the bird killed for the sake of its flesh may also yield valuable feathers and oil.

It ought also to be added that in the balanced order of nature the slaughter of one animal means invariably the increase of another. The unrestricted killing of Seals on the Pribilof Islands, off Alaska, increased the yield of the skins of the Blue Fox, which were valued at £3000 to £4000 annually, but since seal-killing was restricted on the high seas in 1911 and prevented in the Islands since 1912, the output of fox skins has greatly diminished, owing to the lack of carcases upon which the foxes depended for food.

The steps of the decadence of an animal due to deliberate slaughter, or to any other cause, can be traced in stages, first of reduction of numbers, second of curtailment of range, and lastly of extermination.

III. 1

DESTRUCTION FOR SAFETY OF MAN AND OF HIS DOMESTIC STOCK

COEVAL with his search for food began man's active defence against his fellow-dwellers in the wilds, but it was hardly till his wealth and welfare became centred in domestic flocks that his destruction began to tell upon the animal world. Then his energies, directed against marauders, fell heavily upon the beasts and birds of prey, until with increased perfection of weapons, he drove one and then another to extinction within the limits of his homeland. His influence in this respect can best be traced by following the stories of some of the creatures which fell under his ban.

BEASTS OF PREY

THE LYNX

The Northern Lynx (*Lynx lynx*) (cf. Fig. 3, p. 16), once a native of the greater part of Britain, and now confined to the forests of northern Europe although a close relative is found in Asia, makes but one appearance in Scottish history, when it shared with Neolithic man the wilds of western Sutherlandshire. The bones found by Drs Peach and Horne[1] in the Bone Cave of Allt nan Uamh near Inchnadamph, in deposits containing blackened and burnt hearthstones of Neolithic fires, vouch for its presence in the early days, but of its occurrence and disappearance written history makes no mention. It seems to have died out at a far distant period, and the probability is that man, in defence of his flocks, hastened its extermination in Scotland.

---

[1] Identified by Mr E. T. Newton.

THE BROWN BEAR

Few of us can have imagined the possibility of encountering a Brown Bear in the forest glades of Britain, yet our forerunners in the land frequently enjoyed that experience in the far-off days when

> In yon withered bracken's lair
> Slumbered the wolf and shaggy bear.

At the present day in Europe the Brown Bear (*Ursus arctos*) (Figs. 26 and 27) is mainly confined to the forests of Scandinavia, Russia, Hungary and the Pyrenees, but in former days it was a common inhabitant of Great Britain and Ireland. In Scotland, where it lingered longest,

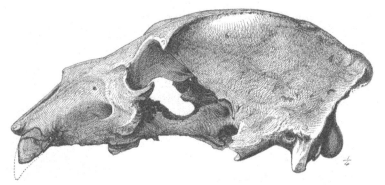

Fig. 26. Skull of Brown Bear from peat-moss in Dumfriesshire. ¼ nat. size.

it ranged over the whole land from Dumfriesshire, where many years ago a well preserved skull and a rib were found in a peat moss at Shaws, to Sutherland and Caithness. The only direct evidence of its association with man in North Britain is afforded by the discovery of a tooth in a broch at Keiss, for the canine tooth found in the bone cave near Inchnadamph in Sutherlandshire lay in a deposit lower than that containing traces of the presence of man. Yet in Yorkshire the Neolithic cave-dwellers of Settle considered the flesh of the Bear a suitable article of food, and its presence, as late as Roman times, is indicated by bones found in refuse-heaps at Richmond in Yorkshire, at Colchester in Essex, and even so far south as London, as well as by the discovery in the Roman camp of Cilurnum or Chesters on

DESTRUCTION FOR SAFETY OF MAN AND STOCK 113

the Tyne-Solway Wall of a tooth of a large Bear, the perforations in which hint that the slayer wore the relic as a badge of prowess.

The assumption is a fair one that if the Bear existed in England at the time of the Roman occupation it was also present in the much wilder country of Scotland, and sidelights of history support this idea. In the years after the Roman occupation of Britain, Caledonian Bears were well known in Rome, whither they were transported over-

Fig. 27. Brown Bear, a former native of Scotland (illustration from individual in Scottish Zoological Park, Edinburgh). $\frac{1}{30}$ nat. size.

seas to make sport in the amphitheatre. Malefactors bound to a cross were exposed to the attacks of these savage denizens from far-off Scotland, as Martial reminds us,

> Hanging on no slim cross, Laureolus
> His naked body to a Caledonian bear
> Thus proffered[1].

That the Bear still survived in Scottish woods after the Roman legions had gone, both archaeology and tradition

---

[1] "Nuda Caledonio sic pectora praebuit urso,
    Non falsa pendens in cruce, Laureolus."

suggest. The canine tooth found with other refuse in the Broch of Keiss in Caithness probably belongs to a period succeeding the Roman conquest, for brochs appear to have been unknown to the historians of Rome. Yet, assuming that it is the relic of a native animal, it indicates the presence of the Bear only in a vague period ranging down to the ninth or tenth century of our era. Tradition is even less definite as to date. Leslie in 1578, speaks of the "Tor-wood" or Caledonian forest as a place

quhair in lyke maner war sa mony wylde bares ['ursos'] that, as the alde wryters make mentione, than being full ['repertissimam'] is now nocht ane (even as our nychbour Inglande has nocht ane wolfe, with quhilkes afore thay war mekle molested and invadet).

And Camden, in his *Britannia* (1607), says of Perthshire:

This Athole is a country fruitful enough, having woody vallies, where once the Caledonian forest (dreadful for its dark intricate windings and for its dens of Bears, and its huge thick-maned bulls) extended itself far and near in these parts.

Almost to our own day, Gaelic tradition has carried the memory of the great *Magh-Ghamhainn*—the "paw-calf"— a "rough dark, grisly monster, the terror of the winter's tale"; and highland legends, such as "The Brown Bear of the Green Glen" recorded by Mr J. F. Campbell, and occasional place names, such as *Ruigh-na-beiste*, the Monster's Brae, and *Toll-nam-biast*, the Hole of the Monsters, may possibly perpetuate the tradition of the last survivors of Scottish Bears.

When did the Bear disappear from amongst the animals of Scotland? Attempts have been made to show that the Clan Forbes owes its name to the slaughter of a Bear by the chieftain, Ochonchar, the founder of the clan, whose surname, bestowed upon him for his prowess, became *Forbear* or *Forbeiste*. Pennant states that a Gordon, on account of his valour in killing a fierce Bear in 1057, was ordered by the King to carry three Bears' heads on his banner. It has even been stated that "in an ancient Gaelic poem ascribed to Ossian, the hero McDiarmid is said to have been killed by a Bear on Beinn Ghielleinn in Perthshire." But in each of these cases a wild Boar and not a Bear was the animal concerned, and the confusion has apparently arisen through the resemblance to "Bear" of the old Scots spelling and pronunciation of Boar, which was "Bare." It is highly probable that a similar confusion of terms led Col. Thornton to state in 1804 that Lord Graham had turned out a few "wild Bears" on the island of Inchmurrin in Loch Lomond, and that the record of Inchmurrin Bears which has found its way into literature is a false one. All that we can say, therefore, is that the Brown Bear seems to have been present in Scotland after the early centuries of our era, and may have existed till the ninth or tenth. Since there were no changes in climate or food supply sufficient to account for its disappearance, the assumption is that man's interference led to its extermination.

The Wolf

Of all the great beasts of prey which harassed man in Scotland, the most troublesome was the Wolf (*Canis lupus*) (Fig. 28, p. 117), which long survived its associates, the Lynx and the Brown Bear. Many bones attributed to Wolves have been found in early settlements, from the Neolithic cave-dwellings of Ayrshire to the times of the brochs and of kitchen-middens of later date, as well as in rock fissures such as that on the Pentland Hills. Yet the story of the Wolf in prehistoric Scotland is an obscure one. For although an odd relic, such as a jaw bone I have examined from a deposit in Ayrshire, probably indicates the presence of the Wolf in the land before the arrival of man, the bones of later ages are apt to be confused with those of domesticated dogs, brought with them by the early Neolithic peoples during their northward wanderings.

Fortunately the obscurity of this period in the Wolf's history in Scotland is of little real significance, since actual remains found in a dozen and more widely scattered English and Welsh counties, show how general its distribution must have been, and the abundant historical evidence of its presence in Scotland in much later times, gives good ground for supposing that in prehistoric days also, it was a common denizen of the woods.

It is impossible here to give in detail the recorded history of the Wolf in Scotland from the legendary times when Dorvadilla, the fourth king of the Scots, who, according to the story, reigned two centuries before the Christian era, "ordaint," according to Boece,

he slayer of ane wolf to have ane ox to his reward. Oure elders persewit this beist with gret hatreut, for the gret murdir of beistis done be the samin.

tThere is the less need for such a detailed history since a very complete account of the Wolf in Scotland appears in Mr J. E. Harting's *British Animals Extinct within Historic Times*, and in Mr R. S. Fittis's *Sports and Pastimes of Scotland*. I shall content myself, therefore, with tracing the main stages of man's interference with this ravager of the flocks.

Throughout all the ages the Wolf was reckoned a grievous pestilence, and even the popularity of a King reflected in some degree his attitude towards this plague. Thus when, as Boece tells, Edeir, a Scottish contemporary of Julius Caesar visited "all the boundis of his realme," his

passaige was the mair plesand to his nobillis, that he was gevin to hunting; for he delitit in no thing more than in chais of wild beistis, with houndis and rachis, and specially of wolffis, for they ar noisum to tame bestiall. This regioun, throw the cauld humouris thairof, ingeneris wolffis of feirs and cruel nature.

The fierceness of Scottish Wolves is attested by many an old story. Witness that of the pursuit of Malcolm II in 1010 in the forest of the Stocket, on the bounds of the city of Aberdeen, when the monarch was saved from a Wolf only by the presence of mind of a younger son of Donald of the Isles, who was rewarded for his timely aid by a present of the neighbouring lands of Skene. Even to the islands the scourge extended, for Arnor, the Earl's skald, tells in the Orkneyinga Saga, that after the Battle of Waterfirth between the invading Norse and the islanders of Skye, in the eleventh century,

> There I saw the grey wolf gaping
> O'er wounded corse of many a man.

Many methods were employed to keep the Wolves in check. It is significant that in the eleventh century, during the reign of King Alexander, when most of the wild creatures were reserved for the royal chase, no one was forbidden to hunt outwith forests and warrens for Wolves; and that from the twelfth century the monks of Melrose were prohibited from hunting and from setting snares in their

DESTRUCTION FOR SAFETY OF MAN AND STOCK 117

lands of Eskdale, except only for the Wolves. Such casual measures of restraining the wolf plague were soon found to be inefficient, and more drastic means had to be enforced. In 1283 King Alexander III made an allowance to his Treasurer for payment of "one hunter of wolves" at Stirling. But even professional hunters proved insufficient, and at last in the fifteenth century the law invoked the aid of the barons an Act of James I in 1428 requiring

that ilk baron within his barony in gaynande time of the year sall chase and seek the quhelpes of Wolves and gar slay them,...and that the barons hunt in their baronies and chase the Woolfes four times a year, and als oft as onie Woolfe beis seen within the barony.

Fig. 28. Wolf—a Scottish scourge exterminated about two centuries ago. $\frac{1}{14}$ nat. size.

Under pain of forfeiting a wedder, the tenants were to "rise with the baron" who was to pay to "the man that slays the Woolfe in his barony and brings his head to the baron, twa shillings."

Even the efforts of the barons were unavailing, and the scope of the hue and cry was broadened not many years later, when in 1457, the inhabitants of the whole countryside were compelled and coaxed by penalties and payments to attend the wolf-hunt at the call of the sheriff, baillies or barons:

item it is ordanyt for the distruccione of wolfes that in ilk cuntre quhar ony is, the sheref or the bailyeis of that cuntre sall gadder the cuntre-folk thre tymes in the yer betwixt sanct marks day [April 25th] and lamess [Lammas—August 1st] for that is the tyme of the quhelpis and quhat evir he be that ryss [rise] not with the sheref or bailye or barone within himself [in his area] he sall pay unforgeuin a wedder as is ordenyt in the aulde act maid thairaponc. And he that slays ane wolf than or ony uthir tyme he sall haif of ilk hous-hald of that parochin that the wolf is slayne within jd [one penny]....And he that slays ane wolf sall bring the hede to the sheref, bailye or barone and he sall be dettor to the slaar for the soume forsaide[1].

A few records of the payment of head money still exist: on the 24th October 1491, the Treasurer of Scotland paid 5s. "to a fellow that brought the King [James IV] two wolves, in Linlithgow."

The necessity of raising a general hue and cry after marauding Wolves led to the general establishment of kennels of wolf-hounds and even to the definition in leases of the duties of tenants on that score. So the monks of Coupar-Angus Abbey in a lease of part of the lands of Innerarity in 1483, bound the occupier to "obey the officers rising in the defences of the country to wolf, thief, and sorners," and many leases enforce the maintenance of "ane leash of good hounds, with ane couple of rachis [wolf hounds] for tod and wolf."

In spite of all such enactments, the effective destruction of the Wolves seems to have rested mainly on the personal idiosyncracy of the landowners, so that, while good service was rendered by an occasional bright spirit such as the Lord Hugh Fraser's lady, "a stout bold woman" as the Wardlaw MS. informs us, who in the latter half of the fifteenth century, "purged Mount Capplach [on the border of the Beauly Firth] of the wolves," yet the plague increased till it reached a climax in the sixteenth century.

Contemporary historians are at one in describing the abundance of Wolves and the terrible devastation wrought amongst flocks and herds by the savage marauders, and in so doing, they picture a Scotland wild beyond the imagination of the present day.

"In all boundis of Scotland," wrote Boece, in effect, in 1527, "except thay partis quhair continewall habitatioun of peple makis impediment thairto,

---

[1] It is interesting to recall that these Scottish Acts "For the distruccione of wolfes" were repealed only in 1906.

is gret plente of haris, hartis, hindis, dayis, rais, wolffis....The wolffis ar richt noisum to the tame bestiall, in all partis of Scotland, except ane part thairof namit Glenmores, in quhilk the tame bestiall gettis litill dammage of wild bestiall."

And half a century later, Bishop Leslie of Rosse found no diminution in the plague:

our nychbour Inglande has nocht ane wolfe, with quhilkes afore thay war mekle molested and invadet, bot we now nocht few, ye contrare, verie monie and maist cruel, cheiflie in our North cuntrey, quhair nocht only invade thay scheip, oxne, ye and horse, bot evin men, specialie women with barne, outragiouslie and fercelie thay ouirthrows.

In spite of the interference of man, the wolf plague had increased beyond the limit of toleration. In her *Book of Highland Minstrelsy*, Mrs D. Ogilvie gives a vivid account of the sufferings to which the natives of north-west Sutherland were subjected.

> The lean and hungry wolf,
>   With his fangs so sharp and white,
> His starveling body pinched
>   By the frost of a northern night,
> And his pitiless eyes that scare the dark
>   With their green and threatening light.
>
> He climbeth the guarding dyke,
>   He leapeth the hurdle bars,
> He steals the sheep from the pen,
>   And the fish from the boat-house spars;
> And he digs the dead from out the sod,
>   And gnaws them under the stars.

And so at last the inhabitants of Ederachillis were compelled to carry their dead across the sea to the lonely and isolated island of Handa, there to lay the poor bodies in peace, far from the reach of the prowlers of the night.

In this sixteenth century, King James V in 1529 and Queen Mary in 1563, witnessed the destruction of numerous Wolves in royal hunts held in the forests of Athole, but there is no mention of Wolves in the account of the royal chase organised by James V in Ettrick Forest in 1528. Apparently Wolves had already been extirpated from the Lowlands. Yet at this very time so dangerous had travelling become in the Highlands that, according to tradition, the great pine woods of Rannoch and Lochaber were almost impassable on account of their savage tenants, and hospices, hospitals, or "spittals" as they were called, were erected on the forest tracks to give refuge to wanderers, caught by nightfall in outlying districts. The site of one of these refuges is commemorated in the name of the "Spittal" of Glenshee, in Perthshire.

The direct interference of man with the Wolf in Scotland failed in its object: hereditary lords and legal governors were as impotent in face of the wolfish breed as were the suffering country people themselves. But where direct slaughter failed, indirect attack banished the Wolves from Scotland, for in the end it became evident that piece-meal slaughter must give way to an extensive destruction of the woodlands in which the Wolves lurked and multiplied in safety. So began that great burning of the forests, the memory of which is still kept fresh in the traditions of many a Highland glen (see p. 318).

And now the Scottish race of Wolves was doomed. It is true a few survivors lingered in woods that were left. In the early years of the seventeenth century, Wolves were hunted in the neighbourhood of Stirling and in Assynt, and in Breadalbane each tenant had to make every year four spears for wolf-hunting; in 1618, Taylor, on his "Pennyles Pilgrimage," saw in the "Brea of Marr," "wolves and such like creatures, which made me doubt that I should never see a house again"; the "Accompt Book" of Sir Robert Gordon, Tutor of Sutherland, mentions, in 1621, "sex poundis threttein [thirteen] shillings four pennies gieven this year to thomas gordoune for the killing of ane wolf, and that conforme to the acts of the countrey," and the same diarist a few years later, specifies the Wolf in his list of the wild creatures of Sutherland. Yet before the end of the century, the Wolf had been all but exterminated. The last Wolf in the north-eastern counties was slain in Kirkmichael Parish, Banffshire, in 1644; the last in the wilds of Perthshire was killed by Sir Ewen Cameron of Lochiel at Killiecrankie in 1680, and about the same time one was killed in Forfarshire. But they probably lingered a little longer in the wilder and more wooded districts, for persistent tradition records that so late as 1743 a Wolf, which had slain two children on the hills by the Findhorn, was tracked and killed by a Highland hunter, Macqueen by name.

So the Wolf disappeared from Scottish hills, though many

place names scattered throughout Scotland, even in the lowlands whence man first banished it, still tell of its former presence and abundance[1].

THE FOX OR TOD

In the old laws the Fox (*Vulpes vulpes*) (Fig. 29, p. 123) keeps disreputable company with its cousin the Wolf, and though both were equally warred against, the smaller animal, as is the rule, has outlived its congener. The general distribution of the Fox throughout Scotland at the present day belies the efforts that have been made to extirpate it, yet its numbers are much reduced from those of former days, when it was reckoned with the Wolf as an evil genius of "tame bestiall."

> There's a tod aye blinkin' when the nicht comes doon,
> Blinkin' wi' his lang een an' keekin' roond an' roon',
> Creepin' by the fairmyaird when gloamin' is to fa',
> And syne there'll be a chicken or a deuk awa'—
> Aye, when the guidwife rises, there's a deuk awa'!
> VIOLET JACOB.

There could be only one end to such iniquity—persistent pursuit and destruction, and for many a long year this has been the fate of the Fox in Scotland. Even in the thirteenth century, special hunters were chosen to keep its ravages in check, for in 1288, in the days of the Maid of Norway, the Court Chamberlain paid 52s. 10d. to two park-keepers and one fox-hunter at Stirling. The fox-hunter long remained an institution in Scotland, but his efforts fell short of the need, and other means had to be taken to keep Foxes in check. One such method was the offering of a reward to a fox-slayer: a Statute of James II enacted, in 1457, that "quha evir he be that slays a fox and brings the hede to the schiref, lorde, barone or bailye, he sall haif vjd [sixpence]." But persuasion

---

[1] Witness a few names gathered at random: in southern counties, such names as Wolf-gill in Dumfries, Wolf-hope, Wolfelee, Wolf-cleugh in Roxburghshire, the Wolf Craigs on Baddingsgill Burn on the southern slopes of the Pentland Hills, Wolfstan in Linlithgow; in more northern districts Wolf-crag on the shoulder of the Ochil Hills, Wolfhill in Perthshire, Wolf-hole in the parish of Birse, Aberdeenshire; and in the Gaelic lands, Toulvaddie or Toll-a'-mhadaidh—the Wolf's Hole, the names of many burns, as Allt-mhadaidh and Allt-a'-choin uidhre—Burn of the Dun Wolf, and lochans such as Loch-a'-mhadaidh and Lochan-a'-mhadaidh-riabhaich—the Loch of the Brindled Wolf.

had to be reinforced by compulsion, and conditions were laid down in many sixteenth century leases similar to those imposed upon David Ogilvy, when in 1552 he received the lands of Glenisla from the Abbot of Cupar, that he should "nurice ane leiche of gud howndis, with ane cuppil of rachis for wolf and tod," and he and other tenants were bound "to be readdy at all tymes quhene we charge thame to pas with ws or our bailyies to the hountis." Every farm of any size had to keep its fox-hounds, and in many districts the fox-hunter became a regular official paid partly by the landlord and partly by the tenants. The latter supplied him with farm produce and entertained him and his dogs during a specified number of nights in the year, according to the extent of land held. In addition, the huntsman received a special fee for every Fox slain by his hounds.

Hector Boece held all such endeavour in disdain and commended a simple recipe learned of the good men of

Glenmores, in quhilk the tame bestiall gettis litill dammage of wild bestiall, speciallie of toddis; for ilk hous of this cuntre, nurisis ane young tod certane dayis, and mengis [mixes] the flesche thairof, eftir that it be slane, with sic meit as thay gif to thair fowlis, or uthir smal beistes. And sa mony as etis of this meit ar preservit twa monethis eftir fra ony dammage of toddis; for toddis will eit na flesche that gustis [tastes] of thair awin kind; and, be thair bot ane beist or fowle that hes nocht gustit of this meit, the tod will cheis it out amang ane thousand.

But the practice of Glenmore apparently did not hold good in other parts, since in the seventeenth and eighteenth centuries annual fox-hunts were the rule throughout the land. To these all the neighbours gathered—that in Strathmore was opportunely convened by the church beadle as the congregation retired from divine service—and for several days Foxes were hunted high and low. In many parts of the country, Sutherlandshire, Aberdeenshire and Dumfriesshire among others, societies for exterminating Foxes were formed, and in the district around Golspie in the closing years of the eighteenth century, according to the Old Statistical Account, "upwards of £100 sterling is yearly expended for the purpose of extirpating that noxious animal that kills young deer and sheep and moorfowl." Such concerted efforts made serious inroads upon the native stock. In five parishes of Aberdeenshire in the district of Braemar 634 were killed in the ten years beginning with 1776; in Sutherlandshire, on the estates

DESTRUCTION FOR SAFETY OF MAN AND STOCK 123

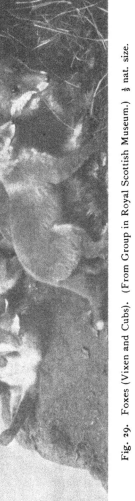

Fig. 29. Foxes (Vixen and Cubs). (From Group in Royal Scottish Museum.) ⅑ nat. size.

of Langwell and Sandside, 546 were killed in seven years from 1819 to 1826, and on the Duchess of Sutherland's estates 193 were killed in the three years from March 1831 to March 1834.

Other inducements beside the protection of stock, have helped to bring Reynard to book, for the value of its skin, especially in the older days, was no mean consideration (see p. 168) and at the present day live Foxes are exported from certain districts of Scotland to stock the coverts of English hunting counties. Yet in spite of all, notwithstanding that his numbers have been sadly diminished, the Fox still holds his ground, and there are few counties which cannot boast many occupied earths. From the islands, however, it has been banished, and none now exists on Mull, where tradition relates that it survived before the middle of the eighteenth century.

The Wild Cat

In the struggle against the wiles of man, the Wild Cat (*Felis silvestris*) (Fig. 30, p. 125) has been less fortunate than the Fox. At a time not very remote it too roamed over the whole of the mainland, and in earlier days even found a home on the islands, as in Bute, whence I have identified bones from the prehistoric settlement of Dunagoil. The increasing cultivation of land and need for more stringent protection of flocks, as well as the growing regard paid to the preservation of the smaller game of the countryside, were certain to tell heavily upon so fierce and persistent a marauder as the Wild Cat. From England man has driven it, as from the Lowlands of Scotland, and even in the Highlands its range is now severely restricted. Thanks to the labours of Dr Harvie-Brown, the steps of its decadence in Scotland can be traced with some precision.

From the great industrial areas and centres of population it first disappeared. No tradition remains of its presence in the busy midland valleys between the Firths of Forth and Clyde: there it has been long extinct. Yet in the wilder lands on either side, it held its ground till recent times. In the counties marching with the Solway and with the English border, it was common at the commencement of the nineteenth century, but about 1830 it had gone from the former, leaving only such place names as Wild Cat Wood and Wild Cat Craigs to commemorate its presence, while in the border counties the last Wild Cat was slain in Berwickshire near Old Cambus in 1849. Here too its former presence is perpetuated in such place names as Wulcatt Yett, Cat-leeburn, and Cat-cleugh in Roxburghshire; indeed there is scarcely a county but has such traditional record of its presence, though this evidence must be accepted with caution on account of the possibility of confusion with the Marten "Cat."

Fig. 30. Wild Cat (from a West Inverness-shire example). ⅛ nat. size.

About the time that it was being exterminated in the Lowlands, the Wild Cat was also hard pressed in the counties bordering the midland valley, and in the more open counties along the east coast. In 1842 it was already extinct in most of the parishes of Stirlingshire, though it still survived in Strathblane. A solitary pair, killed in Glendye in 1850, was the last of the Kincardine breed. In Aberdeenshire, 44 were killed in the ten years succeeding 1776 in five parishes about Braemar, yet in the Don Valley the last was killed near Alford Bridge only in 1862, though in the wilder Glen Tanar it survived till 1875. In the woodlands along the southern shores of the Moray Firth its general disappearance may be placed about 1830, when an individual was killed at Cawdor in Nairn. But in this area also odd individuals may have lingered in the more secluded woods till a later date, for about 1860 the Forest of Dalry near Forres yielded an example.

Even in the wilder counties where one would imagine that abundant shelter might have been found, the same decades saw the gradual disappearance of the Wild Cat. In Perthshire the last in the Athole district was trapped in 1857; in 1863 or 1864 the last south of Glen Dochart was killed upon Ben More; the last loiterer in Glenshee, at Dalnaglar about 1870; and now they have been exterminated throughout the whole of that mountainous county, even to the wilds of Rannoch. In Argyllshire, the shores of Loch Awe saw the last of the race in 1864, although a few miles away, in Glen Orchy, an individual appeared in 1899. Glenmore in Inverness-shire has lost the Wild Cat since 1873, in which year also a solitary example was killed in East Ross, where the species was already extremely rare. An individual, however, is said to have been killed at Edderton on the south of the Dornoch Firth so recently as 1912.

The rapidity of the extermination of the Wild Cat during a comparatively short space of years in country highly adapted for its preservation is a matter for wonder, and says much for the skill and determination with which it has been tracked. In Sutherlandshire on the Duchess of Sutherland's estates, a reward of half a crown was paid for each head, and between March 1831 and March 1834, 901 Wild Cats, Martens and Fulmarts were killed; in the grounds of Dunrobin six Wild Cats were slain between 1873 and 1880; and in the districts of Assynt and Durness, while a keeper killed twenty-four Wild Cats in seven years from 1869 to 1875, in the following five years he obtained only two (see Fig. 42, p. 176), and a colleague who, in the four years 1870 to 1873 killed ten individuals, killed only four in the following seven years to 1880.

Since then the Wild Cat in Scotland has proceeded rapidly upon the path to extermination, and there is little likelihood of its ever regaining lost ground outwith those fastnesses in the forests of western Ross and Sutherland, of Inverness and Argyll to which it has been driven by the hand of man.

Lesser Beasts of Prey

Lesser noxious animals, such as Martens, Polecats, Stoats and Weasels have, like their greater brothers, suffered on account of their habits, but as some have been killed rather for their fur and because of their enmity to game, they will be referred to in the sections which follow.

Birds of Prey

In a land given over to the simple rule of agriculture, birds of prey, of necessity, follow the same hard track of slaughter and extermination trod by beasts of rapine. In the old days Scotland was well plenished with birds of prey that afforded sport for kings, and, because of their service in the amusement of the court, were protected with all the rigour of feudal law. "Of fowlis sic as leiffis of reif [live by rapine]," wrote Boece in the sixteenth century, "ar sindry kindis in Scotland, as ernis [eagles], falconis, goishalkis, sparkalkis, marlyonis, and sik lik fowlis." But of these how few remain in anything like their former abundance, and how many have altogether disappeared. A few examples will illustrate the progressive effect of man's deliberate destruction, which on occasion was insisted upon by law (see p. 217).

The Golden Eagle

Grateful as the presence of the most magnificent of our native birds, the Golden Eagle (*Aquila chrysaëtus*) (Fig. 31, p. 129) may be to the lover of nature, it is little wonder that farmers and crofters of former days waged incessant war upon it. Old records contain many complaints of its destructiveness. "The Eagles [of the islet of Lingay in the Outer Hebrides]," wrote Martin in 1703, "are very destructive to the Fawns and Lambs, especially the black Eagle[1] which is of a lesser size than the other."

The flight of the Eagle was strong and the countryside over which its depredations extended was vast. This also hastened its downfall. The natives of an island adjoining Island Saint or Island-more (Ellan-Shiant or Ellan-Mhuir) in the Outer Hebrides told Martin that the Eagles

---

[1] The universal name for the Golden Eagle in Gaelic-speaking districts is *Iolar-dhub*, the black eagle, or simply *Ant-eun*, the Bird.

never yet killed any Sheep or Lamb in the Island, though the bones of Lambs, of Fawns and Wild-Fowls are frequently found in and about their nests, so that they made their Purchase in the opposite Islands;

and Mr Robert Gray was informed by an observer that he had seen the Eagles of South Uist "coming almost daily from Skye with a young lamb each to their eaglets"—a distance of about twenty-five miles.

Such misdemeanours could not pass unnoticed, so the tale of Dr Patrick Graham regarding southern Perthshire in 1806, might be taken as an epitaph of the eagles throughout the country.

"The black eagle," he says, "has built her eyrie from time immemorial in the cliffs of Benivenow [Ben Venue in Aberfoyle]; but by the exertions of the tenantry, who suffered much loss from her depredations on their flocks, the race is now almost extirpated."

No wonder that the Golden Eagle had all but disappeared in view of the terrible slaughter of its slow-breeding stock. In five Aberdeenshire parishes, clustering about Braemar, 70 Eagles were slain in the ten years from 1776 onwards; on the estates of Langwell and Sandside in Sutherlandshire, 295 old Eagles and 60 young Eagles and eggs were destroyed in the seven years between 1819 and 1826; and rewards of one guinea and ten shillings each respectively brought to book 171 old Eagles and 53 young and eggs on the Duchess of Sutherland's estate in the same county in the three years 1831 to 1834.

The island of Hoy in the Orkneys, to which, about the middle of the seventeenth century, an Eagle is said to have carried unhurt from the mainland a swaddled child, has long been deserted, for a price rested upon the bird's head (see p. 130). So the Golden Eagle, which about two hundred years ago built in Derbyshire, and a hundred years ago had its eyrie amongst the Cheviot Hills, has been banished to the lone islands of the Inner and Outer Hebrides and to the Highlands of the mainland. Fortunately a new sense of the aesthetic value of the Eagle's magnificence has arisen to save this noble bird from extermination, and a wise protection has recently led to a gradual increase of its numbers and extension of its range. At the present time, however, its eastern limit may be traced in the wilds of the Grampians at the head of the Dee, and its southern outposts in the forests of northern Perthshire.

DESTRUCTION FOR SAFETY OF MAN AND STOCK

Fig. 31. Golden Eagle (from a West Highland example). ⅙ nat. size.

WHITE-TAILED OR SEA EAGLE

A harder fate has fallen upon the Sea Eagle or Erne (*Haliaëtus albicilla*) (Fig. 32, p. 131). Greater than its "Golden" cousin, it is no less destructive. Of the Outer Hebrides, Martin said in 1703 "There are Eagles here [Harris] of two sorts, the one is of a large size and gray colour, and these are very destructive to the Fawns, Sheep and Lambs." Of Shetland he wrote:

> There are likewise many Eagles in and about these Isles which are very destructive to the Sheep and Lambs....The Isles of Zetland produce many sheep, which have two or three lambs at a time; they would be much more numerous, did not the Eagles destroy them.

And Brand writing of Shetland about the same time (1701) records that

> There are also many Eagles, which do great prejudice and hurt to the Countrey; for the Lambs they will lift up in their Claws, and take whole to their nests, and falling down upon the Sheep, they fix one foot on the ground and the other on the sheep's back, which they having so apprehended, they do pick out their eyes, and then use the Carcases as they please.

Few of the sufferers were content to save their flocks by the simple charm used by the islanders, who, says Brand,

> when they see the Eagle catching or fleeing away with their prey, use a Charm, by taking a string, whereon they cast some knots, and repeat a form of words, which being done, the Eagle lets her prey fall, tho at a great distance from the Charmer, an instance of which I had from a Minister who told me, that about a month before we came to Zetland, there was an Eagle that flew up with a Cock at *Scalloway*, which one of these Charmers seeing, presently took a string (his garter as was supposed), and casting some knots thereupon with the using the ordinary words, the Eagle did let the Cock fall into the sea, which was recovered by a boat that went out for that end.

So troublesome was this frequenter of the sea-cliffs that from very early times a price rested upon its head in Orkney, witness an act passed at Kirkwall in 1626:

> *Anent Slaying of the Earn*—It is statute and ordained...that whatever persone shall slay the earn or eagle[1] shall have of the Baillie of the parochine where it shall happen him to slay the aigle 8*d.* from every reik [inhabited house] within the parochine, except from cottars that have no sheep, and 20 shill. from ilk persone for ilk earn's nest it shall happen them to herrie; and they shall present them to the Baillie, and the Baillie shall be holden to present the head of the said earn at ilk Head Court.

[1] The words may include both the White-tailed and Golden Eagle.

Fig. 32. White-tailed Eagle or Erne, once common, now practically exterminated in Scotland. (From a Hebridean example.) $\frac{1}{8}$ nat. size.

At a later date, the reward seems to have been commuted into a fixed payment, for Laing in 1806, in his *Voyage to Spitzbergen*, records that "a premium of three shillings and fourpence is obtained for killing one of these eagles ["Earne-eagles"]; and smaller premiums are given for killing less destructive birds."

Nevertheless for a long time the Sea Eagle seems to have held its ground with wonderful tenacity, and it is evidence of the extraordinary rapidity with which a widely distributed and common species may be exterminated that so late as 1871 Robert Gray, in his *Birds of the West of Scotland*, should say of it, "Being a much commoner bird in Scotland than the preceding species [the Golden Eagle], the Sea Eagle has never been at any time in the same danger of extinction." Yet in half a century, man has practically extirpated the Sea Eagle in Scotland. The breeding places on Ailsa Craig and Islay have long been deserted, Orkney is forsaken, and Skye, where under sixty years ago a keeper killed fifty-seven eagles in nine years, now harbours not a single individual. Even in Shetland, where on the tiny island of Vemantry the tenant told Low, about 1774, that he had killed seven in a short time, the death-knell of the Sea Eagle has been rung, for the male of the last pair was killed some years ago, and since then the

old female has returned year in, year out, to the old nest each spring to gaze out over the wide horizon and wait. In the spring of 1916 she was still alive at her post "just hanging about the old place as usual," solitary for the rest of her days[1].

It is possible that the species still breeds in its last out-post, the Outer Hebrides, but at best a few years will probably see the end of the White-tailed Eagle of Scotland.

Vanished and Vanishing Hawks

Hawks should perhaps be regarded as pests and vermin rather than share a place with their nobler kin as destroyers of domestic stock, but for the sake of unifying the treatment of the birds of prey I include them here. In proportion to their number, they have suffered heavily at the hand of man.

[1] Since these words were written this aged White-tailed Eagle of North Roe has disappeared. It ceased to visit its old haunts in the season of 1918, having probably died—the last, it is said, of the Sea Eagles of the Shetlands.

DESTRUCTION FOR SAFETY OF MAN AND STOCK

The GOSHAWK (*Astur palumbarius*), mentioned by Boece in the sixteenth century as a Scottish "fowl of reif," and till the middle of the nineteenth century said to be a regular breeder in the forests of northern Scotland—of Darnaway in Morayshire and Rothiemurchus in Inverness-shire—can no longer be reckoned a native of the British Isles.

Like the Sea Eagle, the KITE or GLED (*Milvus milvus*) (Fig. 33, p. 135) exemplifies how frail is the security of numbers when man sets his hand to interfere. It is almost impossible to believe that a bird once so common that its vast numbers in the streets of London excited the wonder of foreign visitors in the reign of Henry VIII, should have suffered so grievously that in 1905 the few survivors in the British Isles could be counted on the fingers of one hand. Yet so it is, and the love of the "greedy gled" for the poultry yard had much to do with the warfare which has all but exterminated it.

> And other losses too the dames recite,
> Of chick and duck and gosling gone astray,
> All falling preys to the swooping kite:
> And on the story runs from morning, noon and night.
>   CLARE.

It was once a common bird in Scotland, breeding not only in the wilder areas, but even in counties so far south as Stirlingshire and Ayrshire. By the middle of the nineteenth century, however, it had been driven to the solitudes of Perth, Inverness, Banff and Aberdeen shires. From there also it has been banished, though here and there a *Clach-a-chambain* or "Gled Stone," such as that at the head of Glen Brierachan near Pitlochry, marks a well-remembered perching place of the Gled. From Scotland, and from England as well (except in Herefordshire, on the Welsh border), the Kite has been utterly extirpated, and now only a few survivors linger on in Wales, where from the miserable remnant of five birds known to exist in 1905, stringent protection has been fortunate in slowly increasing its numbers.

Less fortunate has been the OSPREY or FISH HAWK (*Pandion haliaëtus*) (Fig. 45, p. 192), which a hundred years ago was so abundant in Scotland that naturalists did not trouble to record its haunts. At the end of the eighteenth century, it probably bred regularly so far south as Dumfriesshire, for in a pamphlet entitled *Observations on Moffat, and its Mineral Waters*, published in 1800, T. Garnett wrote.

> This lake which is called Loch Skeen, is 1100 yards in length, and about 400 in breadth; there is a little island where eagles bring out their young in great safety, as the water is deep, and there is no boat on the lake. The water of this lake abounds with very fine trout[1].

In 1806, Dr Patrick Graham recorded its presence on the southern confines of Perthshire, "The Osprey or Water Eagle, builds her nest in some of the lofty trees in Inchmahoma." It is known also that it frequented the island of Inch Galbraith in Loch Lomond, an islet in Lake Menteith, Loch Awe, Loch Maree and similar places. But these sites have long been deserted.

In the twentieth century its breeding places have been confined to the counties of Inverness and Sutherland, but even here, in spite of all efforts at protection, there has been no security. Loch Askaig has been untenanted since 1911, while Loch-an-Eilein, whose ruined castle, built on an islet, had been a regular nesting-place for a hundred years, has been deserted since 1902 (see p. 192). A pair of Ospreys bred in 1916 in Scotland, in a place that shall be unmentioned, but they are the last of a banished race.

The ranks of other Scottish birds of prey have also been thinned by man, though till now they have escaped the final catastrophe of extinction; but decrease in numbers and limitation of range surely mark the steps of a decadence of which extermination is the end. One cannot think of the persecution which in the case of the PEREGRINE FALCON (*Falco peregrinus*), has replaced the protection of former times, without wondering how this fine hawking bird could survive, were the ranks of Scottish breeding birds not reinforced annually from the Continent. The numbers of the HEN HARRIER (*Circus cyaneus*) have seriously dwindled in Ireland, and in Scotland it has been driven to the Orkneys and Outer Hebrides, and to the fastnesses mainly of the northern mainland. The COMMON BUZZARD (*Buteo buteo*) has disappeared from Ireland, though it still nests in the Inner Hebrides and in the west and central Highlands of Scotland and very rarely in the Outer Hebrides and Orkney. The

---

[1] Situation and name indicate the "Water Eagle"; the little rocky island is a typical nesting-place for the Osprey, as Dr W. Eagle Clarke tells me.

HONEY BUZZARD (*Pernis apivorus*) has been known to breed as far north as Aberdeenshire and Ross-shire, though it can now no longer be reckoned a native of the land.

Fig. 33. Kite or Gled, once common, now exterminated in Scotland. (From a Dumbartonshire example.) ⅛ nat. size.

Even the lesser Hawks have suffered. The records of old vermin lists offer abundant evidence of the slaughter carried on in former days. In the parishes of Braemar, Crathie, Glenmuick, Tulloch and Glengarden, near the head of the Dee in Aberdeenshire, the ten years 1776 to 1786 saw the death of "2520 Hawks and Kites"; the estates of Langwell and Sandside in Sutherlandshire in the seven years 1819 to 1826 yielded 1115 Hawks; and the Sutherland estates in the same county in the three years 1831 to 1834, 1055 Hawks. To what extent such persecution has ultimately affected their numbers, it is not easy to say, though no better proof of their reduction below Nature's standard is needed than the fact that since the Great War has called gamekeepers from forest and moor to the field of battle, many districts have seen such a revival of Sparrow-Hawks, Kestrels and Merlins as has not been known in living memory.

"RAVENOIS FOULLIS"

These birds also, the RAVEN (*Corvus corax*) and its relatives, did much damage in their day to young lambs and the feathered-inhabitants of the farmyard. They too have been subjected to long and steady persecution. One has only to compare their numbers and habits in former days and to-day to realize how much the "ravenois foullis" have suffered at the hand of man.

In the sixteenth century Ravens and others of the race of crows were common in the streets of the large towns, where, as I shall show hereafter, they were encouraged and actively protected for the value of their work as scavengers (p. 224). Even in the streets of Edinburgh and Leith they went about their disagreeable task unmolested, witness Wedderburn's account of the events which followed upon the eclipse of the sun in 1597:

> The peiple with gryt fair fled aff the calsayis [causeways] to houssis mourning and lamenting, and the crawis and corbeis and ravenois foullis fled to houssis to our steple and tolbuith and schip tappis, maist merveulously affrayit.

Now, Carrion Crows and Ravens, once so common, are birds rarely to be seen; from the streets they have been driven to the wilds and to rocky fastnesses in the Lowlands, the Highlands and the Islands of Scotland. "Vermin Lists" again reveal the secret of the disappearance. In the five parishes on Deeside already mentioned, 1347 Ravens and Hooded Crows were killed in the ten years from 1776 to 1786; in two estates (Langwell and Sandside) in Sutherlandshire, 1962 were slain in seven years from 1819 to 1826; and on the estates of the Duchess of Sutherland in the same county, the three years 1831 to 1834 saw the death of 936 Ravens, for each of which the reward of two shillings was paid.

AN INDIRECT RESULT OF DESTRUCTION

The slaughter of beasts and birds of prey for the protection of domestic animals has not been one-sided in its effects, for a natural result has been the indirect protection and inordinate increase of smaller fry whose welfare lay outwith the intentions of the slayers. These smaller beasts and birds have benefited enormously by the disappearance of their natural enemies. This, as well as active protection, may account for the increasing numbers of many small birds, but it is also responsible in part for the multiplication of such pests as Rabbits and Rats. The seriousness of the Vole plague, which in the early nineties of last century ravaged a wide district in the Scottish Lowlands, has also been attributed in some degree to the disappearance of creatures which under natural conditions form an efficient curb to overwhelming multiplication. How will these lesser vermin fare now that war has given their enemies an unforeseen opportunity to increase? Naturalists will be fortunate if the answer leads to a fresh consideration of the influence of birds and beasts of prey upon lesser pests, and to a more reasonable slaughter on the part of the protectors of the poultry yard and the game covert.

III. 2

DESTRUCTION FOR FOOD

ALTHOUGH the destruction of animals for his own safety was a primitive necessity for man, it can scarcely have preceded in time destruction of wild creatures for the sake of their products. The men of the Old Stone Age were hunters by nature. Upon animals they depended in great part for their food, and when clothing was invented, the skins of wild beasts formed a simple and efficient protection from the elements. The bone needles of the later Solutrian deposits of France indicate that in those early (Palaeolithic) days the art of stitching skins was already known.

When the tribes of the much later Azilian culture reached Scotland, the arts of hunting and of the use of skins for clothing were already of long standing, and although refuse heaps show that our earliest settlers subsisted largely upon fish and molluscan shell-fish, there is evidence that the larger animals also fell to their spears. As we shall see, however, the effective interference with animals useful on account of their products, belongs to a much later period of civilization.

At all times animals large and small have been slain for food, but as other motives have entered into the pursuit of many, and especially of the larger creatures—sport in part determining the chase of such as the Red Deer and the Boar, their blubber that of Seals, and their skins that of Hares and Rabbits—these will be mentioned in the sections which follow, leaving for present consideration some of the lesser food creatures. Fortunately, no native of Scotland, unless it be the Garefowl, against which other influences were at work, has suffered the fate of Steller's Sea Cow (*Rhytina stelleri*), which in the course of less than thirty years (between 1741 and 1768) was totally extirpated, simply because it formed a convenient food for the hunters and traders of Bering's Island.

GEESE AND SOME LESSER BIRDS

The reduction in numbers of the Grey Lag Goose has already been referred to (p. 104), and I would only mention here that the process of extermination is even now to be seen in the north of Scotland, its last native breeding resort in the British Isles. Here the native breeders are annually reinforced by large numbers of immigrants, as many as 500 having been seen on the wing at once. But the value of the bird as food, and the damage it causes to the crofters' crops, have combined to reduce its numbers, for when the old birds are moulting and the young are unable to fly, all take readily to the sea and are then easily slain by fishermen, sometimes organized in parties. As a result the number of breeders is decreasing year after year.

The highly specialized method of capturing Geese and Ducks by a system of trap nets or decoys has also accounted for great numbers of these birds. In the thirty-five years following 1833, 95,836 wild fowl were taken from the decoy of Ashby in Lincolnshire and in a single season in the eighteenth century the decoys near Wainfleet captured 31,200 Ducks. The total slaughter caused by decoys and by driving Ducks must have been prodigious before the marshes were reduced by reclamation; but fortunately these deadly devices never gained foothold in Scotland.

Small birds have been and are an easy prey and a favourite food in many countries. To-day we deplore the slaughter of small migrating birds in the European lands bordering the Mediterranean Sea. Quails are netted by the ten thousand when they land on the shores of Europe on their spring migration from Africa—in 1898, 270,000 were sold in the Paris markets—and Larks are killed by the hundred thousand. The War has intensified the slaughter, for in May 1916 it was reported that the peasants of southern Hungary, unable to buy meat at the prices ruling, were killing song-birds, and that the woods were being rapidly denuded of their bird population. Large numbers of lesser birds, Larks, and even Thrushes and Blackbirds, still find their way to Leadenhall market in London, and strings of Starlings are said to be on sale daily in the market at Louth and in other market towns; but the trade in edible songsters has long since disappeared in Scotland. Yet in the old days the birds of the field and moor made a generous contribution to the Scot's table. I need hardly do more in support of this statement than quote, omitting the "tame meat," the provisions wherein an Act passed in 1551 by the Scots Parliament, "statute and ordained" that

the wilde meat and tame meat under written be sauld [sold] in all times cumming of the prices following, that is to say the crane [probably the heron] five shillings; the swan five shillings; the wild guse of the great bind [size] twa shillings; the claik, quink and rute, the price of the peece foure pennies. *Item*, the plover and small mure fowle, price of the peece auchteene pennies; the black cock and grey hen, price of the peece six pennies; the douzaine of powtes twelve pennies. *Item*, the quhaip [curlew] sax pennies.... *Item*, the woode cocke foure pennies. *Item*, the dousane of laverocks [larks] and uthers small birdes, the price of the dousane foure pennies. *Item*, the snipe and quailyie [quails] price of the peece twa pennies....

Bishop Leslie in 1578 also refers to the abundance and utility of Larks: "of Pertrikis in sum cuntreyes [i.e. districts] ar gret abundance, bot of Laferokis [larks] ouer all far gretter, in sa far as that xii. for a frenche sous they cummounlie sell." The "Northumberland Household Book" of 1512 shows that the Percys did not disdain such small fare as "Seegulles," "Styntes," "Snypes," "Redeshankes," "Kyrlewes," "Seepyes" [Oyster catchers], "Knottes," "Dottrells," "Smale Byrdes" and "Larkys," provided always "thay ar in Season."

At a much later date the lesser birds made contribution to feasts, even of Royalty, for at a banquet given by the City of London to George III in 1761, the Second Service included dishes of Ortolans or Buntings, Quails, "Notts," "Wheat Ears," Woodcocks, Teal and Snipes. Moreover it was no insignificant destruction of wild birds that satisfied the table of former days. Glance at the provision made for the great feast at the "intronization" of George Nevelle, Archbishop of York, in 1466, which included

Swannes, cccc.; Geese, MM; Plovers, iiii. C.; Quayles, C. dosen; of the fowles called Rees [Reeves], CC. dosen; In Peacockes, C. iiii.; Mallardes and Teales, iiii. M.; In Cranes, C. iiii.;...Pigeons, iiii. M.;...In Bittors, C. iiii.; Heronshawes, iiii. C.; Fessauntes, CC.; Partriges, v. C.; Woodcockes, iiii. C.; Curlews, C.; Egrittes, M.

Apart from making use of resident native birds our

predecessors did not hesitate to take toll of temporary immigrants. Brand in 1701 tells us that in Caithness "Especially there is a kind of Fowls called Snowflects [Snow Buntings] which resort to this Countrey in great numbers in *February*, they are about the bigness of a Sparrow, but exceeding fat and delicious; they flee in flocks, thousands of them together, many of which the Inhabitants do kill and make use of. They use to *go away in April* and are thought to come from the West Highlands."

CORMORANTS, ROCK-DOVES, AND KITTIWAKES

Larger fry claimed the attention of coastwise dwellers, especially where caves afforded shelter to innumerable hordes of birds. Take for example Macaulay's description of the slaughter of Cormorants (*Phalacrocorax*) and Rock Doves (*Columba livia*) at "Hawskeir," an island near North Uist in the Outer Hebrides:

On the west side of the rock are two remarkably large caves, of a considerable height: To these a vast multitude of sea cormorants retire every evening. Here likewise they lay their eggs and foster their young. The method practised by the islanders for catching fowls of this kind, while secured within such fastnesses, is far from being incurious, though abundantly simple, nor is the pastime at all disagreeable.—A band of young fellows make a party and after having provided themselves with a quantity of straw or heath, creep with great caution to the mouth of the cave which affords the game, armed with poles light enough to be easily wielded: This done, they set fire to the combustible stuff and raise an universal shout; the cormorants, alarmed by the outcry, frightened by a glare so strange, and offended by the smoak, quit their beds and nests with the greatest precipitation, and fly directly towards the light: Here the sportsmen, if alert enough, will knock down a considerable number of them, and together with the cormorants, whole coveys of pigeons.

At Fair Isle, as Dr Eagle Clarke has recorded, a similar method of catching Rock Doves was employed, but there a sail was hung over the mouth of the cave before a lantern was lit within, and the birds making for the entrance flew against the sail, and falling to the ground, were picked up by the hunters. So many were thus slain year after year that, Dr Eagle Clarke tells me, no Rock Dove is now to be seen on the island, although other causes, such as the decrease of cultivated ground and the recent introduction of the gun by the lighthouse-keepers, have contributed to its disappearance.

Even the Kittiwake (*Rhyssa tridactyla*) found its admirers in former times: Sir Robert Sibbald (1684) reckoned it "as good meat as a partridge," and the fishermen and inhabitants of coastwise villages used it largely as food even towards the end of the eighteenth century. At this period Pennant found that it was used in Aberdeenshire, near the Bullers of Buchan, as a whet for the appetite before dinner, and tells a story of a stranger to the custom, who on encountering for the first time this appetiser, as he thought, declared with some warmth after demolishing half a dozen savoury Kittiwakes that he had eaten *sax* and was not a bit hungrier than when he started. But the Kittiwake has survived the gourmets of the Aberdeenshire coast.

Sea-birds formed the staple food of many a coast dweller in the days before travelling facilities had broken down the barriers of isolation. To this we owe the tragedy of the Garefowl in Scotland.

THE TRAGEDY OF THE GAREFOWL OR GREAT AUK

The Garefowl or Great Auk (*Alca impennis*) was a large bird, its flesh was good for food, its fat supplied oil for light, and its feathers were soft and useful. It lived in great colonies, as the Penguins of the southern oceans do to-day, its wings were too small for flight, and the bird was stupidly docile. So it was that when the voyagers to the coast of North America found it in abundance, they made of it an easy prey. At first the Garefowls were knocked on the head with clubs, but the process became too laborious, and finally planks and sails were run ashore and the defenceless birds driven on board the boats by the ton weight, so that the boats were often in danger of being swamped. At first the Garefowls were skinned, their feathers kept, and their bodies salted down like herrings and packed in barrels for food. But such is the ruthlessness of man, that latterly thousands more were captured than could be stored, the valuable feathers were plucked off and the bodies burned for fuel; even when no profit could be made by killing them, the poor birds were tortured and burned alive for the amusement of the barbarian crews bred by European civilization.

Is it any wonder then that this bird, so abundant on the

Atlantic coasts of North America that in 1540 a voyager loaded his two vessels with dead Garefowls in half an hour and had, besides the birds eaten fresh, four or five tons to

Fig. 34. Garefowl or Great Auk (once a native of Scotland, now extinct) with its solitary egg. (From example in Royal Scottish Museum.) ⅛ nat. size.

put in salt, should have been exterminated there in the early years of the nineteenth century?

In Scotland there is no record of such heinous slaughter as darkens the path of the early exploiters of the natural wealth of America. Yet here too, the simplicity and defencelessness of the Garefowl made for its ultimate disappearance. Causes other than its food value operated against it. It was unable to fly and perforce had to nest upon the seashore, where eggs and young lay exposed to the easy attacks of beasts and birds, as well as of man. Its solitary egg told against it, for this slow multiplication offered no chance of recuperation from the destruction which dogged the Garefowl's landward migrations.

All that is known of its history in Scotland marks the stages of decay—reduction in numbers through man's deliberate destruction, curtailment of range, and final extermination. When man first reached Scotland the Garefowl was widely distributed upon the coast of Britain. Its remains have been found in the Cleadon Hills of Durham, and the early immigrants of Azilian culture who settled in Oronsay made use of it for food, as their kitchen-middens clearly show. Even in early historic times it occurred in regions whence it had long disappeared before written history takes up its story; for its bones were discovered in a refuse-heap, probably of the Broch period, in the ancient harbour-mound at Keiss in Caithness.

So long ago as 1684 it was recorded from the Outer Hebrides, and here only and in the Orkney Islands are definite records of its having bred in Scotland. Apart from a few odd references its story is mainly connected with St Kilda, where it used to arrive in considerable numbers in the spring time, seeking the shore from its winter home on the wide sea, in order to lay its solitary egg and hatch its young. One can trace in the accounts of successive visitors to the island its gradual reduction in numbers, for it was slain for its flesh, its oil and its feathers, until it became a rare and occasional visitor and at last disappeared. Sir George M'Kenzie of Tarbat in an account sent to Sir Robert Sibbald, apparently before 1684, merely mentions the "Gare-Fowle" amongst other common sea-birds of St Kilda, as if it merited no special description. It seems still to have been common in Martin's time, since his account of his visit to the island in 1697, faithfully describes the *Gairfowl*, first among sea-birds, "being the stateliest as well as the largest Sort." It is probable that Martin saw many alive, for he says that "it comes without Regard to any Wind, appears the 1st of *May* and goes away about the middle of *June*," and his own visit to St Kilda extended from 1st of June almost to the end of the month, the period when the islanders would be most actively engaged in collecting the birds for their winter stores. Yet in the account of St Kilda in his

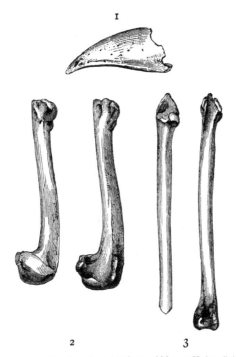

Fig. 35. Bones of the extinct Garefowl from kitchen-midden at Keiss, Caithness. ½ nat. size.
1. Upper portion of beak   2. Right and left wing bones (humeri) (inverted).
3. Right and left leg bones (tibio-tarsi).

Description of the Western Isles of Scotland, published in 1703, Martin mentions only the Solan Goose and the Fulmar as being the most important of all the birds to the inhabitants. Perhaps already the Garefowl had fallen from its rank with these birds, although in Martin's own opinion in 1697 it took its place with them in the island's economy.

A few years later, the minister of St Kilda, Rev. Mr A. Buchan, in an account written between 1708 and 1730, though not published till 1773, mentions the Garefowl as still a visitor to the island; but while he describes the more abundant and useful sea-birds in detail, he passes lightly over the Garefowl, whose importance to the islanders had apparently greatly diminished with its shrinking numbers. Before the next recorded visit to St Kilda was paid, the Garefowl, from being a regular visitor had become a mere straggler to the island, appearing now no longer in May to breed, but as a wanderer in July. In 1758 the Rev. Mr Kenneth Macaulay visited the island and in his description, published in 1764, makes mention of

a very curious fowl sometimes seen upon this coast....The men of *Hirta* call it the Gare-fowl, corruptly, perhaps, instead of Rare-fowl....It makes its appearance in July. The St Kildeans do not receive an annual visit from this strange bird....It keeps at a distance from them, they know not where, for a course of years.

Here the records of any regular sort of visitation of St Kilda by the Garefowl cease, and a few more exceedingly casual appearances complete the story of its existence. A specimen was captured alive off the island in 1821 or 1822, a few years after the last individual had been taken in the Orkneys on Papa Westray in 1813. And with an individual captured on St Kilda in 1840, the history of the Garefowl in Scotland comes to an end. It may have lingered on for a few more years in Iceland or the Faroe Islands, but about 1844 or 1845 the Garefowl disappeared from the world of living things.

THE GANNET AND THE FULMAR

Fortunately not all the sea-birds upon which the inhabitants of the isles depended for food have met the fate of the Garefowl. No bird could well have been more useful to the St Kildans than the Gannet or Solan Goose (*Sula bassana*), whose oil and feathers were of inestimable value, and whose carcases, to the number of over twenty thousand, were preserved annually for winter fare. Of St Kilda and its neighbours, Soay and Boreray, Martin wrote in 1703,

The largest and two lesser Isles...abound with a Prodigious number of Sea-fowl from March till September, the Solan Geese are very numerous here in so much that the Inhabitants commonly keep yearly above twenty thousand young and old in their little stone Houses of which there are some hundreds for preserving their Fowls, Eggs, &c. They use no Salt for preserving their Fowl, the Eggs of the Sea Wild-Fowl are preserved some months in the Ashes of Peats, and are astringent to such as be not accustomed to eat them.

Even on the small area of the Bass Rock as many as 1300 Gannets were slaughtered yearly in the latter half of the eighteenth century, their products being valued at some £120; and the destruction of man has altogether driven this interesting bird from Lundy Island, a former haunt on the coast of Wales. That the Gannets caught on the Bass were widely used for food is indicated by the following advertisement from the *Edinburgh Advertiser* of Aug. 5, 1768:

"SOLAN GOOSE.

"There is to be sold, by JOHN WATSON, Jun. at his Stand at the Poultry, *Edinburgh*, all lawfull days in the week, wind and weather serving, good and fresh *Solan* Geese. Any who have occasion for the same may have them at reasonable rates."

The reasonable rate was about "twenty-pence apiece," but the old birds were said to have had a flavour too rank and fishy for the average palate, so that only young or newly fledged chicks were commonly eaten. They "used to be considered as excellent provocatives."

The Fulmar Petrel (*Fulmarus glacialis*), a bird which is annually increasing its range at the present day, also paid heavy toll at the hands of the people of St Kilda.

"Can the world" said one of the most intelligent inhabitants to the Rev. Mr Macaulay in 1758, "exhibit a more valuable commodity? The Fulmar furnishes oil for the lamp, down for the bed, the most salubrious food, and the most efficacious ointments for healing wounds, besides a thousand other virtues of which he is possessed which I have not time to enumerate. But to say all in one word, deprive us of the Fulmar, and St Kilda is no more."

"Of the fowls themselves," Macaulay tells us, "every family has a great number salted in casks for winter provisions, and the amount of the whole is about twelve barrels." At the present day the Fulmar has entirely replaced the Gannet in the economy of St Kilda.

"BIRD-BUTTER" AND BIRDS' EGGS

The value of the sea-birds was greatly increased by the fat they yielded, which in the hands of the St Kildans was converted into a highly nutritious butter-substitute,—a kind of bird-butter. Martin has described how they manufactured the fat of their sea-fowls into "their great and beloved Catholicon, the Giben, i.e. the fat of their fowls, with which they stuff the stomach of a Solan Goose, in fashion of a pudding."

"This *Giben*," he says, "is by daily Experience found to be a sovereign Remedy for the Healing of Green Wounds....They boil the Sea-plants, *Dulse* and *Slake*, melting the 'Giben' upon them instead of Butter....They use this 'giben' with their Fish, and it is become the common Vehicle that conveys all the Food down their Throats."

In the earlier days, when the Great Auk was abundant, its capacious stomach seems to have been preferred as a repository for the bird-butter, on the same ground that the Greenlanders found it to be the most efficient float for their harpoons.

The destruction of birds' eggs for food has also had some effect upon the bird population. A sixteenth century manuscript in the Advocates' Library in Edinburgh (MS. 31. 2. 6) states of the inhabitants of St Kilda that "thair daily exercitation is maist in delving and labouring the ground, taking of foullis and gaddering their eggis, quharon they leif for the maist pairt of their fude," and Martin calculated that during a three weeks' residence on the island, the members of his own boat's crew and that of the Steward collected "Sixteen Thousand Eggs of Sea-Fowl." The cliff-climbers of Shetland and Orkney, too, were renowned for the success, as for the hardihood of their raids. But the effect on birds which frequent every ledge of a suitable rocky coast in numbers innumerable, is less patent than that upon land birds which occur in more limited numbers. The *Times* of 1871 recorded that so ruthlessly were Lapwings robbed of their eggs, which as a delicacy commanded 3*d.* to 6*d.* each, that, at that time, the bird was almost exterminated in the north of England, and the Statutory Orders protecting Lapwings' eggs at the present day are a sign of the reality of this destruction throughout the country.

FISHERIES

Since the days when the Azilian wanderers from Europe cast on their refuse-heaps in Oronsay the remains of the Wrasse and the Sea-Bream, the Conger Eel, the Spiny Dog-fish and many another, the wealth of our seas has been increasingly purloined on behalf of man. The extent of Scottish sea-fisheries was the constant wonder of early travellers from other lands. "It is impossible to describe the immense quantity of fish. The old proverb says already 'Piscinata Scotia'," wrote Don Pedro de Ayala, ambassador from Ferdinand and Isabella, in 1498; and so also the Italian, Ubaldini, in 1529:

> They have besides...an incredible quantity of fish from all parts of the island and especially when one goes more towards the North,—in such fashion, that the people of the Island being unable to consume so much fish, furnish and load infinite ships every year for France, Flanders, Zeland, Holand, and Germany, and inland even, and even into other and more distant countries, but for the delight of richer, greedy or more gluttonous men.

It is little to be wondered at, therefore, that our own historians found it difficult adequately to describe the resources of our seas. Leslie (1578) says the "Lochis or bosumis of the Sey" are "copious in herring miracolouslie," so that, as the Wardlaw chronicler mentions a hundred years later, "the greatest hearing sold for twopence, at least a penny, the least, two farthings, the hundred. No such penny worth in the world."

> "As tuecheng [touching] vthiris fishes," continues Leslie, "I can nocht tell, gif in ony place in the warlde, athir be mair varietie or mair abundance, of sum kyndes, baith freshe and salt water fishe."

The abundance of fish led to extraordinary slaughter, in which foreign vessels played no little part. It is on record that in one year three thousand busses or small fishing smacks were known to have been employed by the Dutch in the herring fishing in Shetland, beside those fitted out by the Hamburghers, Bremeners and other northern nations; and in recent years the number of fishing craft working round the Scottish coasts has exceeded 10,000, while the amount of fish landed outruns eight million hundredweights.

What the ultimate result of such destruction has been upon the apparently inexhaustible resources of the sea, it is difficult to gauge. The herring shoals which used to frequent many of our sea-lochs seem to have disappeared, and in a limited area, such as the North Sea, there can be little doubt that the stocks of various kinds of fishes have been reduced in numbers, as well as in size of individuals. Many of man's engines of destruction have contributed to this result. Since the introduction of the otter-trawl in the deeper water, the daily catch of Plaice has continually fallen off, and in inshore fisheries such implements as the shrimp-trawl have been responsible for the destruction of hordes of small fishes, the promise of years to come. In half an hour's shrimping in the Mersey, Mr J. T. Cunningham found many young fishes among his 56 pints of Shrimps—10,407 Flounders, 375 Lemon Soles, 169 Hake, 70 Ling, and 12 Soles.

In the more limited fisheries of our rivers, the results of man's destruction are less dubious, although here pollution of rivers and the creation of obstacles to migration have combined with active slaughter to reduce the stock. Think of the massacres which every year overtake Salmon on their migration from the sea. Stoddart gives a vivid description of such a killing, which took place near Melrose in 1846, when

> upwards of three hundred breeding fish writhed and bled on the prongs of a single leister [a type of three-pronged fork, famous in the annals of Border poaching,] and at least six thousand which had escaped the toils of the Berwick fishermen and formed the hope and stay of future seasons of abundance, were cut off by means of the deadly instrument, along the course of the river [Tweed].

But the river slaughter is insignificant compared with the destruction caused by fixed nets and other "engines" which ensnare the Salmon on their journeyings along the coast or in estuaries in search for a suitable stream.

For many hundreds of years the present methods of catching fish immigrating to our rivers from the sea have been practised; and for almost as many years discussions have raged as to the ultimate effect of these methods upon the fish stock of stream and river. As a rule the destruction of migrating fishes in nets and fixed engines at the mouth of a river, has been held to influence adversely the migrant stock, since many of the Salmon on which the bounty of a future season depends fail to reach the spawning grounds.

DESTRUCTION FOR FOOD

So even in the reign of Robert Bruce we find statutes forbidding the use of "fixed engines," and an Act of 1424 (James I) decreed

that all cruives and yaires set in fresche watteris, quhair the sea filles and ebbis, the quhilk destroyis the frie of all fisches be destroyed and put awaie for ever mair.

The Royal Commission on Salmon Fisheries in 1902 reached the same conclusion as to the definite results of nets and fixed engines upon the fisheries of certain rivers. The evidence is vague and contradictory, but it seems to show that there is a gradual decrease in the numbers of Salmon in the upper waters of some netted rivers and in the estuaries of others, the latter point being strikingly indicated by the decline in the number of applications for net fishing licences in certain districts of England and Wales. Besides, there is fairly clear evidence that in some areas the immigrant fishes are unable to reach the upper waters and the spawning grounds until the nets are off, and that the permanent or temporary removal of nets to allow easy access to the river has resulted in improvement of the fisheries.

It is not so much the actual numbers of immigrating fishes destroyed by man that count, though this in itself affects the stock of the river, but that many mature fishes are prevented from reaching the spawning grounds to the prejudice of the stock of future seasons.

A result undoubtedly due, in part at any rate, to man's interference can be seen in the general decline in the catch of grilse, especially marked since the opening of the nineteenth century. At that time the catch of grilse in Tweed alone was occasionally a hundred times greater than the catch for the whole of Scotland to-day, and, wrote Mr W. L. Calderwood in 1916, "in a period of twenty years it never fell below a figure thirty times as great as the present day catch for the whole of Scotland." The economic standing of the Tweed fisheries gives a clear indication of the decline due to destruction, for whereas in 1807 the rents of the Tweed fishings amounted to £15,766, in 1860 their value had fallen to a little over £4000. The accompanying diagram illustrates the results of fifty years fishing at the mouth of the Tweed by the Berwick Salmon Fisheries Company. In the diagram the annual total catch, as well as the catch of Salmon and Trout separately, are shown, the annual average being struck for five-yearly periods. Part of the extraordinary decline, from an average of 109,971 Salmon and Trout in the period 1842 to 1846 (in 1842 the numbers reached 148,930,) to less than 40,000 in the 'seventies, is no doubt due to restrictive legislation controlling the size of mesh of salmon nets, but this still leaves much of the falling off to be explained.

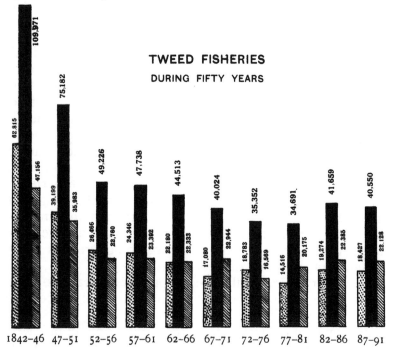

Fig. 36. The Decline of Tweed Fisheries, as shown by the statistics of Trout and Salmon caught during fifty years. Each column indicates the average annual catch in a period of five years. Dotted columns indicate number of Trout caught; lined columns, Salmon: solid columns, total catch of Trout and Salmon.

The figures at tops of columns show actual numbers caught; those at bottom the five yearly periods.

In other countries also, destruction is proceeding apace: in 1913 on the Pacific coast of the United States of America, 140,000,000 Salmon were slain for food, and the gross value of canned salmon packed in North America in 1916 amounted to £8,708,527.

Take again the case of a close relative of the Salmon and Trout—the Char (*Salvelinus alpinus* and its races).

This beautiful fish is a native of our deeper lakes, which it entered, a migrant from the sea, at the close of the Ice Age, and where it has been impounded by subsequent changes in the volume of rivers and in the configuration of the surface of the country. The Char is an excellent food fish, and has long been netted in many lochs. I give only one instance of the destruction of Char in Scottish waters. A few years ago, Mr Campbell, a well-known inhabitant of Kingussie, told me that in his young days (he mentioned about the years 1860 to 1870), Char were so common in Loch Insh, Inverness-shire, that when they came up the river Spey to spawn—a custom peculiar to Loch Insh Char, and probably a relic of the old migratory habit which still characterizes Char in Scandinavia—enormous numbers were netted and snared with a simple ring of brass upon a long stick. He himself with two companions, on one occasion netted in the Spey, in a few hours during the night, as many Char as the three fishers could carry home. The fish were pickled with salt and preserved for winter use. Nowadays Char are by no means so plentiful in Loch Insh, and there can be little doubt that the great slaughter regularly inflicted upon them at spawning time has been a cause of their decline. In several other lakes they have been entirely extirpated within the memory of man.

SHELL-FISHERIES

The extent to which our coasts have been rifled of their molluscan shell-fish, since the times when the products of the sea-shore formed the staple food supply of our Neolithic forerunners, can scarcely be realized. At the present day, taking 1913, the last completed year before the War, as a standard, Mussels, Clams or Scallops, Periwinkles and a few others to the amount of 126,468 hundredweights valued at £16,662, were gathered on the Scottish coast for food or bait, and to these must be added 1,316,100 Oysters valued at £4757. Many of our common shell-fish could successfully withstand much greater depletion, but some have been clearly affected by man's destruction. "Hard by the town [of Leith]," wrote Lowther in 1629, " be oysters dragged which go to Newcastle, Carlisle and all places thereabouts, they be under 3*d*. the 100"; and in 1794 the oyster-beds opposite Prestonpans formed the chief fishery of that port, the ten oyster boats seldom returning without 400 or 500 Oysters each. But, in spite of protection, the oyster-beds of the Forth have long ceased to be worth working and have now almost disappeared. Our forefathers who accumulated the kitchen-middens of past times lived largely on Oysters on several parts of the coast (as in Easter Ross) where Oysters are now scarcely to be found.

In Co. Clare, Ireland, the consumption of the Warty Venus Shell (*Venus verrucosa*) was so great that at one time, according to Mr Damon, the species was almost exterminated there.

Land-shells, because of their more limited numbers, seem to have suffered even more severely than the denizens of the sea. It is on record that in the kitchen-midden of an ancient underground house in N. Ronaldshay in Orkney, great numbers of the Banded Garden Snail (*Helix nemoralis*) were discovered. But the Banded Garden Snail is now extinct on the island. So great did the consumption of the large Roman Snail (*Helix pomatia*) become in France, where during the season 50 tons were sent daily to Paris alone, that its serious depletion in numbers compelled the French Government to protect it by law, and close seasons were instituted when it was illegal to gather or eat the desirable morsel.

Crustacean shell-fish have likewise suffered from the zeal of man. In 1913, Scottish coasts supplied the table with 681,059 Lobsters, and 2,213,866 Crabs, and as a result of constant harrying, and in spite of protecting laws, the numbers, as well as the average size of Lobsters have fallen off in many places.

III. 3

DESTRUCTION FOR SKINS AND OIL

It is only in recent years that the value of fur-bearing animals to mankind has been estimated at its real worth, and that ruthless slaughter has been replaced by endeavours to preserve and encourage the multiplication of creatures whose coats are valuable commercial assets to their native countries. There was no such farsightedness in Scotland of the old days; fashion and profit were the only guides, and the result was a persistent destruction, which, sometimes outpacing natural reproduction, contributed to the curtailment of range or even to the extermination of our few native fur-bearing animals.

In former times Scotland held a European reputation for its skins. What animals contributed to this fame, and suffered on its account? Boece, in the sixteenth century, tells us that

> King Ewin biggit ane othir toon on the river of Nes, quhilk is yit namit Innernes [Inverness], quhair sum time wes gret repair of marchandis, quhilkis come out of Almany to seik riche furringis; as martrikis [Martens], bevaris [Beavers] and siclike skinnis quhilkes aboundis in that regioun.

In another place he mentions the coveted creatures in rather greater detail:

> Beside Lochnes, quhilk is xxiv milis of lenth, and xii of breid, ar mony wild hors; amang thame, ar mony martrikis, bevers, quhitredis [Stoats or Ermines], and toddis [Foxes]; the furringis and skinnis of thaim are coft [bought] with gret price amang uncouth [foreign] marchandis.

And of the people of Orkney, Leslie said in 1578 that part of "thair riches consistes...in the skinis of wilde beistes."

Further light is thrown upon the traffic in Scottish skins by an analysis of the customs duties levied upon exports. The Ayr manuscript, written in the days of King Robert Bruce, in a chapter on "*Peloure* or *Peltry*," enumerates along with the commoner skins of Tod, Whitret, Mertrick and Cat, those of Beaver and Sable, perhaps foreign skins re-exported, as well as hides and Deer skins. And at the Port of Leith in 1482, among the taxed articles are mentioned skins of Calf, Goat, Kid, Rabbit, Polecat, Otter and Badger. Many varieties of skins of domestic animals contributed to the exports—fleeces of sheep, skins of "shorlings," lambs and "futfallis" (lambs that die just after birth), goat skins and calf skins, kid skins and salt hides, but such skins scarcely entailed slaughter other than food requirements made necessary. Of the skins of wild animals the chief annual exports in the early years of the seventeenth century, as shown in an important paper from the charter chest of the Earl of Mar and Kellie, were

hairt hyddis [Red Deer skins] 91 daicker[1] extending, at £20 the daicker, to £1830[2]; rea [Roe Deer] skynnis, 240, at 16s. the pece, £186[2]; tod skynnis, 1012, at 40s. the pece, £2024; otter skynnis, 44, at 40s. the pece £88; and cuneing [Rabbit] skinnis, 53,234, at £6 the hundreth, £3194.

Apart from this considerable export of Scottish skins, many changed hands within the country at local fairs, of which the annual "Fur Market" of Dumfries was typical. Here there was on sale every February, the year's produce of Dumfriesshire, of the Shire or Stewartry of Galloway, of the counties of Ayr, Lanark, Peebles, Selkirk, Roxburgh, and even of Cumberland and Northumberland; Hare skins, sometimes to the number of 70,000; Rabbit skins, in one year as many as 200,000; Fitches, Foumarts or Polecats, on one occasion 600; and skins of Otters, Badgers, Foxes and Cats as the supply offered.

The constant drain upon the wild inhabitants of the country for the sake of their skins, cannot but have told upon their numbers. Yet so complicated is the influence of man that the ultimate effects are not always easily to be traced. On the one hand, new opportunities of increase in numbers, afforded by the development of cultivation of the soil, more than compensated for the destruction of Hares and Rabbits, while on the other hand, their destructiveness, real or fancied, hastened the decrease of "vermin" such as the Marten and Polecat, the Otter, Badger and Fox Nevertheless an account of the histories of a few typical fur-bearing animals in Scotland will give some indication of the effects of continued persecution.

[1] A daicker or daker—Lat. *decuria*, from *decem*—comprised 10 hides.
[2] I trust that the charter-writer's statistics are more reliable than his arithmetic in these cases.

THE BEAVER

Rarest and most interesting of all Scotland's fur-bearers was the Beaver (*Castor fiber*) (Fig. 37). Of its presence throughout the country in days long gone by, there is indubitable evidence, for its remains have been found in the deposits of ancient lakes in which it disported itself before man's advent to North Britain. From the Solway to Perthshire in these days the Beaver was common. Its

Fig. 37. European Beaver—exterminated in Scotland. $\frac{1}{18}$ nat. size.

remains have been found in Dumfriesshire, and in a peat moss at Kimmerghame in the parish of Edrom, Berwickshire, many bones were discovered in 1818 together with a well preserved skull. Linton Loch in Roxburghshire has also yielded a Beaver's skull, preserved in Kelso Museum, and in the marl of the loch of Marlee, in Kinloch parish, Perthshire, a skeleton was discovered of which the skull and one of the haunch bones were presented to the Society of Antiquaries of Scotland in 1788. Of its continued presence in the Lowlands during man's tenancy of the country the evidence is equally clear, for in two prehistoric settlements in Ayrshire—a shell-mound at Ardrossan and a cave shelter at Cleaves Cove, Dalry—the bones of the Beaver have been discovered amongst the miscellaneous contents of the kitchen-middens. In historical times there are few references to its presence, yet we cannot but believe that it existed in the less frequented rivers almost till the middle ages. There are Gaelic traditions telling of the presence of *Dobhran-losleathan*—the Broad-tailed Otter—in many parts of the Highlands, and it is said to have been plentiful at one time in the district of Lochaber in Inverness-shire.

In the twelfth century, Giraldus Cambrensis, who found Beavers in Wales, recorded that he had been informed that they still existed in one river in Scotland, and in the early half of the same century "*Beveris*" are included in a list of animals whose skins were subject to export duty in the reign of David I (1084-1153). I have already quoted passages in which Boece mentions it as occurring in Loch Ness (p. 155), and though it is possible, as has been suggested, that Boece was recording only some vague tradition that had reached his ears, yet many of his statements, it seems to me, have met with unnecessary scepticism, and I see no reason why the Beaver may not have lingered on in the wilds of Inverness-shire even to the sixteenth century, since many others of our decadent creatures found there a safe retreat. If so, it could not long have survived the date of Boece's record; and in the light of its history here as in other countries, it is safe to attribute its extermination to the destructiveness of man. I know no case which illustrates more clearly how a single whim of fashion can affect the creatures of a land even far distant, than the history of the Beavers of North America (*Castor canadensis*). So long as beaver fur was used on a large scale for the making of hats, the Beaver was so keenly hunted that it was threatened with extinction, but the invention of the silk hat, in ousting the 'beaver,' resulted in an immediate increase in the numbers of American Beavers, and in their reappearance in places they had long forsaken.

THE MARTEN OR MERTRICK

Once a common denizen of Scottish woods and wilds, the Pine Marten (*Martes martes*) (Fig. 38), or Mertrick in the Old Scots tongue, was in former times persistently slaughtered for its "costly furrings." In the old laws and ledgers reference is often made to the Marten's skin. It was one of the valuable items of Scottish export, and a regulation made in the reign of David II (1324-1371),

Fig. 38. Pine Marten—approaching extinction in Scotland. (From individual killed at Kintail, West Ross, in 1886.) ⅓ nat. size.

imposes a "custom of 4*d*. to be paid on each timmer[1] of mertrick skins at the outpassing." In 1424, the duty was raised to 6*d*. on each Mertrick skin exported.

It was possibly the fact that Martens had become rarer, owing to the continual drain upon their numbers, that suggested the use of their increasingly desirable skins as

[1] *Timmer* or *Timber*, a merchant's term used to denote, according to kind, a number of skins. A timber of such as Martens, Polecats, and Ermines, contained 30 skins, of other creatures, 130 skins.

a caste distinction in the fifteenth century; seeing that in 1457, the Scottish Parliament ordained "burgesses (unless bailies or councillors) and their wives and daughters, and clerks (unless dignitaries of cathedrals) not to wear furrings of mertricks except on holiday."

The result of such slaughter as the Marten was subjected to could be in no doubt, and the history of the Marten in Harris in the Outer Hebrides, as given by successive visitors, well illustrates its progressive effects. In 1549 Monro wrote "In this countrey of Harrey [is]...*infinite slaughter*[1] of otters and mactickes [probably a mis-spelling for mertricks]." A century and a half later, in 1703, Martin says "they are *pretty numerous* in this Isle; they have a fine Skin, which is smooth as any Fur, and of a brown Colour." Little more than a century passed and in 1830 Macgillivray found that it was "*not very uncommon*"; and less than fifty years later, Harvie-Brown regarded it as *almost exterminated* there.

On the mainland the same process of extirpation proceeded, intensified by the fact that the Marten's depredations in the poultry yard and amongst game, and its occasional forays against lambs and even grown sheep, led to its enrolment in the class of "vermin." The disappearance of the Marten has been more rapid than that of the Wild Cat, for, notwithstanding that at a late date it covered a wider area of Scotland, it is now the rarer of the two creatures. Dr Harvie-Brown has given a comprehensive account of its history in Scotland up to 1881, and later information has added little to the general truths his facts brought to light. It is true that the track of its disappearance is often obscured by the sporadic occurrence, in districts far from their birth place, of wandering individuals or pairs, which have little chance of setting up successful new colonies. Yet the main drift of the disappearance of the Marten is clear.

The cultivated districts were earliest forsaken, an indication that pest rather than pelt determined its disappearance, and indeed, vermin or rabbit traps have accounted for most of the numbers slain during the nineteenth century. A few pairs may still lurk in the Cheviots, but the Lowlands of Scotland are now practically deserted, although it was common so far south as Kirkcudbrightshire in 1796, and

---

[1] I have italicised the significant words in each passage.

odd examples were killed in Ayrshire in 1876 (Maybole) and in 1878–9 (Minnoch Water). From the midlands too it has gone. It was "rare" in Stirlingshire at the end of the eighteenth century, and the last wanderer to Dumbartonshire was killed at Arrochar in 1882, the last to the kingdom of Fife near Dunfermline in 1873. It has been driven even from the wilds of southern Perthshire, which ceased to be an important breeding ground about the 'fifties and 'sixties of last century, although odd stragglers were seen up to 1880, when one was killed at Balquhidder.

The east coast also has been deserted. The last recorded example in Forfarshire was slain about 1860. In Aberdeenshire the Marten has been driven to the wilds of Strathdee and Strathdon, for although a straggler was found at Gourdas in Fyvie in 1894, the last individual on the coast was killed at Ellon in 1874. The southern border of the Moray Firth was abandoned many years previous to the appearance of a pair of wanderers near Burghead about 1868.

In the more northern counties the Marten still retains a hold, but in reduced numbers which find sanctuary in the protected wilds of the deer forests. The last remnant of a once universally distributed and flourishing race has been pressed backwards and ever backwards by the persecution of mankind, till it now finds itself concentrated in the forests and moors of the central Highlands, in Sutherland, Ross and Inverness and perhaps in Aberdeen, Perth and Argyll. From the wilder breeding centres of these districts the Marten still ranges occasionally into new territory (an individual appeared at Colintraive on the Kyles of Bute in 1914), but such wanderers invariably meet a fate unworthy of their venturesomeness, and the Marten, first pursued for its skin and later for its transgressions, has already trodden far upon the path to extinction.

THE POLECAT OR FOUMART

The history of the Polecat (*Mustela putorius*) (Fig. 39, p. 163) or Foumart (this Old English and Scottish equivalent meaning Foul Marten, from the atrocious smell of the creature) runs parallel with that of the Marten. Once an abundant and universally distributed denizen of Scottish wilds, it was hunted first for its skin, and later on account of its love of poultry and game. This, and especially its fondness for Rabbits, has led to its undoing.

In the old days its skin was an export of some value, the export duty of 4*d.* on each timmer[1] of Ferret skins imposed in the reign of David II (1324–1371) being raised, in 1424, to "8*d.* to be paid on each ten fulmart's skins, called fethokis, exported." In the country's fur markets, too, it held an important place, and the influence upon its welfare of man's interference can be clearly traced in the dwindling numbers of skins offered for sale, as well as in the rising prices paid for skins by traders at local fairs. At the annual Dumfries Fur Fair, the contemporary records of which, from 1816 till 1874 when the Fair ceased, were collected by Mr R. Service, there were exposed for sale in 1829 400 Foumarts' skins, in 1831 600 and in the following year, they were, as the contemporary account puts it "a drug on the market." Yet in 1856 the numbers had fallen to 240, in 1860 to 168, in 1866 to 12 and from 1869 till the Fair ceased there were "no foumart skins on offer."

The diagram (Fig. 40, p. 165) shows graphically the effect of constant slaughter on the numbers of Foumarts collected throughout the Lowlands of Scotland, and, equally instructive, the gradually increasing price which the skins commanded, for although the trend of fashion and demand may account for minor fluctuations, there can be little doubt that in the main the rising value is to be associated with the growing scarcity of the animal. The price is reckoned upon "the furrier's dozen" which consisted of

twelve very best full-sized skins, or a greater number of small-sized or secondary quality, or torn skins, so that a "dozen" sometimes really consisted of twenty or thirty or more of inferior skins.

The fixing of prices according to the "furrier's dozen," has the advantage, from our point of view, of practically eliminating price fluctuations due to the quality of the skins. The general rising tendency of the values, as plotted in the graph, is apparent; before 1850 twice only did the price reach 20*s.* and sometimes it fell to 12*s.* a dozen, after 1850 it fell beneath the 20*s.* standard only twice, and in one year ranged from 42*s.* to 45*s.* The steady and gradual rise is still more marked

[1] See footnote p. 159.

DESTRUCTION FOR SKINS AND OIL

Fig. 39. Polecat and young—approaching extinction in Scotland. ⅛ nat. size.

when the extremes are eliminated by the simple method of taking the average of five consecutive prices.

The slaughter, begun on account of the value of its pelt, was continued, in part by design, because of its thieving tendencies, and in part by accident; for after the Rabbit had been introduced and encouraged by man, the Polecat found in it an easy prey, and, congregating where Rabbits most abounded, fell an easy victim to the steel traps of the rabbit-catchers.

How, in face of such misfortunes, does the Polecat stand in Scotland to-day? The records of its occurrences up to 1881 have been collected by Dr Harvie-Brown.

In the south of Scotland it is almost if not quite extinct. Its general disappearance from Berwickshire may be placed about the 'sixties of last century, though one was seen near Linhope in 1880; the last Roxburgh example on record was caught in Liddesdale in the winter of 1879–80; in Dumfries one appeared on the lands of Glenlee in 1892, the only one heard of in the county for upwards of twenty-five years; an example killed in Troqueer parish in 1880 was probably the last of the Kirkcudbrightshire race. In Ayrshire, the Polecat was regarded as almost extinct in 1881; in Lanark none have been heard of since about 1860; and the last Renfrew example on record was killed in 1868.

In the midlands it has fared no better. In Haddington the last example recorded was shot near North Berwick about 1860: in Midlothian none have been seen since "a number of years prior to 1880" (Fala Hill); a stray wanderer to Kinneil in 1886 completed the Linlithgow tale; in Stirling and Dumbarton, where they were once so plentiful that one could be caught at any time, a solitary survivor was seen at Garden in the winter of 1879–80. Kinross has been deserted since 1860, Fife since 1880, when one was seen at Falkland. Southern Perthshire and the Loch Awe district of Argyll have been forsaken since 1860, though in the wilder parts of the latter county, in Sunart and Ardnamurchan, 50 were killed between 1870 and 1880.

Only a rare straggler now occurs along the east coast. There have been none in Forfarshire since 1860. Except for a casual wanderer such as that killed in a glen near Peterhead in 1894, and notwithstanding the fact that in two years about 1863 and 1864, one keeper killed 30 on a single estate (Littlewood) on Donside, and that 57 were killed in five parishes on upper Deeside in ten years from 1776, Aberdeenshire has ceased to be tenanted since about 1870.

The same tale has to be told of the southern border of the Moray Firth. A Polecat appeared at Whitewreath, four miles south of Elgin in 1898, but its fellows had been exterminated before 1880. In all but the wilder parts of Inverness it is extinct, though less than a hundred years ago it was so common that in three years (1837–40) Glengarry alone furnished 109 individuals.

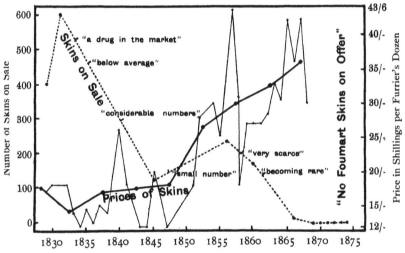

Fig. 40. THE DECLINE OF THE POLECAT; as shown by the dwindling numbers of skins and rising prices at Dumfries Fur Fair. Numbers of skins are indicated on the left, prices per furrier's dozen on the right. Broken line indicates *skins on sale*; thin unbroken line, *prices recorded*; thick unbroken line, *five-yearly averages of prices*.

The graph of skins is based on available numbers only, the remarks inserted are from the Market Reports of the Fair.

Even in Ross-shire it is now seldom seen, the example last recorded having occurred at Leckmelm in 1902. In the wilds of Sutherlandshire the Polecat has longer kept a footing: in 1912 one was seen at Lairg near Loch Shin, and at Inchnadamph the years between 1880 and 1889 yielded seven, and 1890 to 1899 eighteen individuals. It is hardly necessary to add that the islands have been long deserted, though Monro wrote in 1549 "Oronsay......quhair there is mayne laiche [low] land, full of hairs and foumarts."

The conclusion is clear that the curve of the decadence of the Polecat shown for the Lowlands of Scotland, applies with slight shiftings of dates to the country as a whole, and that this creature, once a universal native of Scotland, and now a rare dweller in the wilds of Ross, Sutherland, and Inverness, is balanced on the brink of extinction, to which it has been driven by the persecution of man.

Of Scottish skins there were none that ranked in value with those of the Beaver, the Marten and the Foumart, yet several other creatures suffered in less degree for their pelts. Among these were Rabbits and Hares, and, in order of descending importance, the Fox, the Otter, and the Badger; but the destructiveness of these creatures, as well as the demand for their pelts, accounts in great measure for their persecution.

RABBITS AND HARES

In a future chapter (p. 247) I shall discuss the introduction of the Rabbit (*Oryctolagus cuniculus*) to Scotland, but it may not be out of place to remind the reader that this native of south-western Europe was brought to Scotland and planted in warrens throughout the country, mainly on account of the value of its skin. This in former times far exceeded the value of its flesh, and commanded a price varying from half-a-crown to three shillings, according to its quality and size. In early times there was considerable foreign demand for rabbit skins, so that a Scottish law of 1424 imposed a duty of 12*d*. on every 100 "cuning" skins exported. In the beginning of the seventeenth century the Scottish export exceeded 53,000 a year. In the same century a great stimulus was given to the use of rabbit and hare skins when, in 1621, the wearing of "castor" or beaver hats was forbidden by law, except to the highest in the land to whom the special privilege of wearing "beaver" was granted in 1672, and when, following upon this prohibition, a further statute, of 1695, granted authority for making hats of rabbit and hare skins. For this purpose the underwool or down alone was used, but the whole pelt also was made into muffs and tippets, and was used for lining robes. So great was the demand for skins that in the middle of the nineteenth century a Parliamentary Report estimated the annual consumption of rabbit skins in England at 30,000,000.

In Scotland a great trade in the fur of Hares[1] and Rabbits was carried on at the local fur markets, Rabbits in southern Scotland being classed, in order of decreasing value of skin, as "warreners, parkers, hedgehogs and sweethearts," according as they lived in warrens, pleasure parks, had no fixed abode, or were tame. At the Dumfries Fur Market as many as 70,000 hare skins and 200,000 rabbit skins have been on sale in single years, but even such figures give no idea of the constant drain upon the numbers of these animals caused by the steady demand for fur.

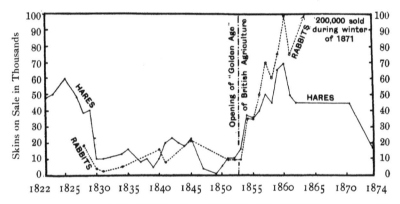

Fig. 41. THE DESTRUCTION OF RABBITS AND HARES; as indicated by the numbers of skins on sale at Dumfries Fur Fair.

Numbers of skins shown on right and left; years below; dotted line, *Rabbit skins*; unbroken line, *Hare skins*.

The position of the commencement of Prothero's "Golden Age" of British Agriculture is indicated.

In the above diagram I have represented graphically the numbers of hare and rabbit skins offered at the Dumfries Fair from the early years of the nineteenth century till the market came to an end about 1874.

The graph clearly shows that during these years Rabbits in the Lowlands did not suffer ultimately in number through the persecution of man, as one would have expected, but that on the contrary, while Hares showed a gradual decline till 1849 and then a steady rise till 1860, the

[1] Almost altogether skins of the Common Hare (*Lepus europaeus*), but perhaps also a few skins of *L. timidus*, the Mountain Hare.

numbers of Rabbits increased more than fourfold, an extraordinary contrast to the rapid decline which marked the history of the Polecat in the same area. The increase of Rabbits is especially striking when one recollects that at the beginning of the century they were comparatively few in the south-western Lowlands, for the *Agricultural Survey of Dumfriesshire* states that in 1812 "a few rabbits are to be found, but hardly worth mentioning. There is no regular warren." The increase of these grass-eaters is to be definitely associated with the increasing supply of food afforded them during a period which, from 1837 onwards but especially after 1853, was marked by steadily increasing agricultural activity throughout Scotland (see also p. 390).

It is worth noting also that the graph may afford some evidence of that antipathy between Rabbits and Hares to which I shall allude elsewhere (p. 253); for it is known that locally in Scotland, as in New South Wales, where exceptional increase of Rabbits has taken place, the Hares have been driven out. The close agreement, for the greater part, between the motions of the two curves may indicate that only very large numbers of Rabbits adversely affect the welfare of Hares.

THE FOX OR TOD

Of the destruction of the Fox (*Vulpes vulpes*) on account of its evil ways, I have already spoken, but it is well to remember that the value of its skin acted like a price upon its head in encouraging more strenuous pursuit. In the fourteenth century an export duty of fourpence was levied on each timmer of skins (see footnote, p. 159) and this was raised, in 1424, to sixpence on every ten "tod" skins exported. In the beginning of the seventeenth century the skins exported (valued at 40s. apiece) numbered 1012 a year, no inconsiderable slaughter when there is added to it the number traded within the country, and when the limited extent of the native stock is taken into account.

THE OTTER

The demand for the skins of Scottish Otters (*Lutra lutra*) has been long on the downgrade. The Otter skins which were exported in the early seventeenth century—the annual export averaged only some forty-four skins—fetched "40s. the pece" but thereafter the price seems gradually to have fallen. About 1800, according to the *Dumfries Courier* of February 21st, 1829, a Dumfries dealer who purchased *sixty* otter skins from a single individual, paid close on 30s. each for them; but the statistics of the Dumfries Fur Market show that from 1829 to 1869, when otter skins ceased to be forthcoming, the price averaged rather under 10s. a skin, and although it rose in 1840 to 13s., it frequently fell so low as 5s. and 6s., and touched its lowest ebb of 3s. to 6s. in 1866.

Nevertheless in the early days of trading the export was sufficiently great to warrant the imposition of a customs duty, which was modified from one halfpenny "on ilk otyr" in the fourteenth century, to sixpence on every ten otter skins exported in the fifteenth. In later times Scottish otter skins were mostly forwarded to London, where they were manufactured into gaudily decked purses for export to Africa, but apart from this, the demand was chiefly a local one.

The destruction of Otters for their fur, because of their raids upon Salmon and Trout, and for sport, has made inroads upon their numbers, occasionally attributed to other causes.

"The Ottars, also Seals or Selches, and other such Sea-creatures," wrote Brand in 1701 concerning Orkney, "are very numerous but now their number is so much diminished, that not one of Twenty is to be seen, and they have found several of them lying dead upon the Shore; some hence observing that the Judgments of GOD as to scarcity of suitable Provisions to these Creatures are upon the Waters also."

The Dumfries Fur Market gives clearer evidence of their gradual decline, for while 1829 saw 50 skins on sale, and 1831, 226, thereafter the greatest number recorded was 36 (1863), and after falling to six in 1866, otter skins ceased to appear three years later. As the prices, which during these forty years remain wonderfully steady despite an occasional large rise or fall, give no indication that the falling off was due to lack of demand, it is reasonable to conclude that it may represent a real decline in the numbers of Otters in the Scottish Lowlands, whence the supply of the market was drawn.

The tale of a single "vermin" list will indicate the penalty paid by the Otter for its depredations in fishing streams. On the Duchess of Sutherland's estates in Sutherland, during the three years from March 1831 to March 1834, a reward of five shillings offered for each head was paid on 263 Otters.

THE BADGER OR BROCK

In Scotland the decadence of the Badger (*Miles miles*) is to be accounted for by a multitude of influences which told severally and directly against it. It was hunted for sport, it was caught for baiting, it was destroyed for its destructiveness, it was killed for food, and its skins were a marketable commodity. Of all these direct influences the value of its skin was probably that which least influenced its welfare, for in Scotland the skins never created any great demand, and I mention the Badger here simply because the skins, which sold at some 5s. or 6s. each, made an occasional appearance at the Dumfries Fur Fair up to about the middle of the nineteenth century.

The case of the Badger, however, is typical of most of the fur-bearing natives of Scotland, and in concluding an account of the influence of the trade in skins upon the Scottish fauna, I would emphasize again that skins alone seldom formed the whole object of the persecution to which their possessors were subjected, but that other and varied objects, and especially the protection of game and minor domestic stock, combined to intensify the pursuit of most of the fur-bearers. In Scotland there is no instance of that single-mindedness and intensity of destruction which led to the slaughter by single ship's crews of 57,000 Fur Seals in 1800 during the short season in South Georgia, and of at least 74,000 in Australia in 1804, and which on account of their skins has almost exterminated Fur Seals in the Southern Hemisphere. In Scotland motives have been more complex, but the results if less striking have been no less fatal to the races of fur-bearing animals.

DESTRUCTION FOR OIL

In discussing the creatures destroyed for their fat or oil, we are reminded once more that a rigid classification of the motives which have led to the slaughter of animals conveys only part of the truth. While the fat of many a creature was manufactured into oil,—during centuries a necessity for light and heat to the dwellers on the more remote parts of the coast,—the flesh was often used as food. In particular the Solan Goose and Fulmar were invaluable in supplying both food and oil to the natives of their haunts, especially in the isolated outer islands. The inhabitants of the Bass Rock, in the beginning of the nineteenth century, were accustomed to obtain ten gallons Scots of oil drawn from the fat of the Solan.

For the most telling illustrations of ruthless and deadly destruction of oil-bearing creatures, the enquirer must look beyond the coasts of Scotland. In 1868 a few American whaling ships turned their attention to the Walrus (*Odobenus rosmarus*), which yields about 20 gallons of oil, and finding it an easy prey, each ship accounted for from 200 to 600 individuals. The result was that in succeeding years more ships followed suit, so that in 1870 the American whaling fleet is believed to have destroyed not fewer than 50,000 Walruses[1]. At the same time American traders were pursuing with easy energy the Penguins of the southern Ocean, so that small vessels specially fitted out for the work returned after a six weeks' cruise with 25,000 to 30,000 gallons of oil, and this, since eleven birds yield only a gallon, represented a slaughter of some 300,000 birds to each ship. Or take the case of the Turtle: Bates, in his *Naturalist on the Amazon*, records how its eggs are collected for the oil they contain, and estimates the destruction at 48,000,000 a year.

Although Scotland can instance no such appalling destruction of life, nevertheless the demand for oil proved a constant drain upon her more slender resources.

SEALS

So long ago as the days of St Columba, the Monastery of Iona reserved for itself a small island lying off the coast, whereon a colony of Seals was protected in order that the monks should be furnished with food and with oil to lighten the dark days of winter (see p. 222). On other parts of the coast, also, regular seal fisheries were engaged in in former times, for we learn, from a charter of David I to the

[1] This destruction was reflected in our fauna, for since the middle of the nineteenth century the occasional appearance of the Walrus on the coasts of Scotland has ceased.

Monastery of Dunfermline, that on the east coast Seals were objects of trade even in the twelfth century. Both the Common Seal (*Phoca vitulina*) and the Grey Seal (*Halichœrus grypus*) were objects of pursuit, but as a rule fisheries on the east coast were concerned with the former, and on the west coast and amongst the islands most often with the latter. At a later date, in the neighbourhood of North Uist, the slaughter was reckoned sometimes at 320 individuals a year. Here as elsewhere, the flesh of the Seal as well as its oil was used, the former being preserved for winter food. Martin tells how the men of "Heiskir" caught the Seals in a narrow channel between that and a neighbouring island, by means of a net of horse hair ropes, "contracted at one end like a Purse," and gives a detailed account of a seal-hunt in the Outer Islands, as he observed it about the opening of the eighteenth century.

On the western coast of this Island [of Heiskir] lyes the Rock Cousmil, about a quarter of a mile in circumference, and it is still famous for the yearly fishing of seals there in the end of October....These Farmers [probably the men of N. Uist] man their Boat with a competent number for the business, and they always imbarque with a contrary wind, for their security against being driven away by the Ocean, and likewise to prevent them from being discovered by the Seals, who are apt to smell the scent of them, and presently run to sea.

When this Crew is quietly landed, they surround the passes, and then the signal for the general attack is given from the Boat, and so they beat them down with big staves. The Seals at this On-set make towards the Sea, with all speed, and often force their passage over the necks of the Stoutest assailants who aim always at the Forehead of the Seals, giving many blows before they be killed, and if they be not hit exactly, on the Front, they contract a Lump on the forehead, which makes them look very fierce, and if they get hold of the staff with their Teeth, they carry it along to sea with them. Those that are in the Boat shoot at them as they run to Sea, but few are catched that way. The Natives told me that several of the biggest seals lose their Lives by endeavouring to save their young ones, whom they tumble before them towards the Sea. I was told also that 320 Seals, Young and Old, have been killed at one time in this Place....The Natives Salt the Seals with the ashes of burnt Sea Ware and say they are good Food, the vulgar eat them commonly in the Spring time with a long pointed Stick instead of a Fork to prevent the strong smell which their hands would otherwise have for several Hours after. The Flesh and Broth of fresh Young Seals, is by experience known to be Pectoral, the Meat is Astringent and used as an effectual remedy against the *Diarrhoea* and *Dysenteria*; the Liver of a Seal being dry'd and pulverized and afterwards a little of it drunk with Milk, *Aquavita*, or Red Wine is also good against Fluxes.

DESTRUCTION FOR SKINS AND OIL

Some of the Natives wear a Girdle of the *Seals Skin* about their middle to remove the *Sciatica*, as those of the Shire of ABERDEEN wear it to remove the *Chin-cough*....

The Seal, tho' esteemed fit only for the Vulgar, is also eaten by Persons of Distinction, though under a different name, to wit, *Hamm*.

Sixty years later, Macaulay, describing the same sealing ground, adds

that the fat of the Seals, is by the people, to whose share that perquisite falls, converted now into oil and sent to market. But in that writer's [Martin's] time, and for ages immemorial before, this, together with the flesh of these animals, was eaten either fresh or salted.

And in 1830 Macgillivray could still write of "Gaskir [or Haskeir] twelve miles from Harris" that "great numbers are killed upon it annually, upwards of a hundred and twenty having been destroyed in one day."

Close on two hundred years before Martin wrote of the Outer Hebrides, "Jo. Ben." (perhaps the Bellenden who translated Boece's *History*) described, in a Latin manuscript, a very similar seal-hunt which in the sixteenth century took place annually at "Selchsskerry" in the Orkneys, and even in 1795, Low in his *Fauna Orcadensis* relates that "a ship commonly goes from this place once a-year to Soliskerry [Suleskerry], and seldom returns without 200 or 300 Seals."

Now the significance of the number of Seals killed depends not upon its intrinsic greatness, but upon its relation to the annual increase of the stock which inhabits the hunting-grounds. There can be no doubt that the slaughter of former days in Scotland exceeded the natural increase and trespassed upon the breeding stock of Seals. The result has been that the Seals of Scotland have been greatly reduced in number, a result especially evident in the case of the larger and more valuable species, the Grey Seal. This fine creature, the object of the seal-hunts at Haskeir in the Outer Hebrides, and amongst the Orkney and Shetland Islands, is thought to have been at one time the commonest seal on the east coast also. Its bones have been found in a kitchen-midden on Inchkeith, and at an early Christian settlement in Constantine's Cave near St Andrews. But its numbers have been so reduced that it is now seldom seen on the east coast, and the stock throughout the whole of Scotland where it was once very numerous has been estimated to have fallen to less than 500 strong. So threatening did the outlook for the unprotected Grey Seal seem to be that in 1914 a Parliamentary Bill became law, making it illegal to destroy the Grey Seal between October 1 and December 15, a period covering the breeding time of the species, under penalty for every offence of £5 for the slayer and £10 for the owner of the boat employed.

Whales

Five hundred years ago, Whales were abundant in every ocean. Even the Mediterranean Sea furnished a regular fishery which has long since died out for lack of Whales to slay. And although the actual hunting of the larger and more valuable species, the Greenland or Right Whale (*Balæna mysticetus*) and the Sperm Whale (*Physeter macrocephalus*), was never carried on systematically in the immediate neighbourhood of the Scottish coasts, the Scottish fauna is the poorer for the destruction that was visited upon these wanderers of the ocean in distant parts.

In this destruction Scottish vessels and Scottish mariners played their part. At Peterhead the whale fishery was started in 1788 and in less than a hundred years, to 1879, had accounted for the capture of 4195 Whales yielding 30,975 tons of oil (a ton measuring 252 gallons) and 1549 tons of whalebone, apart from a total of 1,673,052 Seals, yielding another 20,913 tons of oil. At Dundee the fishery commenced in 1790, and up to 1879 there were captured 4220 Whales, yielding 32,774 tons of oil, and 1640 tons of whalebone, together with 917,278 Seals, yielding 10,464 tons of oil.

And this is no more than a drop in the bucket. In the thirty-eight years from 1835 to 1872, the American whaling fleet is credited with having captured or destroyed 292,714 Whales. The story of the more valuable Whales can be read with no uncertainty in the statistics of the whaling industry: a gradually decreasing catch and following upon this a reduction in the numbers of ships that set out for the whaling grounds, so that the United States fleet, which numbered some 730 vessels in 1846, had fallen off to 218 in 1872—such facts tell of the pitiful decline of the great ocean wanderers.

Now the Sperm Whale and the Whalebone Whales are so scarce that they no longer offer a profitable fishery in our northern waters, and attention has been turned to species which formerly were passed by in contempt.

On the coasts of Scotland, before the war, the whaling stations of the Outer Hebrides and of the Shetlands, manned by Norwegian fishermen, accounted for hundreds of Rorquals (*Balænoptera*) a year; and here as well as in the Southern Oceans, where in the Falkland Islands group, 9429 Whales were slain in the season of 1913–14, the slaughter is bound to bring even these flourishing species of the Whale stock near extinction, if legislation does not speedily protect them by close seasons or other devices. Rorquals or Finners are slain entirely for their blubber, and in South Georgia the carcases are set adrift to rot in the sea, so that huge decaying masses are said to lie for miles round the different stations. But in these lean days the attention of the Food Controller might with profit be drawn to the fact that in the seventeenth century the people of the Western Isles of Scotland found by experience that Whales supplied nourishing food, for this Martin was assured of, "particularly by some poor meagre people who became plump and lusty by this Food in the space of a week; they call it Sea Pork." Their method of capturing the "Whales," probably the Pilot or Ca'aing Whale—the Round-headed Porpoise (*Globicephalus melas*)—was of the simplest. A school having been sighted

the Natives employ many boats together in Pursuit of the Whales chasing them up into the Bays, till they wound one of them mortally and then it runs ashore, and they say that all the rest follow the track of its Blood, and run themselves also on shore in like manner; by which means many of them are killed. About five years ago there were fifty young whales killed in this manner, and most of them eaten by the common People.

Of Whales it may be said that inordinate destruction compelled by the greed of gain, has far outrun the natural increase of the race, so that all the oceans of the world have been impoverished, and the seas of Scotland have shared in the loss, being deprived of many a visitor such as, in the old days, made a chance pilgrimage from the northern ocean.

III. 4

THE DESTRUCTION OF VERMIN AND PESTS

IN the gamekeeper's "larder" or "museum," that miscellaneous collection of fresh bodies and dried skins nailed round the girth of a tree or tacked to a doorpost, we have a tangible epitome of the destruction of vermin. The assortment of Stoats, Magpies, Hooded Crows, Jackdaws, Weasels, and alas, Owls, must be so familiar to every lover of the country, as to make it unnecessary here to enter into the

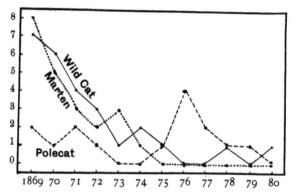

Fig. 42. Decadence of "Vermin"—Wild Cat, Marten and Polecat—through twelve years' work of one gamekeeper. The numbers are indicated on the left, the years beneath.

details of the slaughter. Many of the creatures mentioned in other connections in this account of man's destructiveness have been allotted a place on the vermin lists—Eagles, Foxes, Wild Cats, Martens, Polecats, Ravens and Hawks,— and it cannot but be that, where keepers are employed mainly with the view of destroying vermin for the sake of game, a serious falling off in the original stock is likely to follow. The result of a single keeper's efforts offers a summary of the whole, and, as represented in the diagram above, shows how rapidly regular trapping may lead to the reduction or disappearance of our native animals. In twelve years the Marten may be regarded as having been exterminated in this keeper's beat, in Assynt in Sutherlandshire, the Wild Cat was reduced to the verge of extinction, and even the Polecat showed an ominous uncertainty of tenure.

Of lesser vermin the slaughter has at all times been great since universal game-preserving became the fashion. On the estates of Langwell and Sandside in Sutherlandshire, in seven years from 1819 to 1826, Carrion Crows and Magpies to the number of 2647 were slain, in addition to 1799 Rooks and Jackdaws; while in three years from 1831 to 1834 on the Duchess of Sutherland's estates in Sutherland, 1739 Carrion Crows and Magpies were destroyed.

ROOKS AND CHOUGHS

Some of the lesser vermin are miscreants steeped in crime. The Rook was no less vehemently denounced by the farmer of the fifteenth century than by his successor to-day. Even the Scottish Parliament took up the cudgels against it, and by a curious method sought to keep this thief in check. For seeing that "ruks bigande in kirke yards, orchards, or treis does gret skaithe apone cornis," a statute of 1424,

> ordanyt that thai that sik treis pertenys to, suffer on na wyse that thai birds fle away. And whar it beis tayntit [known] that thai bige and the birds be flowin [flown], and the nests be fundyn [found] in the treis at Beltane, the treis sal be forfaltit [forfeited] to the King.

A later law, of 1457, also provided for the destruction of Rooks, Crows and other birds of prey which injure corn and game; and at the present day every country district is familiar with the annual "crow-shoot" whereby an endeavour is made to limit the numbers and destructiveness of the Rook.

The Chough also has suffered from the zeal of the vermin-killer. At one time it seems to have been widely distributed even in inland districts. Leslie makes undoubted reference to it, when, as Dalrymple translates his Latin, he says:

> Sche is said to be fund in ane only Ile, in the sey cost besyde Cornwale foranent [over against] the Realme of France, bot with ws [us] this fowle [Lat. *corniculam*, little crow] may be seine with neb and feit of purpur hew, nocht only in ane place, that only is thocht to be fund in Cornwale of sum. [The Latin adds, "whence we give it its name," *i.e.* Cornish Crow.]

In inland localities, which it has now altogether forsaken, the Chough nested down to the eighteenth and even to the nineteenth century. The following facts are summarized from Mr J. H. Buchanan's account of the Chough in Scotland. In the midlands it occurred in Glenlyon in Perthshire in 1769, in 1795 there was a pair or two on the Campsie Fells and records exist of its presence on the Ochil Hills and on the Clova Hills of Forfarshire. In the Lowlands it frequented the Corra Linn Fall on the Clyde about 1770, and the last individual from an inland breeding-place was shot at Crawfordjohn in Lanarkshire in 1834.

On the cliffs of the coast as a rule it held its own to a later date, but from most of these also it has disappeared. On the east coast it has been found at Dunrobin in Sutherlandshire and at St Abbs in Berwickshire, but even in 1851 all but a single pair had forsaken the fastnesses of the latter neighbourhood. On the west coast its former haunts are better known, for in the secluded places of that wild shore it is making its last stand in Scotland. It has gone from the parishes of Kilbrandon and Kilchattan, and of Giga and Cara in Argyllshire, where the writers of the Old Statistical Account knew it. On the island of Lismore on Loch Linnhe, where flocks existed at the opening of the nineteenth century, it is extinct. The majority of the islands have been deserted: Skye, Raasay, the Long Island, where it still existed about 1830, Tyree, Rum, Mull, Colonsay, Iona and Arran (where the last pair was shot in 1863), on all of which it once had harbourage, know it in its numbers no more. On a few islands of the Inner Hebrides, especially on Islay, and on the coast near the boundary of Ayrshire and Wigtownshire, it retains its last feeble hold in Scotland, but without generous protection its race in the northern kingdom is doomed, a result in great part due to the exertions of the game-preserver, and a penalty ill-becoming one of the most interesting of " vermin."

Other creatures are slaughtered in immense numbers on account of their harmfulness, though the fact that many remain pests, indicates that their numbers are not seriously on the wane. Most of the following belong to this category.

RATS, MOLES AND SPARROWS

Rats have been slain by the hundred thousand in Indian and American cities on account of the damage they do and of the bubonic plague which they carry; and in our own land single farms and country villages have on occasion accounted for many thousands of these notorious thieves. As, however, this aspect of its multiplication will be considered in the discussion of the Rat's introduction to Scotland (see p. 431), the subject need not be more than mentioned here.

The damage caused by Moles, since cultivation by increasing their food multiplied their numbers, must be great and their destruction by professional mole-catchers on a corresponding scale. I have no means of estimating the annual slaughter in Scotland, but Mr O. H. Wild tells me that in April 1918 he saw the dead bodies of some 750 individuals hung along a hundred yards of fence at Aberlady railway station—the result of three weeks' trapping on farms in the neighbourhood by a single mole-catcher. To this collection bodies were still being added at the rate of twenty-five a day.

The Common Sparrow has been slain in its thousands on account of its devastation in cornfields and gardens. Sparrow Clubs have been formed in many districts, especially in England, for the reduction of the pest, and during the War the Government issued special orders for its destruction. In three years the Tring Sparrow Club accounted for 39,058 individuals; the Ixworth Sparrow Club slew 14,669 in 1915, and the Slimford Rat and Sparrow Club killed in a few years 84,590 vermin and destroyed 17,201 Sparrow's eggs.

RABBITS AND HARES

Just as man has created the sparrow pest by the increasing perfection of his tillage, so in Scotland he has also created the hare and rabbit nuisance. Since the Rabbit escaped from the warrens to which it was introduced, and took to living wheresoe'er it pleased, and upon whatsoever of the farmer's crops it could most readily obtain, it has become an unmitigated pest, against which, in these days of increased demand for home-grown food, emergency legislation has done well to set its hand (see p. 215).

The history of the Hare as a pest is of more than ordinary interest, because at common law the Hare is regarded as ground game, and as such was at one time protected—so strictly that in 1707 the shooting of Hares by any person whatsoever was prohibited under a penalty of 20 pounds Scots. Such a law at so late a date can mean nothing else than that the Hare was an animal comparatively rare. The gradual development of the misdeeds of the Hare can be traced from successive statutes, which point to an extraordinary increase in numbers contemporaneous with the agricultural activity of the nineteenth century, which began about 1837. The first law aimed against the Hare was passed in 1848 and permitted "all Persons at present having a Right to kill Hares in *Scotland*, to do so themselves, or by Persons authorized by them, without being required to take out a Game Certificate." Judging from the statistics of the Dumfries Fur Market (see Fig. 41, p. 167), this permission, coming at a time when agriculture was about to reach its highest development in the nineteenth century (1853 to 1862), seems to have resulted in an intensified slaughter of Hares, which finally caused a falling off in the stock. Still the Hare remained a burden greater than the farmer could bear, and for his benefit there was passed in 1880 "in the interests of good husbandry" the Ground Game Act which gave the tenant equal rights with the proprietor to kill and take ground game on his holding. The immediate result of this breaching of the privileges of the Game Laws is in no doubt—the Hare suffered so greatly in numbers that in the Lowlands of Scotland it was threatened with extinction, and over all, the slaughter was so excessive that in 1892 the Hares' Preservation Act was passed, granting the persecuted creature a measure of protection by instituting a close season from March till July.

THE DESTRUCTION OF VERMIN AND PESTS

RED DEER, SQUIRRELS AND OTHERS

Other creatures, beginning their course innocent of serious evil, have, through excessive increase in numbers, developed into formidable pests. The Red Deer with the Rabbit and Hare have been the objects of emergency legislation during the years of the War (see pp. 212 and 215), and the Squirrel shares with them unenviable notoriety. In another place (p. 295) I have indicated the rapidity of the Squirrel's spread and multiplication in the northern counties, following upon its introduction at Beaufort Castle on the Beauly Firth in 1844, but of recent years its increase there has been phenomenal. The Highland Squirrel Club was formed in 1903 to counter the devastation wrought in the woods of eastern Ross-shire, part of Sutherland, and that portion of Inverness north of the Caledonian Canal. The results of its activities are astounding, when it is recollected that three-quarters of a century ago the Squirrel was unknown in the district. During the fifteen years up to the end of 1917, 60,450 Squirrels had been killed; in 1903 alone 4640 were destroyed, in 1907 6628, and 1909 provided a record of 7199 individuals[1]. The price paid for tails by the Club varied from 3d. to 4d.

With such nuisances no sympathy can be felt, although the destruction of Owls, which themselves are pest destroyers, and of Stoats and Weasels, which probably do almost as much good by their enmity to rats and mice as they do harm, is regrettable; but there is one so-called pest regarding the destruction of which the naturalist can have nothing but regret—the little Dipper or Water Ouzel.

---

[1] On account of the interest of this warfare against the Squirrel in the North, I give the annual numbers of squirrels killed since the Club was started, from lists kindly sent me by Mr A. H. Duncan, the Secretary:

1903......4,640	1907......6,628	1911......3,056	1915......2,601
1904......3,988	1908......3,197	1912......3,679	1916......2,692
1905......3,431	1909......7,199	1913......3,283	1917......3,998
1906......4,007	1910......4,235	1914......3,816	

THE DIPPER OR WATER OUZEL (*Cinclus cinclus*)

For years this lively little bird, whose presence brightens the stretches of many a dull stream and river, has been an object of persecution on the ground that it destroys the spawn of salmon and other fish. On the Spey the slaughter of a Water Ouzel during the fishing season used to be rewarded by bestowing upon the slayer the right to fish salmon with the rod during the close season—an iniquitous provision which has dropped into desuetude, but not before the number of Dippers had been seriously reduced. On the Sutherland estates of the Duchess of Sutherland a reward of 6*d.* used to be paid for each Dipper slain, with the result that from March 1831 to March 1834 "548 King's Fishers" were slaughtered, "King Fisher" being a local name for the Water Ouzel; and for the six years between 1873 and 1879 the vermin list of the Reay country, also in Sutherland, included "368 Water Ouzels." Regret at the slaughter of an interesting bird is intensified by the knowledge that its actual food is not so much the spawn of fishes as the larvae of dragon flies and water beetles, which themselves commit serious havoc among the spawning beds. A little knowledge of Natural History is a dangerous thing.

It is no simple matter to reach an estimate of the influence of man upon the numbers of vermin and pests, but every dweller in the country and especially in game-preserving districts knows that the destruction is no light one. That the enemies of game, be they real or fancied, have been reduced below the standard which Nature sets, is shown by the unwonted increase of such "vermin" during the War years, when gamekeepers have been called from their wonted beats. Will the increase of the game-preserver's "vermin" affect some of the smaller pests? Can there be any doubt that there is a close relationship between the decrease of the beasts and birds of prey and the increase of the creatures upon which they were accustomed to feed? Has the game-preserver not been too ready to shoot and too reluctant to spare; has he not brought to earth such pest-destroyers as the Honey Buzzard, the Kestrel and the Owls, simply on the hearsay of a bad name?

But new ideas of the value of a native fauna have taken root and already bear promising fruit. The success of a game-drive is not everything. There must be balanced against the bag of grouse and the head of pheasants and partridges, the majesty of the eagle's flight, the grace of the kestrel's hover, the lithe movements of the marten in the forest and of the ermine in the coverts. And who will say that the beauty of nature is to be filched from the countryside for the sake of the bag of game?

III. 5

DESTRUCTION FOR SPORT

"The Scottes" wrote Holinshed, "sette all their delighte in hunting and fowling, using about the same to go armed in jackes and light iesternes with bowe and arrows, no otherwise than if it had been in open warre, for in this exercise they placed all the hope of the defence of their possessions, lands and liberties."

From very early times the law, as well as the national taste, encouraged the chase in Scotland, and we should therefore expect to find here clear evidences of the effect of sport upon animal life.

Destruction for sport is, however, less inimical to animal life on the whole than one might suppose, for over-destruction must mean in the long run the death of the sport itself. Consequently the risk of over-destruction is modified by the introduction of various modes of protecting the animals hunted, a subject which I have discussed in another chapter. Yet, in spite of protection, sport has resulted in the decrease in numbers and even in the total or local extermination of some members of the fauna.

THE WILD BOAR

The occurrence and the decline of the Wild Boar have already been mentioned (p. 89), and I would only recall that in spite of the fact that its chase was strictly preserved to the royal court and the great landed nobles, it was exterminated in Scotland probably in the early years of the seventeenth century.

THE BROCK OR BADGER

The Brock (*Meles meles*) has long been ranked with the Wolf as an object worthy of the chase.

"Throuch thir woddis" Dalrymple translated Leslie, "the gretter parte of the nobilitie hes thair maist recreatione in hunting with the sluthe-hundes, for that, this recreatione hes our countrey men ather in the feildes to hunte the hair and the fox, or in the sandes and water brayes the Brok, or in the mountanis the Wolfe, or the Wilkatt."

Yet the Badger seems to have maintained its ground with great persistence in spite of hunting and harrying, for

Fig. 43. "Catching the Badger"—from a coloured plate, dated 1820, after H. Alken.

even during the nineteenth century it was so common that many a parish had its "brockair" or brock-hunter, and in several counties it was even sought after for food, Badger hams being preserved for winter use. Dr Campbell wrote in 1774 that the Badger is

hunted and destroyed whenever found; and being by Nature an inactive and indolent Creature, is commonly fat, and therefore they make his hind Quarters into Hams in North Britain and Wales.

In the *Spectator* of Sept. 29, 1917, a correspondent gives many recipes for cooking the Badger, which are said to have been familiar to his mother as practised in the Outer Hebrides[1].

The sportsman, hunter, baiter and the ham factor must have reduced its numbers, but the Badger is a shy nocturnal creature whose earth is not always readily found. Indirectly rather than directly man has wrought its doom, for a great blow to its existence in Scotland seems to have fallen with the extension of cultivation in the latter half of the nineteenth century, and especially with the feverish cutting of woodland and breaking in of waste ground which accompanied the agricultural boom of the 'fifties and 'sixties. These processes destroyed many a safe refuge, and exposing the Badger to the attacks of man, exterminated it from many a district where it was once common. There is probably no county in Scotland where it is not much rarer nowadays than it was a century ago, and although at the present day it still occurs in the wilder parts of the Highlands, in the Lowlands it has become exceedingly scarce as a native of the soil, many of the examples now recorded being no more than escapes from confinement or their descendants.

GAME BEASTS AND BIRDS

The history of the more important game creatures has been traced in the chapter dealing with their protection, and since the stringency of protection is a measure of the intensity of destruction, the account there given illustrates sufficiently the trend of the slaughter due to sport.

[1] While the recipes may be correct, the locality seems to be at fault, for, according to Alston and Harvie-Brown, Badgers are unknown in the Outer Hebrides.

It must be remembered, however, that influences are complex and that while a widespread influence, like the destruction of woodland, was the main factor in causing the decline or extermination in Scotland of such as the Red Deer and the Reindeer, yet sport also contributed to their disappearance, for we have the evidence of the Orkneyinga Saga that even in the twelfth century journeys were made from distant parts expressly for the purpose of hunting the Red Deer and the Reindeer in the uplands of Caithness and Sutherland. The results of excessive sport have been shown nowhere more clearly than in the United States of America, where, before protecting laws had been introduced, the native carnivores, deer and game birds of many States had been brought to the verge of extinction.

So also causes other than sport told hardly upon certain of the game creatures. The Great Bustard and the Bittern were chased and hunted and are exterminated, yet the evidence seems to show that sport had much less to do with their disappearance in Scotland than had the cultivation of waste places in the case of the former (see p. 366) and the reclamation of marshes in the case of the latter (see p. 374). Such at least I take to be the indication of the unsuccessful attempts to re-establish the Great Bustard in Britain, now that all question of active destruction is eliminated.

The use of game birds as food further combined with the destruction of sport to accentuate the reduction of their numbers, witness the declaration of the Statute of 1600, heralding the imposition of new and heavier penalties against illegal dealing in game—one hundred pounds to be paid by both buyer and seller if they were "responsal in gudes," and if they were not that they be "scourged throw the burgh or town where they shall be apprehended." Regarding the profiteers of the seventeenth century, this Act alleges that

diverse and sundry persons, having greater regard of their gaine and commodity whilk they purches by the selling of the said wyld-fowle to sik persons wha prefers their awne inordinat appetite and gluttony either to the obedience of the said lawes or to the recreation that may be had by the direct slaying of the samine, hes used all the saids indirect meanes in slaying of the saids wyld fowles and beastes, whereby *this country being so plentifully furnished of before is become altogether scarce of sik waires*[1].

[1] My italics.

Whether one ought to include the great slaughter of Ducks, Geese and Swans, which in earlier days and to a lesser extent to-day, is carried out by such murderous weapons as punt guns—which still survive in Scotland in the Moray Firth and in the Solway—as mainly destruction for food or as destruction for the sake of sport, I leave my readers to decide. In either case the result from the point of view of animal life is the same—the reduction in number of the immigrant Ducks, Geese and Swans which seek temporary harbourage in our sheltered creeks.

III. 6

DESTRUCTION FOR PLEASURE OR LUXURY

THERE is a type of animal slaughter, widespread in most countries, which has nothing of the spirit of sport, and is not animated by the utilitarian element that has led to most of man's destructiveness. It is notoriously exhibited in the yearly massacres of birds of gay plumage, in order that, at the behest of fashion, the human race may flaunt their stolen plumes. What this massacre means few can realize. In 1913, the last completed year before the War, there were offered in the London salerooms 12,850 oz. of "osprey" feathers, representing a slaughter of about 77,000 individuals of the harmless heron-like Egrets (*Ardea egretta* and *A. candidissima*), as well as 19,125 Osprey wing quills; 16,211 White Crane wing quills; quills of Bustard, 10,800, and of Condor, 48,321; the skins of Emus, 1233, of White Terns, 5321, of Crowned Pigeons, 11,478, and of Smyrnian Kingfishers 162,750.

The destruction for pleasure and luxury is no faddist's hallucination, and it pays no regard to the rarity, interest or usefulness of its quarry.

In Scotland such revolting butchery as the tropical forests witness has had no place, yet in less obvious, but scarcely less deadly ways, the same craving after pleasure or luxury or possession has affected some of our native animals. Three types will illustrate the point—the collector, the bird-catcher, and the pearl-fisher.

THE SINS OF "COLLECTORS"

It is regrettable, but necessary, that reference should have to be made to a type of destruction which has taken place under the guise of the study of Nature, the fruit of an inordinate desire for possession, which sets numbers before real nature study. Yet we cannot close our eyes to the fact that the collector of specimens has caused or has hastened the extermination of several members of the old fauna.

Upon the head of the entomologist lies the guilt of the extermination of the English "Large Copper" Butterfly (*Chrysophanus dispar*), whose head-quarters were in the Whittlesea Meer of Huntingdonshire. There persistent collecting of the caterpillars by persons young and old, abetted by an unusual flood, exterminated about 1850 a creature so abundant twenty years before that visitors who had seen it never dreamed of its extinction.

The methods and occasional intensity of the entomologist's pursuit are well illustrated in the case of the Artaxerxes Butterfly (now regarded as the variety *artaxerxes* of *Polyommatus agrestis*), which was first discovered and described from Arthur's Seat near Edinburgh—for long the only known locality. Here the Artaxerxes (Fig. 44) was so common that collectors flocked from all parts to plenish their collections, with the result that in 1844 it was said that "all the English cabinets and the principal foreign ones, are now abundantly supplied from that locality." But collecting did not stop when moderate demands had been met. Still collectors flocked to the crags above Duddingston Loch, as Mr William Evans has informed me, from the season's beginning to the season's end, not only capturing the butterfly on the wing, but collecting pupae and caterpillars for subsequent development, and even plucking those leaves of the Rock Rose on which the eggs had been laid, towards the same end of stocking the collections of Europe and of the world with Artaxerxes when the imago appeared. The result of such persistence could have been in no doubt in the case of so local a form. The last specimen was found in 1868—Artaxerxes was exterminated from its world-renowned home on Arthur's Seat. It was not due to the foresight of collectors that other localities in Scotland have since been discovered which carry on the tradition of the existence of the Artaxerxes Butterfly.

Birds have suffered even more than Butterflies. The St Kilda Wren (*Troglodytes hirtensis*) was so persecuted after the discovery of the unique character of this inhabitant of the lonely island, that in a most destructive raid in the spring of 1903 it was believed, according to Sir Herbert

Maxwell, that "every egg was taken on the principal island, and many of the parent birds were shot," to satisfy the brisk trade at high prices created by their rarity. So a special law had to be enacted in 1904, protecting it from the collector's gun and the nest-harrier. In England the extinction of other birds—threatened by the increase of cultivation—has been ensured by the zeal of the collector. Amongst such we may safely reckon the Red Night Reeler or Savi's Warbler (*Locustella luscinioides*) of Norfolk and Lincolnshire, exterminated about 1849, the Black-tailed Godwit (*Limosa limosa*) and the Black Tern (*Hydrochelidon nigra*) of the

Fig. 44. Artaxerxes Butterfly (upper and under sides)—exterminated on Arthur's Seat. Natural size.

Norfolk Broads, the former of which finally disappeared between 1829 and 1835, and the latter in 1858.

It is an unfortunate paradox, which must remain from the nature of the case, that just when the creatures he envies require protection most, the collector's enthusiasm rises and his efforts redouble. He seldom exterminates a species from the beginning, but when other causes have threatened the existence of a creature, he puts a seal upon its doom. It was only when the Kite became a rare bird that he turned his attention to it and its eggs, and now the most careful protection can scarcely outwit the fate he has decreed for its last British survivors in Wales.

192 DELIBERATE DESTRUCTION OF ANIMAL LIFE

So too with the Osprey in Scotland. The brunt of its disappearance also must be borne by the collector; for even when it was reduced to a few pairs he harried its nest time and again, until one site after another was forsaken. The later history of the Loch-an-Eilan Ospreys in Inverness-shire is a tale of ruthless persecution. There are many blanks in the records, but it is surprising that where the

Fig. 45. Osprey or Fish Hawk—practically exterminated in Scotland. ¼ nat. size.

Ospreys were actively protected by the proprietor, and where secrecy was a first aim of the harrier, so many of the records which have survived should reveal the latter's nefarious deeds. In each of the successive years 1848, 1849, 1850, 1851 and 1852, the eggs were taken, and in the latter year the nest was twice harried. Is it any wonder that in the following year the Ospreys forsook for the time being their old site in the Loch?

Wherever the birds sought harbourage, whether they

DESTRUCTION FOR PLEASURE OR LUXURY

moved to Loch Morlich or Loch Gamhna, the egg-stealer pursued them. From 1843 to 1899, when the last pair built at Loch-an-Eilan (although individuals were seen up to 1902), there are definite records that the Ospreys nested in twenty-four years in that or the other two Lochs mentioned, and there are records as definite that the nests were harried and the eggs taken on fifteen occasions! The wonder surely is that the Speyside Ospreys survived this heartless persecution so long.

BIRD-CATCHING

The barbarous habit of keeping wild birds in cages for the beauty of their plumage or the sweetness of their song must also be pilloried from the point of view of interest in our native fauna. The toll taken by professional bird-catchers is common knowledge to the frequenter of bird-shops, but it is perhaps not realized how seriously their depredations affect the numbers of certain kinds of wild birds. It is supposed that in recent years bird-catching has been on the decrease, but the lists of cases tried in court and resulting in proof of guilt, show that the practice is still very common, and the numbers of bird-catchers who escape the law, as the naturalist who is familiar with favoured bird re sorts in Scotland knows, is out of all proportion greater than the numbers convicted.

Take only one or two cases in illustration of the results of this trade in living flesh and blood. It is on record that in January 1895, when a late migration brought great numbers of Twites, Linnets, and Red-polls to the shores of eastern England, one bird-catcher netted, in four successive days, 70, 130, 220, and 330 Linnets. And in the winter of 1900 another expert caught 140 Siskins one morning before breakfast, on a decayed lettuce patch. In recent years the London market has considered a supply of 400 dozen Linnets a week in October a small average, and 600 dozen has been reached.

The case of the Goldfinch is even more pitiful. These graceful and beautiful birds frequent, in flocks, patches of teazle or thistles, and remain attached to a patch until all the food afforded by the seed-heads has been consumed. It is a practice of some bird-catchers to plant teazle in their gardens for the sole purpose of attracting Goldfinches, and of capturing one bird after another until the constancy of the flock is rewarded by its total extinction. In 1860, the annual average of Goldfinches caught in the neighbourhood of Worthing is said to have been 1154 dozen, but the supply —and how can one wonder—has fallen off, and to replace it Goldfinches have been imported in large numbers from Germany.

The pity of it all is that not only are the birds removed from surroundings and localities they would adorn with their wild grace and beauty, but that the cost of the process in bird life is so appalling, for the female birds are in many cases killed as they are netted, since they do not sing, and of birds consigned to the market, fewer than half live even to delight the ears and eyes of man, since it is said that 60 per cent. die miserably within a fortnight after they have traversed the short distance between the bird-catcher and the trader of the bird-shop.

SCOTTISH PEARL FISHING

As a last example of the influence of fashionable pleasure upon Scottish animals, take the case of the insignificant Freshwater Pearl Mussel (*Margaritifera margaritifer*). Through many centuries the fame of British pearls has spread beyond the limits of our islands, for Suetonius alleges that the hope of enriching himself with them helped to induce Julius Caesar to venture across the Channel; and in the times of the Venerable Bede (673–735) they were a valuable British commodity. In the twelfth century there was a European market for Scottish pearls, and in the sixteenth Bishop Leslie says of them:

in Laudien Land [the Lothians] farther, and lykewyse in vthir prouinces with ws ar funde Gemis,...to wit...the Margarite in gret number...the Margarite is baith welthie [*lit.* abundant] and of a noble price. Thay indeed schawe a schyneng brichtnes, notwithstanding mair obscuir than thay quhilkes ar brocht in frome the Eist. In freshe water buckies nocht pleisand to the mouth, na lesse than in salt water buckies growis the Margarite.

In 1560 "large handsome pearls" were exported from Scotland to Antwerp, and in 1620 there was found in the Kelly Burn, a tributary of the Ythan in Aberdeenshire, a fine pearl which was carried to King James VI by the Provost of Aberdeen, and became one of the Crown jewels. The Provost was rewarded for his trouble by being presented with "twelve to fourteen chalder of victuals about Dunfermline, and the Customs of Merchant's goods in Aberdeen during his life"; the pearl-fishers were rewarded by an order of the Privy Council passed in 1621, proclaiming that Pearls found within the Realm belonged to the Crown, and appointing conservators of pearl-fisheries in several counties, including Aberdeen, Ross and Sutherland. The Aberdeenshire conservator was specially commended in that "he hath not only taken diverse pearls of good value, but hath found some in waters where none were expected."

The Scottish pearl-fishery caused an appreciable reduction of Pearl-mussels in the rivers. In his "Tour" (1771) Pennant records of the Tay and Isla that

there has been in these parts a very great fishery of pearls got out of fresh water muscles. From 1761 to 1764, £10,000 worth were sent to London and sold from 10s. to £1. 6s. per ounce....But this fishery is at present exhausted, for the avarice of the undertakers. It once extended as far as Loch Tay.

The destruction of the Pearl-mussel brought to an end for the time being the pearl-fisheries of Scotland; and they were not revived till about 1860, when a German merchant, Moritz Unger, travelled through Ayrshire, Perthshire, and Aberdeenshire, buying up all the pearls that could be found. With the valuable stock he then acquired, he reawakened the demand for Scottish pearls, and created afresh the pearl-fishing industry. I have often spoken with an old fisher on the Don in Aberdeenshire, who, first encouraged by Unger and later by two other traders, Selig and Aaron, spent the summer, year after year, when the river was at its lowest, "at the pearls." The prices paid by these itinerant pearl-dealers ranged from 5s. to 7s. a grain. Now, owing to its rarity as well as to the increased appreciation of the liquid beauty of the Scottish pearl, the value has reached from £1 to £10 a grain, according to quality and size. A small collection of half a dozen Tay pearls bought for the Royal Scottish Museum from Unger in 1859 for £12 is to-day valued at over £50.

The result of the new demand was that the more accessible shallows of pearl rivers were soon denuded of their

Mussels, for the methods employed were of the most ruthless kind, all the Mussels found being opened and destroyed on the chance that a pearl might lurk within. The shallows have become almost entirely unproductive, and of recent years attention has been turned to the deeper waters. The climax of destruction was reached in 1913, when two bands of fishers set out with motor bicycles and collapsible boats, and, touring the country, harried rivers far and near. The financial result repaid their exertions: one large and fine pearl obtained was valued at £300, and the total proceeds covered the cost of bicycles and transport and secured an ample margin of profit. But the outcome is that many a river and burn has been almost cleared of its Pearl-mussels. A few such raids and the famous Pearl-mussels of Scotland, already sadly reduced in numbers, will be almost exterminated and the Scottish pearl-fishery doomed.

CHAPTER IV

PROTECTION OF ANIMAL LIFE

> To birds man gives his woods,
> To beasts his pastures, and to fish his floods.
> For some his int'rest prompts him to provide,
> For more his pleasure, yet for more his pride.
>   POPE.

IT is fitting that an account of man's destructiveness in the realm of animal life should be followed by the kindlier theme of his protection of the lower creatures, for the two are inseparably united. In the world of living things, as in our physical environment, the natural law holds, that every action produces an equal and opposite reaction; and so destruction of one animal as a rule entails the protection of another, just as protection of one species involves the destruction of its enemies. The sportsman aims at protecting his game, and, as we have seen, reaches his end by destroying the beasts and birds of prey; but his action is more comprehensive than his intention, for he protects not only the objects of his sport, but the myriads of smaller vermin which also contribute to the food of the creatures he destroys.

It is less my intention to discuss here, however, such undesigned results of the protection of animal life than to trace the more direct consequences which have followed upon a deliberate policy of protection. These direct consequences, it need hardly be said, are as a rule, an increase in numbers of the creatures concerned, and frequently an extension of their range to new areas. They may involve so great a multiplication of animals desirable from one point of view, say that of the sportsman, that from another point of view, say that of the farmer, the protected creature becomes a nuisance, and protection has to give way to destruction. But in most cases protection does little more than preserve its sheltered favourites in numbers approximating to their natural proportion in the fauna: it does little more than compensate for the destructiveness of man.

Broadly speaking three influences tend to foster animal life: the premeditated protection of the law, the gentle sway of popular favour, and the solicitude bred of superstition. Of these the first has proved to be the most potent.

THE PROTECTION OF THE LAW AND ITS EFFECTS

In Scotland, as elsewhere in countries long settled by man, protective legislation on behalf of animals has been a matter of much experiment, and of a slow and gradual evolution which has progressed step by step with the advance of political and social ideals. Indeed the laws relating to animals were for long entirely social in their bearing, and had as little thought for the animals themselves as they had much concern for their assumed lords and masters. The development of a practical love for nature is a national acquisition of very recent date.

So we find three broad but distinct stages in the evolution of protective legislation. At first the king only and his nobles were the principals concerned, and the laws were feudal laws, the main object of which was to ensure the entertainment of the Court. So arose the Game Laws—that great bulk of legislation regulating the preservation of sport animals. With the decay of feudalism and the extension of power amongst the landed classes, the great lairds and lesser lairds came to be included in the magic circle for whose benefit the Game Laws exist.

Further, with the decay of feudalism and the growing importance of the economic aspect of the country's welfare, a second class of law arose—that regulating the protection of animals whose significance lay in their economic value.

And lastly, only in recent times, there arose, with the spread of power amongst the people at large and the development of a democratic instinct, laws which endeavour to preserve the rarity and beauty of the countryside for the people —the beginnings of an aesthetic code.

The influence on Scottish animal life of these various bodies of law, protecting creatures for the sake of sport, for utility, or for aesthetic reasons, can best be shown by a discussion of the bearing and scope of each, and this is attempted in the following three sections.

IV. 1

PROTECTION OF ANIMALS FOR SPORT

HAWKING AND HAWKS

THE sport of hawking and the possession of hawks were recognized from early times in Britain as badges of nobility. So significant was the presence of trained hawks that members of the Court and the nobles seldom rode without them, and seldom dispensed with them on their travels or even on the field of battle, where the surrender of a hawk was accounted the surrender of honour. It is little wonder, then, that from early times care was given to the preservation and rearing of hawks. Tradition goes to show that even during the period of the raids of the Norsemen, about the tenth century, hawking was practised by the noblemen of Scotland; but the earliest historical reference to protection occurs in the reign of William the Lion (1165–1214), when Robert of Avenel, in granting his lands in Eskdale to the Abbey of Melrose, reserved the right to the eyries of Falcons, and to tercels or male hawks. Even the tree whereon a hawk had built in one year was safe from the axe of the woodsman, until succeeding years made clear that it had been deserted as a nesting site, a wise provision depending upon an intimate knowledge of the limited choice of sites available for a hawk's nest and nest flight. Similar reservations hedged about the eyries on estates in Ayrshire granted about the same time to the same Abbey by the Stewarts.

In the thirteenth century hawking was a recognized privilege of the Court, and Alexander III kept Falcons at Forres and at Dunipace in Stirlingshire, and in 1263, so the accounts of the year show, paid for eight and a half chalders of corn consumed by William de Hamyl during his twenty-nine weeks' stay at Forfar with the Kings' Falcons. King Robert the Bruce had his falcon-house at Cardross in Dumbartonshire repaired shortly before his death, and in 1343 his successor, David II, granted to John of the Isles many islands and lands "*cum aeriis falconum.*" About this time the names of both Goshawks and Sparrowhawks appear in the public accounts, but from the fifteenth century onwards fashion tended to make the Peregrine Falcon the hawk *par excellence*.

During the reign of James III, the law made the protection of hawks general, ordaining in 1474 that no one should take trained or wild hawks or their eggs without leave of the owner of the ground.

It was probably owing to the careful protection enforced on account of hawking, that birds of prey, including even the Goshawk, which has long ceased to breed in Britain, were common in Scotland in the sixteenth century, for Hector Boece tells us that "of fowlis, sic as leiffis of reif [live by rapine] ar sindry kindis in Scotland, as ernis [eagles], falconis, goishalkis, sparhalkis, merlyonis, and sik like fowlis." They appear to have been widely distributed, for Rogers mentions that "in 1496 the King's falconers were recompensed for procuring hawks in the Forest of Athole also in Orkney and Shetland." Hawks, and it is evident that the Peregrine, now a somewhat rare breeder in Scotland, is the bird referred to, had their eyries at the Abbey Crag near Stirling, and also upon a summit of the Ochils, where the birds were preserved. But the most remarkable Falcons were obtained in the northern counties. Falcons from the eyries of Caithness were sent by James V as gifts to the King of France, to the Dauphin, and to the Duke of Guise. The extraordinary value which attached to well-trained birds must also have tended to keep the breeding-places under strict protection. It is on record that James IV paid £189 to the Earl of Angus for a single bird, and that in the reign of James VI, a pair of Falcons were valued at £1000.

Through many centuries Orkney and Shetland were specially favoured in the quality of their Peregrines, of which frequent mention is made in old charters and deeds. When, in 1539, the "Channonis of the Cathedrall Kyrk of Orknaye, under ane reverend fader in God Robert be the mercye of God bischop of Orknaye and Zetland" made a deed of "all and haill our lands lyand in Zetland" in favour of "our weil belovit brothir and freynd Schyr David Fallusdell," they specifically included the "halkings and huntings." In the following century it was stipulated that Sir John Buchanan,

PROTECTION OF ANIMALS FOR SPORT

to whom the Lordship of the Orkneys and Shetlands was granted by Charles I, should pay £235. 13s. 4d. Scots a year to his Majesty's falconers "for thair chairgis in vplifting his Majesties haulks thair yeirlie"; and even in the beginning of the eighteenth century, Brand records that

Fig. 46. Peregrine Falcon—formerly protected for hawking. ¼ nat. size.

the King's Falconer used to go every year to the Isles [Orkney and Shetland], taking the young Hawks and Falcons to breed, and every House in the Country is obliged to give him a Hen or a Dog, except such as are exempted,

these perquisites being originally designed for the maintenance of the King's hawks. Still later, in 1707, the Scots Parliament in making a grant of the lands of Orkney and Zetland, reserved the hawks and the falconers' salaries as pertaining to the crown.

The greatness of the effort made in Scotland to preserve

hawks, and this as a rule meant the Peregrine and the Sparrowhawk, may be judged from a Statute passed in 1621 which raised the "unlaw" or fine for stealing a hawk from £10 to £100, and another which, in 1685, condemned the stealer of a hawk from nest or eyrie or of a vervel (the equivalent of a modern marking ring) from a hawk, to a fine of 500 merks.

Yet even with all the protection afforded them, it is doubtful if hawks did much more than remain stationary in numbers in these favoured times, for it is well known that each pair reserves for its hunting a patrol area, within which no other pair can breed, and suitable breeding places, especially for Peregrines, are limited in number. So long ago as the twelfth century, Giraldus Cambrensis observed that, in spite of the care taken of the breeding places of Falcons and Sparrowhawks in Ireland, their nests did not become more numerous and their numbers did not increase. Doubtless the same was also true of Scotland.

The decay of the noble sport of hawking and the rise of modern game-preserving have reversed the judgments of old, and the birds which were once preserved to supply man's sport, are now destroyed lest his sport should be interfered with.

HAWKING AND THE QUARRY

The institution of hawking entailed the protection of the wild birds that formed the objects of the chase. This protection was, as a rule, an absolute protection, unlike most of the regulations that refer to game valued mainly as food, in which case protection during part of the year was thought sufficient. There is further evidence of the stringency of the protection in the severity of the penalties imposed, as befitted measures providing for the amusement of the highest in the land.

As early as 1493 the Heron, "royal game," was placed under the law for three years, and persons protecting it for the King's pleasure were to be rewarded. In 1551 in order to preserve the sport of hawking, killing game with guns was prohibited under pain of death, and no one was allowed to kill game for three years, except gentlemen with hawks. In later years the extreme penalty was modified, for in 1567 the offender who slew, with gun or bow, Herons or "fowls of the revar" was to forfeit his moveable goods, and, if he were a vagabond, to be imprisoned for 40 days for a first offence, and for a second to have his right hand cut off; although at a still later date a third offence could be expiated only by hanging to the death.

The most complete list of the "wylde foulys" which hawking brought under protection appears in an interesting Statute of James VI passed in 1600, which specifies

partridges, moore-fowles[1], black-cokes, aithehenis[2], termigants[3], wyld-dukes, teillis[4], atteils[5], goldings[6], mortyms[7], schidderenis[8], skaildraikes[9], herron, butter[10] or any sik kynde of fowles commonly used to be chased with halkes.

Woodcocks, Plovers and Wild Geese were specially excepted. The attempt at protection in the case of the birds mentioned was exceptionally thorough, for not only was the offender who shot at the birds liable to a fine of £100, but the buyer or seller of any of them was held equal in guilt.

So we leave consideration of special laws, which, though primarily intended for the benefit of the Court, could not but have had a beneficial effect upon the wild fauna of the country, and pass to a short discussion of the legislation relating to "wylde foulys" in general.

"WYLDE FOULYS" IN GENERAL

The greater number of the sporting wild-fowl of Scotland is included in the prohibition of 1600 just quoted, which was specially designed to encourage the sport of hawking, and which reminds us that

by the common consuetude of all countries, special prohibition is made to all sorts of persons to slay wild fowl, hare, or venison...*in respect the samine as well has been created for the recreation of mankinde as for their sustentation*[11].

Yet a study of the old Scots law leads me to think that a broad distinction was drawn between preservation for sport

---

[1] Grouse. [2] Grey-hens. [3] Ptarmigan. [4] Teal.
[5] Probably the Wigeon. [6] Perhaps the Golden-eye Duck.
[7] Supposed to be Martins.
[8] Also "schiwerines" and "schildernes," i.e. shield-nosed, probably *Spatula clypeata*, the Shoveller or "Spoonbill" Duck.
[9] Sheldrakes. [10] Bittern. [11] My italics.

royal and general wild-fowl protection. In the former case, the protection was, as might be expected, an absolute protection enforced by the severest penalties, while in the latter, less stringent measures were considered sufficient. These measures took different forms, and seem to have been graded according to the scarcity or otherwise of particular birds at particular periods.

At one time nests were protected, and the eggs of Partridges or Wild Ducks were forbidden to be taken, under a penalty of 40s. (1474). At another time the young birds, "pouts" of Partridges or of Moor-fowl (Grouse), were exempt from destruction (1599, 1600, 1685), or the old birds were to be spared during the moulting season (1457). But perhaps the method most frequently adopted was that now universal in the case of game birds, the fixing of definite "close seasons," when wild-fowl could neither be killed, sold, bought or eaten without incurring penalties of various degree, from a paltry 40s. to a prohibitive fine of £100.

It is hard to estimate the actual effect of these measures upon the numbers of wild-fowl, but the constant repetition of the old, and the appearance of new laws throughout the centuries, leads one to suspect that the Statutes were honoured as much in the breach as in the observance. Even after two centuries and more of legal protection, Moor-fowls or Grouse, a Statute of 1682 states, were so much destroyed that there was fear of their total decay, and they were accordingly prohibited from being bought or sold for seven years under a penalty of £100 to be incurred as well by the buyer as the seller. Recourse was again had to this expedient a few years later when, owing to Moor-fowls being so "decayed," they were in 1698 absolutely protected for another seven years. Nevertheless, if in spite of their general inefficiency, the laws saved some of the birds of the moors and marshes from the sorry fate of extinction which overtook the Bittern and the Bustard in Scotland, they did good service on behalf of our native fauna.

MODERN GAME BIRDS

The measures adopted in more recent times for the protection of game birds in Scotland differ in one important respect from those of early days. Formerly the wildness of an animal was a guarantee of its freedom from proprietary rights—it was no man's beast until it was slain. Only "royal game" was reserved for the king. But the development of a fresh body of legislation had the effect of vesting property in wild animals in the owner of the land on which they were found. This important restriction of the common rights was carried out by means of laws actually restricting, through licences, the right of killing and selling "game" and by others, such as the Night Poaching Act of George IV and the Day Poaching Act of William IV, making trespass, in itself merely a civil misdemeanour, a criminal offence to be expiated by heavy penalties. In other respects the mode of protection nowadays is similar in nature to one formerly adopted—the institution of definite "close seasons," during which particular birds may not be shot.

The variety of birds protected by these measures is somewhat less than in the days of Scotland's independence, for "game birds" are variously defined as "partridges, pheasants, muirfowl, tarmagans, heathfowl, snipes and quails" (13 Geo. III), and as "pheasants, partridges, grouse, heath or muir game, black game, and bustards" (9 Geo. IV), although in addition to "game," "woodcocks, snipes, quails, landrails, wild ducks, and conies" are protected in the Day Poaching Act of 1832.

The effect upon the fauna of Scotland of these statutes has been very great, for the practical granting of proprietary rights in game birds, has led to a great increase in the numbers of the more favoured species. Pheasants, imported aliens (see p. 264), have been reared with the greatest care and at enormous expense, Partridges have been encouraged by the strictest preservation, and Grouse and Black-game have had enormous tracts of country set apart for their use. So that in most places where they are found, these species occur in numbers beyond their natural proportion in the native fauna, for in addition to protection from man, they have benefited by the deliberate destruction of beasts and birds of prey—their natural enemies.

The result of the preservation of game birds has been strikingly shown in the course of the War. In the absence of the annual battues on the extensive scale of pre-war days, the birds have so increased in number as to become a serious nuisance to farmers and a menace to home-grown food supplies, and to cause special measures to be taken for their destruction under the Defence of the Realm Regulations, "with a view to preventing or reducing injuries to crops by game birds" (27 February, 1917).

As it is difficult to indicate, except in general terms, the effects of the Game Laws upon the numbers of the protected in Scotland, I may be excused for turning to an American source for a statistical account of the effects of game legislation, especially as the instance in question has the merit of illustrating the differences in method between the slowly evolved empirical laws of an old country and the scientifically based laws of a new. In 1906, according to Mr J. B. Burnham, in Monroe County, New York, game of all kinds had become so scarce that sportsmen had abandoned shooting. During the six years prior to 1904, 135 Chinese Ring-necked Pheasants were distributed for stocking purposes, and the shooting of them was totally prohibited. In 1908 they had so increased in numbers that a short shooting season was opened, *but for cock birds only.* Since that time, till the present day, shooting has gone on annually, and in some years more than 6000 Pheasants have been killed. Yet owing to the protection of hens, the supply is still increasing at a great rate, and sportsmen are attracted to the county in great numbers and from long distances.

Apart from the direct and apparent results of the protection of game birds, there are more subtle but not less important effects which follow in chains of circumstances. Pheasants and Partridges, Black-game and the young of Grouse are largely insect-eaters, and as such, benefit their protectors to no small degree—the crop of an Argyllshire Pheasant has been found by Mr P. H. Grimshaw to contain close on 2800 recognizable insects—the staple food at the time having been a Two-winged Fly, *Bibio lepidus* (the maggots of which live in the ground upon roots of plants), 2286 specimens, and the Heather Beetle, *Lochmæa suturalis,* 508 specimens. Black-game also are formidable enemies of the latter insect, the grubs of which destroy annually hundreds if not thousands of acres of heather. Much more subtle are the connecting links in the series of events which, through the protection of Pheasants and the consequent encouraging of Black-headed Gulls in order that their eggs might supply the game-birds with food, led to the alteration of a local flora, and the replacement of the old by a new association of animals. To this interesting succession of results more detailed reference will be made elsewhere (see p. 501).

THE DEER FOREST

"The Scottes," said Holinshed, "sette all their delighte in hunting and fowling"; so it came to be that, as the numbers of wild animals diminished, the beasts of the chase fell under a system of protective laws as strict as those which governed the fowling quarry. Chief of the hunted was the Red Deer:

> The best of chase, the tall and lusty Red,
> The stag for goodly shape and statelinesse of head,
> Is fitt'st to hunt at force.

It is not too much to say that but for the care which has been taken of it by the law, the Red Deer in Scotland would probably long ere now have followed to extinction the Scottish Reindeer and the Elk.

The system of protection dates from a very early period, for in former days the Red Deer was royal game, *"inter regalia,"* and could be hunted only by the King himself or by those to whom he had given a grant of forestry. Yet in the *Fragmenta Collecta* of Scottish law, of uncertain date, a curious rule is made for deciding property in Deer: Deer were to be regarded as wild by nature, but in forests they were "thine as long as they have a desire to come to thee, and when they have no desire to come again they are not thine."

RED DEER IN THE LOWLANDS

It is interesting to note how frequently from the eleventh down to the sixteenth century the nobles of Scotland pursued the Red Deer in areas whence they have long since disappeared. David I (1124–1153) had a hunting house at Crail in Eastern Fifeshire; Walter the Stewart, in founding the Abbey of Paisley in 1160, granted the monks a tithe of his hunting, with the skins of the deer slain in his forest of Ferenze in Renfrewshire. Robert the Bruce (1306–1329) was repeatedly baulked by a white deer which he started among the Pentland Hills; and in a charter of 1328 he conferred on the monks of Newbattle an annual donation of five harts at the feast of St Cuthbert's Translation, to be taken from his forest of Selkirk. There, too, Bruce's successor, David II, hunted; in 1330, his Chamberlain paid 24s. for "a chalder of large salt for salting the king's venison at Selkirk," and in the following year, 16s. for salt for the venison at Ettrick Forest. In 1334 David renewed the grant made by his father to the Priory of Newbattle.

The history of the Red Deer in the Lowlands is carried down to the sixteenth century by Lindsay of Pitscottie who records that

> the second day of June the King [James V, the year being 1528] passed out of Edinburgh to the hunting with many of the Nobles and Gentlemen of Scotland with him to the number of twelve thousand Men; and then past to Meggitland, Crammat, Popert Law, St Mary's Laws, Carlaverick, Chapel, Ewindoores, and Longhope. I heard say he slew in these Bounds, eighteen score of Harts.

Although deer were evidently still plentiful in 1528, by the middle of the century constant slaughter and, more important, the development of farming and of the practice of pasturing upon the Lowland hills large flocks of sheep, sometimes ten thousand in number, as Bishop Leslie tells us, led to such a decrease in the numbers of Red Deer that vigorous efforts were made to protect the relics of the herds. In 1551 in the reign of Queen Mary, a Statute was passed complaining that "deare, rae, or uther wild beasts and wild fowles are clean exiled and banished by schutting with half-hag, culvering, and pistolet," and enjoining "that nane of our soveraine ladies lieges, of quhatsumever degree hee be of, take upon hande to schutte" at these animals "under the paine of death and confiscation of all their gudes for their contemplation." Notwithstanding, no measure could check the downward way of the lowland Red Deer, and with the troubles and lawless years of the seventeenth century, and the steady growth of agriculture, they were finally banished from the uplands and forests of southern Scotland.

Modes of Deer Protection

From a general point of view the modes adopted by Scots law in protecting deer range round several well recognized expedients. In the first place deer were not to be slain but by properly qualified persons: in the earliest days by the King and his Court, or by persons to whom he had deputed or granted rights of forestry; in later days by the great landowners on whose ground the deer roamed, for no one dare slay "Deare or Raes in utheris closes or Parkes....but [except with] special licence of the owners, under the paine of dittay [criminal prosecution], and to be punished as thieft" (1474). The "unlaw" or fine for such misdemeanour, at first "X punds," was raised in 1579 to "ten pundes" for a first offence, "twentie pundes" for a second, and "fourtie pundes" for a third, but if the wrongdoer was "not responsall in guddes," he was to "be put in the stokkes, prison or irones auct [eight] dayes on bread and water" for the first fault, "fifteene dayes" for the second, and for the third was to suffer "hanging to the death."

In the second place, since deer were preserved for hunting, and hunting in the early days meant the "clamorous hunt" with trained deer-hounds, all other methods were prohibited, such as shooting or slaying with such noxious inventions as "hag buttes [arquebus], hand gunnes, croce bowes and pistolettes, and taking of them with girnes [snares] and nettes" (1597 and earlier Acts).

In the third place, deer were not to be slain until they were able to take care of themselves, for it was provided in 1474, "that na man slaie onie of their Kiddes quhill [i.e. until] they be ane yeir auld, under the paine of X punds. And it to be a point of dittay."

In the fourth place, a sort of "close season" (instituted in 1400) was confirmed by the Act of 1474, since it was ordained "that na man slaie Daes nor Raes, nor Deare in time of storme or snaw....under the paine of X punds."

And lastly a kind of total prohibition was tried, for Parliament ruled that from June 1682 venison be not bought or sold for seven years.

Effects of Deer Legislation

The importance attached to the protection of these objects of the chase and the great efforts made to preserve them, may be judged from the fact that, in the sixteenth century alone, no fewer than eleven Acts are concerned with the penalties prescribed for illegal slaughter of deer. That these laws effected an increase of deer in suitable localities, such as the Highlands of Scotland, there can be little doubt, judging by the numbers seen and slain in some of the great hunts of former times. Take for example that famous hunt, in the summer of 1529, when King James V "togidder with his mother and ane ambassadour of the Paipis [Pope's] wha was in Scotland for the tyme, went all togidder to Atholl to the huntis." The quaint description by Lindsay of Pitscottie may be quoted, as much to show the pomp and circumstance which attached to a royal hunt as to illustrate the significance set upon the protection of deer, and the efficiency of legal preservation.

Preparations for the hunt were made long in advance, armies of beaters being sent to the hills, weeks or even months before the date of the final massacre. During the intervening time the beaters concentrated from all quarters upon the place chosen for the last act, driving before them the herds of deer from valley and mountain. To the rendezvous the King was conducted and housed in state; but let Pitscottie carry on the narrative in his own inimitable language.

> This noble Earl of Atholl caused mak ane curious pallace to the King, his mother, and the ambassadour, quhairby they were als weill eased as if they had been in ony pallace either of Scotland or England, and equivalent for the tyme of thair hunting; quhilk was biggit in the midle of ane greine meadow, and the wallis thereof was of greine timber, wovin with birkis, and biggit in four quarteris, as if it had been ane pallace, and in everie quarter ane round lik ane block-house, quhilkis was loftit and jeasted thrie hous hicht; the floore was laid with grein earthe, and strowed with sick floures as grew in the meadow, that no man knew quhairon he yead, bot as he had beine in ane greine gardeine. Fardder, thair was tuo great roundis on everie syd of the yett, and ane great portcullis of trie falling down, as if it had beine ane barrace yett, with ane gritt draw bridge, and ane foussie of sixteine fute deip, and thrittie fute broad, of watter. This pallace was hung with fyne tapestrie within, and weill lighted in all necessair pairts with glassin windowis.
>
> The king was verrie weill entertained in this wildernes the space of three dayes with all sick delicious and sumptuous meattis as was to be hade in Scotland, for fleschis, fischis and all kindis of fyne wyne and spycis requisit for ane Prince. Fardder, thair was no fischis that could leive in fresh watters but war thair swimming in the foussie about the pallace. It is said by the space of thir three dayes that his Grace was thair, the Earle of Atholl was everie day ane thousand pundis of expenss. This Pope's ambassadour, sieing so great ane triumph in ane wildernes quhair their was no toun neir be twentie myllis, he thought it ane great marvell that

sich ane thing sould be in Scotland, that is, so court-lyk and delicious entertainment in the Highlandis of Scotland, quhair he saw nothing bot woodis and wildernes. Bot, most of all, this ambassadour, when the King was cuming back from the huntis, marvelled to sie the Highlanderis sett all this pallace on fire that the King and the ambassadour might sie it. Then the ambassadour said to the King, I marvell Sir yea latt burne yon pallace quhair yea war so weill eased. The King answered, 'It is the use of our Highland men that they be nevir so weill lodged all the night, they will burne the same on the morne.' This being done, the King returned to Dunkell that night, and on the morne to St Johnstoun. It is said at this tyme, in Atholl and Stratherdaill boundis, thair was slaine threttie scoir [600] of hart and hynd, with other small beastis, sick as roe and roe-buck, woulff, fox, and wyld cattis.

I have already referred to another hunting of James V, when, in 1528, 360 harts were slain in the forest county of Selkirkshire and the adjoining counties. In 1563 Queen Mary was present at a Highland hunt when, in the course of two months "tenchel" driving, two thousand Scottish Highlanders collected in the wilds of Athole, Badenoch, Mar and Moray a huge herd of "more than two thousand deer." Many of these, owing to a sudden stampede of the herd, broke bounds and escaped, but notwithstanding, "there were killed that day 360 deer." Taylor, the Water Poet, relates that, in his presence, in 1618, in the "Brea of Marr"—Braemar in Aberdeenshire—"in the space of two hours, fourscore fat deere were slaine," and the Wardlaw Chronicles tell how, in 1655, in the "Forrest of Monnair" on the borders of Inverness, Ross and Cromarty,

We travelled through Strathglaish and Glenstrafarrar to Loch Monnair.... Next day we got sight of 6 or 700 deere, and sportt off hunting fitter for kings than country gentlemen.

These facts go to show that, while legislation was unable to save the deer of the Lowlands, where deer preservation came into violent conflict with agricultural progress, it did result in keeping up the numbers in the wilder areas of the Scottish Highlands.

Deer Protection at the Present Day

In our day also, a similar result follows upon the protection afforded to Red Deer. So efficiently are they preserved that there can be little doubt that in many areas the number exceeds what the ground could naturally bear. Why should it be necessary to hand-feed deer in many forests during the winter, were it not that the "forest" does not afford sufficient sustenance for the numbers of deer upon it? And how else can we account for the great mortality that occurs from natural causes every season? In the deer-forests of Jura, a count extending over ten years revealed the fact that, over and above the slaughter due to sport, an average of more than one hundred deer died every year. The conclusion that protection has increased the stock beyond the capacity of the country, is confirmed by the fact that while the total number of deer in Jura is now over 2,000, in Martin's day, in the closing years of the seventeenth century, the hills ordinarily had "about three hundred Deer grazing on them," and even then they were protected, for they were "not to be hunted by any, without the Steward's License."

The War has further emphasized the tendency to overprotection. During the past thirty years, the area of Scottish deer forests has increased by many thousands if not by millions of acres. Deer have now become so numerous, partly owing to the absence of sportsmen and keepers on the trail of greater game, that they have overrun great stretches of the best sheep-grazings and are said to have destroyed fields of valuable crops within twelve miles of Glasgow. So serious a menace have they become to the interests of agriculture and to the food-supply of the nation, that the protective legislation which has held sway from time immemorial has been rescinded for the time being. An order of the Board of Agriculture for Scotland issued in February 1917 authorizes the occupier of any agricultural holding in Scotland to kill, by any means available, deer that are trespassing on his grazings, or causing injury to his crops; and this with only a gun licence and without even a licence to kill game—surely a measure and an expedient which might well raise from their last resting-places the Scots law-makers of old, with their mutilations and their hangings.

Nevertheless, in spite of the severity of the old laws of the deer-forest and their one-sided point of view, naturalists owe them a debt of gratitude; for, taking account of the influence of changed conditions upon the deer themselves (see under Destruction of Forests, p. 335), of the present necessity for artificial feeding in many areas, of the annual tale of natural deaths, of the history of the lowland Red

Deer, of the disappearance of the Reindeer and the Elk, and of the destructiveness of man, I would unhesitatingly say that, but for the protection afforded by the law, Red Deer would long since have ceased to exist in Scotland as the wild denizens of our mountains and highland moors.

ROE AND FALLOW DEER

Roe and Fallow Deer were held by the law in much the same light as Red Deer; in many of the Statutes all three are specified together and on equal terms. As Fallow Deer were as a rule "Parked Deare," the property of the owner of the enclosure, interference with them was a simple crime of theft; and as Roe Deer were of less value for purposes of the hunt, the protection of them was on the whole less stringent than that of the "best of chase, the tall and lusty Red."

LESSER GAME—HARES

Hares did not enter the desirable regions of protection at so early a date as Deer, for the "tim'rous hare," "a weak, harmless flying creature," was scarcely to be reckoned *inter regalia*—sport for a king. Indeed in the twelfth century, during the reign of King Alexander, there was no prohibition against hunting hares except they were in forests and warrens, where they were private property. Yet in many ways the protection of hares ran parallel with that of deer, although it was as a whole less comprehensive.

In the fifteenth century a close season of a kind was instituted, for hares might not be slain in time of snow under a penalty of 6s. 8d. (1400). In the sixteenth century (1567) a much more strict preservation was enforced under pain of forfeiture of all the offender's moveable goods, or if he had no goods, of imprisonment for 40 days for a first offence and, for a second, loss of his right hand. As hares, like deer, were to be reserved for the chase, shooting with "hag buttes, hand gunnes, croce bowes and pistolettes" as well as snaring and netting were forbidden; in 1579 the "slaying of Haires" was included in the comprehensive Act which threatened offenders for a third time, with "hanging to the death"; and in 1685, on account of the decay of game in the Lowlands, the shooting and selling of hares at any time were forbidden. A curious penalty attaches to a decree of 1707, made during our conflict with Louis XIV of France, for it was ordained that no one shoot hares without a licence from the proprietor of the ground, under penalty of being sent abroad as a recruit.

What the actual effect of such protection was upon the stock of hares in the country I have no means of estimating. I imagine that it must have kept the number up to a level that could not have been attained had hares been slaughtered at all times by all and sundry; and this supposition is confirmed by the reduction of numbers which followed upon the setting aside of the older Game Laws by the Ground Game Act of 1880[1]. Yet in the old days hares do not seem to have been overplentiful, judging from the slight evidence I have gathered. To Mr A. O. Curle I am indebted for an extract from the manuscript of the Moray Papers, which I quote here by permission of the Right Hon. the Earl of Moray. It appears in a letter, dated 22 January 1582, from John Guthrie in Castle Campbell to the Countess of Argyll, relating a libel against Argyll made to the King by the Prior of Pluscarden, accusing him of

> the foullest and greatest slauchter of hares that ever he saw, felling them in thair setts and lowsing of 10 or 12 leish of dogs by [? forbye] ane great number of raches at ane hare and so wald slay in ane day 12, 16, or 20,—

no very great "slauchter," as things go in our day of sporting guns and ammunition, though with dogs a very fair bag. In the middle of the eighteenth century the number of hares and leverets in Great Britain was estimated, according to Dr John Campbell, I do not know on what basis, at a stock of twenty-four thousand, and an annual breed of twelve thousand.

As in the case of Deer and Game Birds, the War has made more apparent than ever before, the influence upon the hare's numbers of even the modicum of protection now afforded it. Owing to the absence of sportsmen and gamekeepers, the annual destruction of hares has been lessened, and the effect of preservation, no longer obscured by compensating slaughter, stands out clearly in an enormous

---

[1] See p. 180, where also the Hares' Preservation Act of 1892 is referred to.

increase of numbers. So great has become the damage done by this unwonted increase that "with a view to preventing or reducing injury to crops or wastage of pasturage," special Orders have recently been issued (17 April 1917) by the Board of Agriculture for Scotland and the corresponding bodies in England and Ireland, authorizing and encouraging

> the killing and taking, the sale and purchase, and the possession of any birds or hares or rabbits at any time when the killing and taking, the sale and purchase, or the possession thereof would otherwise be unlawful,

and granting these liberties to persons who have no game-licences and no rights according to the laws of game protection.

IV. 2

PROTECTION OF ANIMALS FOR UTILITY

I HAVE suggested that, in a general way, the development of laws protecting animals because of their usefulness, succeeded in time the development of laws protecting the objects of sport. Under the feudal system the king and his court held first place, and so their pleasure was more important than the nation's profit, since the nation, as we think of it, consisted but of underlords and vassals. Thus it was only with the gradual decay of the feudal idea and the growth of a new appraising of values that the utilitarian aspect of wild animals forced itself into notice. Even so, the growth of a true economic protection has been a slow one, for the earliest efforts at the legal protection of useful animals, were mainly aimed at protecting such for the nobles and the proprietors of land, and from the masses of the people. With a few exceptions, it is only within comparatively recent years that there has been passed utilitarian legislation concerning animals, which affects the greater part of the population.

The laws protecting useful animals have had several distinct ends in view. Some animals have been protected because they could be used as food; others because of the profit to be gained from their fur or skins; a few because of their value as scavengers; and a larger number because of the special services they render to the agriculturist and the nation at large.

PROTECTION OF FOOD ANIMALS

Beasts and Birds

History furnishes unpleasantly frequent records of the occurrence in Scotland of times of great scarcity, when prices of common articles of food rose to a prohibitive height (so that, as in 1551, Acts establishing fixed prices of food-animals had to be passed), when the poorer inhabitants were forced to the sea-shore there to live upon the shell-fish gathered on the rocks and from the sands, and when, in spite of all efforts, many died of want. Such recurrences probably first turned the attention of our legislators to the need of supervising the wild stock of the country with a view to future food supplies. Thus alongside protection of game birds from the sporting point of view grew protection of wild fowl from a utilitarian standpoint.

In an Act passed in 1457, in the reign of James II, provision was specifically made for such wild fowl "as ganis to eit for the sustentacion of man, as pertrykes, pluvars, wilde dukes and sik lik fowlys." Of these neither the nests nor eggs were to be taken or destroyed, nor were the birds themselves to be killed in moulting time. Further, that the stock of wild fowl should be still more augmented, the same Act provided for the destruction of "foulys of reif," or birds of prey in a wide sense, such being "ruikes, crawes, eirnes[1], bissettes[2], gleddes[3], mittalles[4], the whilk destroyes beast, corne, and wilde foulys." Great efforts were to be made to keep these destroyers in check, for they "sall utterly be destroyed by all manner of men, be all engine[5] of all manner of crafts that may be founden" since "the slaughter of them sall cause great multitude of divers kind of wilde foulys for man's sustentacion."

In the sixteenth century there were also passed Acts for "staunching dearth" which provided for the preservation of "hart, hinde, dae, rae, haires, cunninges, and utheris beasts." The many statutes protecting the inhabitants of the dovecot, the "dows" and their "dowcattes," of which a summary has already been given (p. 98) were also clearly devised for the increase of food supply, although the supply was primarily intended for the table of the laird.

Fishes of Fresh Waters

Ranking almost in importance with the preservation of game is the protection which has been given to the more valuable fishes of our fresh waters, Salmon and Trout. The details of the protective measures vary for almost every great river, and it would serve no useful purpose to discuss them here, but in the main they follow several wide principles.

[1] Eagles.   [2] Buzzards.   [3] Kites.   [4] "A kind of halk"—Jamieson.
[5] Ingenuity.

Originally the right to catch Salmon was a common right, but the value of the fish soon led to interference with the privileges of the people at large, and at an early date property in salmon fishings became vested in the Crown. By grants and charters the Crown bestowed the rights in these valuable properties to private individuals, often independently of the land in which the fishing lay, so that the taking of Salmon became a right limited to a favoured few of the population, for whom it was protected by statutes innumerable.

Yet from the time of Robert the Bruce, the law made many endeavours to preserve the fishes themselves, and these regulations follow well-recognized lines common to all legislation protecting animals—the protection of adults, especially at breeding times, and the safeguarding of the young.

Modes of catching the fish were severely restricted. Fixed engines, stake-nets and bag-nets were prohibited in rivers and in tidal estuaries, and it was enacted in the reign of James I "that all cruives and yaires set in freshe watteris, quhair the sea fillis and ebbis, the quhilk destroyis the frie of all fisches, be destroyed and put awaie for ever mair" (1424). The general effect of such restrictions, which are of wide application, is to give adult fishes, returning from the sea to fresh water, a sporting chance of reaching their spawning beds in the upper reaches of the rivers.

The safety of breeding fish was sought by prohibiting the taking of "baggit" Salmon, that is, Salmon about to spawn, and by the creation of close seasons, which were first instituted in the Act of James I just quoted, the close time being there designated as from

the Feast of the Assumption of our Ladie quhill [until] the Feast of St Andrew in winter....Quha sa ever be convict of slauchter of salmonde in time forbidden be the law, he sall pay fourtie shillings for the unlaw, and at the third time, gif he be convict of sik trespasse, he sall tyne [lose] his life or then [else] bye it [i.e. pay its value as ransom].

What the ransom was for a life in 1424, I do not know, unless it was that of a law attributed to a much earlier date, which stipulated "For the life of ane man nine time twentie kye," and proceeds to put a mercenary value upon various acts of violence—

For ane fute, ane marke. For ane tuth, twelve pennies...For ane strake with steiked neif [clenched fist], twelve pennies...For ane strake with the fute, fortie pennies. For the blude shed out of the head of ane Earle, nine kye.

In Scotland at the present time the annual close season for Salmon differs according to local conditions in individual rivers, but in no case must it be less than 168 days. A weekly close time for nets from 6 a.m. on Saturday till 6 a.m. on Monday, the "weekly slap," gives much needed facility for migrating Salmon to pass beyond the estuaries up the rivers.

Adult Salmon are further protected by comparatively recent legislation (1862) prohibiting the taking of fish which, having spawned, have not yet completed their seaward migration—foul fish or kelts, and unseasonable fish.

Several measures have been adopted for the preservation of young Salmon: spawning beds may not be disturbed, even banks suitable for spawning must not be interfered with, the taking of Salmon parr is totally prohibited, and as early as 1457, it was decreed that "na man in smolt time set veschellis, crelyis, weris, or any uthir ingyne, to let [prevent] the smoltis to pass to the see."

Trout are protected by restricting to private individuals what were at first public rights, by restrictions as to the modes of fishing, and by an annual close season extending from the 15th of October to the 28th of February.

So the "multiplicatioune of fische, salmonde, grilsis and trowtis" has been encouraged in Scotland; but the direct protection has given rise to another and important means of "multiplicatioune," for the institution of private property in fishing has encouraged the artificial rearing of Salmon and Trout fry in specially designed hatcheries. There they are brought through the early stages of life, free from the innumerable dangers from physical accident and natural enemies to which they are exposed in open rivers, and are launched annually in their thousands on suitable feeding-grounds in their own or in distant waters. Some notion of the significance of fish hatcheries may be gathered from the facts that in the season 1915–16 the 224 breeding stations in Switzerland hatched 157,971,000 eggs of fresh-water fishes, and that in 1915 the hatcheries of the United States of

America liberated over 535 million fry and over 18 million fingerlings, yearlings, or adults in streams debouching on the Pacific, while since the inception of the work of hatching, they have distributed 6291 millions of fry.

Sea Fisheries

The history of the protection of sea-fisheries is complicated by infinite variety of limitations, most of which have now fallen into desuetude. At the present time, white fish and Herrings are protected mainly by regulations fixing a minimum size of mesh in nets used for their capture, and by the institution of reserves—the three-mile limit and certain special areas such as the Moray Firth and the Firth of Clyde—within which specific modes of fishing, for example trawling, are prohibited.

Molluscan and Crustacean Shell Fish

A few species of molluscan and crustacean shell-fish have been protected with a view to increasing the stock available for food. Oyster beds and Mussel scalps are regarded as private property and are protected as such. Oysters are further subject to a close season extending from 14th May or 15th June till 14th August, according to whether they occur on inshore or deep-sea beds.

Of Crustacean shell-fish, Crabs and Lobsters are by far the most important from an economic point of view (see p. 153). The protective regulations vary in different Fishery Districts but a minimum size of $4\frac{1}{2}$ inches across the back for Edible Crabs and of 8 inches length for Lobsters is enforced, and as a rule, the taking of crabs with spawn and "hen lobsters in berry" is prohibited.

Such regulations, while they have had, in most cases, no appreciable effect in increasing beyond their original numbers the stocks of the chief food animals of the country, have had the negative but highly important effect of preventing a serious depletion of these stocks, a result which would inevitably have followed promiscuous destruction of creatures of such great economic value.

PROTECTION OF FUR-BEARING ANIMALS

When one thinks of the extent and of the content of the great fur countries in the northern territories of the Old and the New World, exporting annually their hundreds of thousands of valuable pelts, it seems as if the limited fauna of Scotland could never have ventured to take part in the trading of furs. Yet it is not so, for long before those northern regions had been tapped, Scotland, as we have already seen (p. 155), was much resorted to by the merchants of the Continent on account of the rich furs it exported. Skins of the mertrick (marten), foulmart (polecat), beaver, otter, tod (fox), whitret (weasel, perhaps also stoat or ermine), and cunning (rabbit), were exported in quantity. In view of the importance of this trade in furs, it seems extraordinary that, apart from a few regulations limiting the wearing of the more valuable kinds to men of high degree, and a few enactments levying a toll upon exported skins, the law made no attempt to foster so valuable an asset by endeavouring to protect the animals themselves, except in the solitary case of the rabbit.

The Rabbit

In another place I have discussed the introduction and establishment of the Rabbit in Scotland (p. 247), so that here I would do no more than indicate its standing as a fur-bearer and the protection which has been awarded it on this account. Although of no mean value as a food animal, the Rabbit owes its early preservation rather to the quality of its skin. Witness the statement of Dr John Campbell in his *Political Survey of Great Britain*, published in 1774:

> Their Flesh at a proper Age, and in proper Seasons, is thought equally wholesome and delicate. But this, though in some Degree an Object of Profit, did not so much recommend or render them so valuable, as their Skins, which are now much reduced in their Price from a Variety of Causes, and though thus reduced they are still of no despicable Value.

He goes on to say that in his time "it hath been computed, that Skins included, the annual Produce of Rabbits within the Bills of Mortality comes to about Forty Thousand Pounds."

The price of a rabbit for food was fixed in Scotland in

1551 at from 1s. to 2s., but the value of the skin was much greater, for Campbell says

the Skins of large well chosen Rabbits would produce Half a Crown, or even Three Shillings a Skin, being then used in lining Robes, in Muffs, in Tippets, &c. The Down was employed in making Hats, and in both ours [an apparent printer's error for *colours*] was highly esteemed in France, especially the jet Black, and such as had only a Sprinkling of White amongst the Black, and was very much preferred to their own.

A great fillip was given to the use of rabbit skins for hat-making in Scotland when, in 1621 the wearing of castor or beaver hats was forbidden (although in 1672 the privilege was allowed afresh to noblemen), and when, in 1695, authority was given by law for the manufacture of hats from rabbit and hare skins.

In view of the value attaching to rabbit skins, it is little wonder that "cunninges" and "cunningaires" or rabbit-warrens, were protected with a jealousy which in our day of a rabbit pest seems remarkable. A Statute of James VI, passed in 1579, was particularly severe on the "breakirs of cunningaires" who, should they be found "unresponsall in gudes," that is, unable to pay the fine of "fourtie pundes" demanded on a third offence, were to suffer "hanging to the death." This value, too, is accountable for the rapid spread of the rabbit in Scotland and for the creation of innumerable warrens, many of which, as that on the links at Aberdeen, were commonties, the property of the neighbouring town or village. An index to the esteem in which the skins were held in countries outwith our own boundaries is furnished by an interesting entry in a Charter of the Earl of Mar and Kellie, which gives the yearly average export between 1611 and 1614: "Of cuneing skinnis 53,234 at £6 the hundreth, £3194."

Seals

Abundant though Seals were in former days on the coasts of Scotland, and valuable as their carcases must have been in furnishing food and oil as well as in supplying skins for clothing, no general attempt was ever made to protect them. Yet an interesting story of St Columba shows that at one time and in one place breeding colonies of seals were jealously guarded. Adamnan, in his *Life of St Columba*, tells how the Saint despatched from Iona to Mull two brethren empowered to catch a robber from Colonsay who, having come over in his boat at night to a small island where the young seals were brought forth and reared, killed and stole many of them. The seals were the property of the settlement in Iona and were carefully protected and encouraged to multiply as a reserve for supplying food, clothing and oil to the monks.

Simple as was this oldest protection of the seal in Scotland in the sixth century, it is still the latest method; for it was instituted in 1912 in the Pribilof Islands, off Alaska. The prohibition of killing in that famous breeding-ground has increased the stock of Fur Seals (*Arctocephalus ursinus*), according to the enumeration and calculations of the United States Bureau of Fisheries, from 363,872 in 1915 to 417,329 in 1916, a gain of over 53,000 in a single year.

The protection of even a single breeding-ground on a small island on the west of Scotland must in that area have had no little effect in keeping up the numbers of seals. Otherwise, under the constant demands of man they must have followed that course of decline which elsewhere has invariably accompanied unrestricted slaughter. This very slaughter, to which I referred in the preceding chapter (p. 172), rendered necessary the protection of the Grey Seal (*Halichœrus grypus*) in British waters, and this was granted in 1914 by an Act of Parliament which instituted an annual close season from the 1st of October till the 15th of December.

PROTECTION OF ANIMALS AS SCAVENGERS

The encouragement and protection of such animals as feed upon carrion and garbage is a well established custom in many and especially tropical lands where sanitation is in its infancy, and where festering heat hastens decay and noisomeness. In such countries, Jackals by night and Pariah Dogs by day are tolerated with a benevolent neutrality on account of their services to cleanliness, while the sacredness of the Vultures of India and the sufferance accorded to their relatives in the warmer regions of Europe, Asia and Africa can be traced to their efficiency as cleansers of the earth.

In Scotland in early days and down to comparatively recent times the sanitary condition of towns and villages was not very different from that of many modern sites of habitation in the East. Imagine the state of the streets of Edinburgh, for example, when, even so late as 1730, the frequent cry of "Gardy Loo" (*Gardez l'eau*) heralded a deluge of household slops and filth from the window of an upper storey upon the causeway, or when middens plentifully bestrewed the main avenues of traffic, to the hindrance of the lieges, as Lindsay (*d.* 1557) relates:

> Marie ! cummand throw the Schogait,
> Bot thair hes bene ane great debait
> Betwix me and ane sow.
> The sow cryit guff, and I to ga,
> Throw speid of fute I gat awa
> But in the midst of the cawsa[1]
> I fell into ane midding.

In such conditions many animals prone to nose in garbage were afforded a mild sort of protection. Swine were to be found, especially at night, roaming from one heap of offal to another, and it was no unusual chance that led the hero of the verses quoted to meet one in a thoroughfare. Nor was it a fortuitous juxtaposition that occurred when the Assize of Haddington in October 1543 ordained "Item all muk to be put off the Gait [street] and all swyne to be put off the Towne." The foragings of dogs and cats also were so much encouraged that their presence became a public nuisance, and the Haddington Assize was compelled to ordain "that the hangman sall escheit to hymself all swyne doggs & catts at [that] he fyndis one [on] the gait fra this nycht furcht."

Birds of Carrion

Many carrion-feeding birds were encouraged to aid in the work of scavenging; else how can we account for the numbers which frequented the streets of Edinburgh and Leith? A casual reference in Wedderburn's Accompt Book indicates the presence of these visitors, whose appearance in the streets called for no remark on ordinary occasions, although during an eclipse of the sun in 1597,

the peiple with gryt fair fled aff the calsayis [causeways] to houssis mourning and lamenting, and the crawis and corbeis and ravenois foullis fled to houssis to our steple and tolbuith and schip tappis, maist merveulously affrayit.

[1] Causeway.

PROTECTION OF ANIMALS FOR UTILITY

I think it highly probable that till a comparatively late date, Rooks, Carrion Crows and Ravens were afforded a real and active protection in our towns and villages on account of their services to cleanliness, but that, as with so many habits and customs now curious to us, the very ordinariness of their presence in former days led to its being passed over unremarked by contemporary writers. We must remember, however, that, according to Clusius, vast numbers of Kites were in constant attendance in the streets of London during the reign of Henry VIII, on the look-out for the offal with which the thoroughfares were polluted.

It has been left to a German nobleman, soldier and traveller, Von Wedel, to record as strange to his eyes, what our own historians passed over in silence, the presence and protection of Ravens as scavengers in our towns.

"On the 6th [of September 1584]" he wrote in effect, "we rode to Belfart [Belford], twelve miles, and from thence twelve miles again with fresh post to Barwick [Berwick-on-Tweed]....The houses in the town are mean and thatched with straw....There are many ravens in this town which it is forbidden to shoot, upon pain of a crown's payment, for they are considered to drive away bad air."

Here is obviously a reference to their removal of evil-smelling garbage and the pestilence it engendered; but the significance of this protection of scavengers is made quite clear by one Capello, a Venetian ambassador, who, after spending the winter of 1496–7 in England, wrote

Nor do they dislike what we so much abominate, i.e. crows, rooks and jackdaws; and the raven may croak at his pleasure, for no one cares for the omen; there is even a penalty attached to destroying them, as they say that they will keep the streets of the town free from all filth. It is the same case with kites, which are so tame, that they often take out of the hands of little children, the bread smeared with butter, in the Flemish fashion, given to them by their mothers.

The protection of the Raven and its kind as scavengers was no new feature of sixteenth century life in Britain, for even in Roman times, as the results of excavations indicate (although excavators and commentators have missed the significance of their presence) these birds were welcomed in the settlements and cities.

"The most common birds [found in the excavations of the Roman station of Calleva, now Silchester, in Hampshire]," says Mr H. Jones in *Archaeologia* in 1892, "after those of the domestic fowl, have been identified as those of the Raven...The numerous bones found amongst the Roman remains would almost point to its having lived there in a semi-domestic state."

And the excavations of the following year showed that "The Raven and the Crow, especially the former, seem to have been very plentiful, and gave the largest number of identifiable bones." A commentator has surmised that the abundance of the Ravens at Calleva may have been due to the fact that they were hung in cages at the entrance to the houses, as the Magpie was in Rome, to keep guard against intruders and to salute those who were invited to a villa. But such surmises are unnecessary: the numbers of their remains, the analogies of the municipal laws of Berwick and of the bird frequenters of the streets of Leith and London, make it perfectly plain that the Ravens and Crows were encouraged and protected in the Roman settlements on account of their value as disposers of garbage.

The result of such protection was that for many centuries "the birds obscene that croak and jar," which now, in numbers miserable, are banished to the wildest crags, were not only common throughout the land, but were the constant companions of the traffic of the streets even in our largest and busiest seaports and towns. Strange, wild times when the Sparrow scavengers of modern thoroughfares were replaced in numbers and in assurance by Rooks, Carrion Crows and Ravens!

It is not necessary to suppose, however, that these Raven hordes built and nested in the near neighbourhood of the towns they frequented, for it is well known that they, in common with all birds of carrion, gather from afar to the feast. Macgillivray instances such a congregation on the Island of Pabbay in the Outer Hebrides, when as many as two hundred Ravens gathered over a stranded herd of Grampuses or "Killer" Whales; and Dr B. N. Peach tells me that on a similar occasion, when five "Killers" came ashore in Weisdale Voe on the mainland of Shetland, he witnessed the congregation of a flock of Ravens of which he estimated the number at five hundred, many of which, in his opinion, had gathered from the outlying islands.

PROTECTION OF THE FARMER'S FRIENDS

Since everyone knows that many of our wild birds in working out their own life-stories, perform deeds without which the fruitful tilling of the soil could not long survive, it seems strange that so little care should have been given to the welfare of these friends of man. What has man done to encourage the useful bird?

> "The stingy farmer," wrote Michelet, "has not a grain for the creature, which, during the rains of winter, hunts the future insect, finds out the nests of the larvae, examines, turns over every leaf, and destroys every day, thousands of incipient caterpillars. But sacks of corn for the mature insect, whole fields for the grasshoppers, which the birds would have made war upon. With eyes fixed upon his furrow, upon the present moment only, without seeing and without foreseeing, blind to the great harmony which is never broken with impunity, he has everywhere demanded or approved laws for the extermination of that necessary ally of his toil—the insectivorous bird."

And Michelet's condemnation could be applied with force to Britain, for when he wrote *L'Oiseau* in 1856, the law had made no effort to preserve any such bird for the welfare of the country at large. Legislation for the protection of game, there was in plenty, but the democratic idea of preserving birds for their services to the people had not yet been born.

Since Michelet wrote, some progress has been made in other countries as in our own. Many of the United States of America, like West Virginia, specify in their protection laws, birds "that promote agriculture and horticulture by feeding on noxious insects and worms"; and in 1902 there was signed in Paris by many of the European countries (*not* including Great Britain) a "Convention pour la Protection des Oiseaux utiles à l'Agriculture," which forbade at all times and in all ways the killing of birds useful to agriculture, especially insectivorous birds, and the destruction of their nests and eggs.

In Britain insectivorous or useful birds have not been singled out for special protection, but they are, of course, included in the general protection granted to all wild birds during a close season instituted by the Wild Birds' Protection Act of 1880, and extending in each year between 1st March and 1st August, the breeding season. Such birds as are mentioned in a Schedule attached to the Act are absolutely protected during this period, while the others remain at the mercy of the occupier or owner of land. Many also have been protected under subsequent Acts (especially of 1894 and 1896) providing for the preservation of the eggs of particular wild birds, for the adding of new birds to the Schedule, for the extension of the close season in particular cases, and for the creation of areas where all wild birds are protected all the year round.

The result of these enactments has been on the whole a marked increase of insectivorous birds. Indeed so great has been the increase in the case of certain species that they have been forced by the competition of numbers to turn more and more from their staple food of insects to the grain of the farmer and fruit of the gardener. On this account many demands have been made that certain of the species which have most multiplied under protection should be removed from the guardianship of the law. Although these demands must be carefully tested in the light of unbiassed observation of feeding habits throughout the year, there is a consensus of opinion, based upon field observations and the examination of the food found in the crops of thousands of individuals, that in certain areas the numbers of such insect-eating birds as Rooks, Starlings, Gulls, especially Black-headed, and Blackbirds, could be limited without detriment to their useful labours, and to the great advantage of agriculture generally.

I conclude this reference to the "farmer's friends," with an account of a curious type of special protection which used to hold in Shetland. There the Great Skua (*Stercorarius skua*) was preserved and encouraged because of its services to the farmer in driving the Sea-Eagle from the island. Low in his *Tour through Orkney and Shetland*, which took place in the latter half of the eighteenth century, wrote

> In Foula there is a privileged bird, no man will nor dare shoot it, under the penalty of 16s. 8d. Ster., nor destroy its eggs; when they meet it at sea, whatever fish they have in the boat Skua always gets a share, and all this out of gratitude for beating off the Eagle, who dares not venture to prey on the island during the whole of the breeding season....Skua is not so strong as the Eagle, but much more nimble; strikes at him without mercy, with such effect that he makes the other scream aloud, and his retreat is so sudden as to avoid all danger from the Eagle.

IV. 3

PROTECTION OF ANIMALS FOR AESTHETIC REASONS

In the slow progress of the protection of animal life by law, it seems somehow natural to suppose that the idea of preserving creatures for no other reason than their rarity, their beauty of plumage, their sweetness of song, or the interest of their habits, should have been the last to have borne fruit. Probably the reason lies partly in the fact that the general interest in nature for its own sake is a development of very recent growth, partly in the truth of the old saw which says that "Ilka man's business is nae man's concern"—in other words, that because of their lack of expression, the concerns of democracy have been stifled by the clamour of sections more powerful and more immediately interested.

So it happens that the protection of creatures "attractive in appearance or cheerful in song," as the Nebraskan law puts it, is a matter of our own day, for we can hardly regard the Scots laws of James I and Charles II (1621 and 1672), forbidding the wearing of feathers, as having been instigated by any concern for the "plume question," as it is understood to-day.

Even yet, so far as I know, the law takes no cognizance of our rare or interesting animals other than birds. The numbers of the Badger may dwindle on our hill-sides, the Wild Cat and Marten may become extinct in our highland woods, the Large Copper and the Artaxerxes Butterflies may be exterminated, but the law says never a word.

Birds stand in happier case. Our birds of fine plumage and sweet song, our rare birds and interesting birds shared with all others the benefits of the close season of the Act of 1880. Since then many of our rarer birds have been afforded special protection, as have also their eggs, and a special Act was passed in 1904 for the protection of wild birds on the Island of St Kilda, notably the St Kilda Wren, which was threatened with extinction owing to the regardless enthusiasm of collectors.

There can be no doubt that the numbers of many of our interesting birds have increased greatly during the past thirty years mainly owing to the protection afforded them. Year after year the reports on Scottish bird life in the *Annals of Scottish Natural History* and its successor *The Scottish Naturalist*, tell of increasing numbers and of extensions of range, and to these I would refer the reader, contenting myself here with a few typical examples, mainly extracted from the pages of that magazine. Rare birds like the Red-necked Phalarope (*Phalaropus lobatus*) "seem to be increasing every season"; the Great Crested Grebe (*Podiceps cristatus*) "has been increasing of late years as a breeding bird in Scotland," and since its first recorded appearance about 1877 has extended its range to the faunal areas of Solway, Tweed, Forth, Clyde, Tay and Moray; the Scottish Crested Tit (*Parus cristatus scoticus*) (Fig. 47)—a bird exclusively Scottish—has colonized many new areas in Moray, Nairn, Inverness and Ross during the last ten years, despite the persecutions of collectors; and the Hawfinch (*Coccothraustes coccothraustes*) has continued its general extension in Scotland, in spite of the ill-will of growers of peas. Birds of fine plumage, like the Goldfinch, the Jay and the Kingfisher are steadily increasing in numbers notwithstanding the attentions of birdcatchers, gamekeepers, and preservers of fisheries. Song-birds, such as Linnets, Sedge Warblers, Wood Warblers, Thrushes and Blackbirds are multiplying in several or many areas. And of birds of general interest, the Golden Eagle, the Raven, the Magpie, the Gannet and the Fulmar Petrel are a few which have increased and have settled in new breeding grounds.

In many cases, it is interesting to note, these benefits have accrued to the naturalist and to the people in general, in face of the vested interests of sport: the Golden Eagle multiplies in spite of the toll it levies upon grouse and ptarmigan; the Kingfisher in spite of its depredations on young trout. And this is due in part to a new sense of aesthetic value which has led landowners and owners of fisheries to balance against the game bag and the basket of trout, the beauties of undisturbed nature.

So successful has been the legal protection of wild birds that several species have increased beyond reasonable limits, and have come to be recognized as pests. Amongst these beneficiaries under the Acts and their supplementary Orders are the Gulls, especially Blackheaded Gulls, which have turned their attention to grain, turnips and potatoes;

Fig. 47. Scottish Crested Tit—increasing in numbers under protection. (From a Speyside example.) About ½ natural size.

the Merganser, Goosander and Heron, whose numbers are said seriously to deplete freshwater fisheries; the Starling, Thrush, Blackbird and Bullfinch, which have become the bane of the fruitgrower; and those doughty thieves, the Rook and the Wood-pigeon. In the interests of the home grown food supply of the country, the importance of which the War has emphasized, such birds, in districts where satisfactory evidence of their harmfulness has been produced, ought to be temporarily excepted from the modes of protection awarded them, so that, by the reduction of their numbers, good may come to the country at large. But with birds as with men evil is most often tempered with good, and only after long observation and careful balancing of facts can a final and fair judgment be reached. Even then, I am convinced, little harm would befall were the final judgment to be tempered with mercy.

IV. 4

PROTECTION OF ANIMALS THROUGH POPULAR FAVOUR AND SUPERSTITION

Man has so much to answer for as a destroyer and exterminator of his co-dwellers on the earth that there is something pleasing in the thought that on occasion his goodwill has been extended to animals, apart from the thought of future sport or of gains to come. This goodwill has affected a favoured few of our native animals, and as a rule has been determined by some special interest attaching to the creatures it shields, in view of which minor delinquencies on their part have come to be overlooked.

The Otter is a serious menace to freshwater fisheries, yet it has probably been saved from extermination because of a tender thought in the heart of many a proprietor of a trout stream for this "lord of the stream" who surveys "all the finny shoals his own," whose destructiveness is compensated for by the interest and beauty it adds to many an angler's haunt. But perhaps the thought of a day with the otter hounds has also had its influence. The Badger, too, however much it may have been persecuted in former times, has now gained a place in popular favour because of its interest and rarity, and there is little likelihood that the extermination which once threatened it, will be carried into effect. May it be taken as a gratifying sign of the growing regard for our wild animals, that a colony of Badgers has inhabited for many years, without disturbance, a site on Corstorphine Hill almost within the boundaries of the city of Edinburgh? But if its size and clumsiness add interest to the Badger, for it is one of the largest of our surviving wild animals, diminutiveness and grace make equal claims upon the affection of man, and with these the Dormouse, a native of southern England, has made an appeal which has not been unanswered.

It is to the Birds, however, that the goodwill of man has most frequently turned. Where their song has charmed his ear, or their confidence his heart, he has extended to them a protection more sympathetic and more potent than that of the law. Poets without number have sung their praises, and even the schoolboy stays his harrying hand with ancient lilts born of long-time appreciation.

It is little wonder that in England the Nightingale, "night's sweet bird," should have claimed the reverence of men when even

> Highest oakes stoop downe to heare,
> And list'ning elders prick the eare.

And to a less extent the songsters in our northern land, especially the Blackbird and Thrush—Merle and Mavis—despite their peccadillos in the orchard, have gained a patronage which their glad music well merits.

What is one to say of the favour in which the "foolish Cuckoo" is held? It has no wonderful song to sing, its habits are unattractive, yet its "plaintive roundelay," its "two old notes," awaken a sympathetic thrill as every spring comes round. Perhaps Montgomery was right,

> Why art thou always welcome, lonely bird?
> —The heart grows young again when I am heard;
> Nor in my double note the music lies,
> But in the fields and woods, the streams and skies.

As little claim upon our regard has the Turtle Dove as a songster, yet there can be no doubt of the favour in which poets have held the "plaintive moan," which with generous interpretation they have regarded as symbolic of tenderness and peace, of constancy and truth. To what extent the "too saint-like turtle" has been protected by the praises of the poets, which, as a rule, echo the feelings of the people, it is difficult to say; but there can be no doubt of the efficiency of the protection which has shielded the Robin, on account of the affection in which it has been held through many ages and countries.

Probably the confidence with which man looks upon the Redbreast was in the first place simply a reflex of the confidence with which the Robin treated man. But legend and fairy tale have created a sacred and a sentimental halo round the little bird, and have raised it to a first place in man's esteem. The breadth and depth of this popularity has been indicated by Wordsworth—

> Thou the bird whom men love best,
> The pious bird with the scarlet breast,
> Our little English Robin;
> The bird that comes about our doors
> When autumn's winds are sobbing.
> Art thou the Peter of Norway boors,
> Their Thomas in Finland
> And Russia far inland?
> The bird who, by some name or other,
> All men who know thee call their brother,
> The darling of children and men.

Protection by favour extends even to the lower tribes of the animal world. There are few insects that are spared at the hand of youth, but the Lady-bird[1] is one of them, and this not only in our own country, but in Germany as well, where it is known as *Gotteskäfer*—God's Beetle. Compare also the rhymes with which youthful Britons and Teutons set free on fresh journeys those little beetles, the Lady-birds and *Maikäfer* or Cockchafers. The British schoolboy chants to the former:

> Lady-bird, Lady-bird,
> Fly away home.
> Your house is on fire,
> Your children all gone.

While his German equivalent, with the same root idea, repeats to the Cockchafer,

> Maikäfer, flieg'! Maikäfer, flieg'!
> Dein Vater ist im Krieg'
> Deine Mutter ist im Pommerland,
> Und Pommerland ist abgebrannt.
> Maikäfer, flieg'![2]

The Scottish version applied to Lady-birds is different from either:—

> Lady, Lady Landers,
> Lady, Lady Landers,
> Tak' up yer cots[3] aboot yer heid,
> An' flee awa' t' Flanders.

---

[1] Beetles of family *Coccinellidæ*.

[2] Fly! fly! Beetle of May,
Thy Father is gone to the war away.
Thy Mother rests her in Pomerane,
And fire has burnt that fertile plain.
Fly, Beetle of May!

[3] Petticoats—an amusing and striking description of the raising of the wing-cases.

In whatever way this widespread custom may have originated, its effects, so far as Lady-birds are concerned, can be none but good, for the Greenfly and Scale-insect pests of our gardens have no more persistent nor voracious enemies than the innocent Lady-bird and its children.

ANIMAL SANCTUARIES

This account of the workings of popular favour on behalf of animals would be incomplete were no reference made to a development of recent date which promises to do much for the preservation of native faunas—the establishment of animal sanctuaries.

It is true that animal sanctuaries have been in existence in Britain for long centuries, for in 1125 Malmesbury drew a doleful picture of the New Forest, created by William the Conqueror, as a place appropriated, from the use of man, for the nurture and refuge of wild beasts. The protected areas of the royal and ancient "forests" of Scotland, too, must have done something for the preservation of wild life, as indeed still do the wild deer-forests of to-day. But in such cases, sport is the object in view, and the protection of any but a few creatures is no more than an associated accident. That areas should be set aside for the sake of the animals themselves is a development of a newly awakened love of nature.

The foremost examples of modern animal reserves are the Yellowstone Park and Mount Rainier National Park, which, with other great reserves in the United States, cover an area of more than 70,000 square miles, and wherein representative sections of the wild life of North America have been gathered together in security, free from all outside interference, for the benefit of present and future generations.

In our own country the formation of such sanctuaries was encouraged by the Wild Birds' Protection Act of 1896, and since that time, thanks especially to the efforts of the Royal Society for the Protection of Birds, many small areas in Britain have been set aside and carefully guarded so that within them birds of the rarer kinds may nest and multiply in safety. One of the most interesting of Scottish sanctuaries, since it is designed to protect all kinds of wild animal life, promises to be that at Taradale in Ross and Cromarty, the birthplace of Sir Roderick Murchison, where the grounds have recently been bequeathed as part of the Murchison of Taradale Memorial "to form a sanctuary or reserve for the preservation of the wild life of the Highlands."

PROTECTION THROUGH SUPERSTITION

Since the earliest days of humanity, man has regarded certain animals with particular veneration and such he has spared and protected. The ancient Egyptians held sacred the Crocodile, the Cat, the Ibis and other creatures, whose bodies, even after death, they preserved and reverenced. Many nations have raised animals to the level of gods, and many religions have seen in the animal creation symbols of their highest faith—we think of the Lamb and the Fish, symbolic to the early Christians of Christ Himself. Some primitive peoples have, by identifying their tribes or their families with particular animals in the cult of totemism, created thereby a special protection for the totem animal, for its blood became as their blood and its life their life, so that on no occasion could it be slain, except as a sacrifice to their god. Buddhism spares all living things, and partly as a result, there are in India 40,000 deaths a year from snake-bite.

Religious symbolism of such a type, it is hardly necessary to say, has had no place in later Scottish life, though the animal symbols upon the early Christian monuments of the northern and eastern counties hint at a time when it played a part in Scottish history. Yet to-day there are many superstitions regarding animals, which still act as protectors of the lower creatures.

I have already mentioned, in passing, the legends that have hallowed the Robin Redbreast—his care for the bodies of the dead, how

> Cov'ring with moss the dead's unclosèd eye,
> The little redbreast teacheth charitie.

But more striking are the stories that account for his ruddy breast: the story of his pitiful service on Calvary:—

> Thou from out His crown, didst tear
> The thorns to lighten the distress
> And ease the pain that He must bear,
> When pendent from thy tiny beak
> The gory points thy bosom pressed,
> And crimsoned with thy Saviour's blood
> The sober brownness of thy breast![1]

Or that tale of his journeys to "the fiery pit":—

> He brings cool dew in his little bill,
> And lets it fall on the souls of sin:
> You can see the marks on his red breast still
> Of fires that scorch as he drops it in[2].

I do not doubt that these tales, carried through the ages, have much to do with the origin of the schoolboy superstition that groups the Robin with the birds which bring ill-luck to their harriers:

> The laverock[3] an' the lintie[4],
> The robin and the wren:
> Gin[5] ye harry their nests
> Ye'll ne'er thrive again.

It is strange that while in England, Ireland and the Isle of Man, the "hunting of the wren" was a common and vicious custom of Christmas Day or the day after—St Stephen's Day—in Scotland the Wren was regarded almost as a holy bird and was exempted from the ordinary rifling of nests.

> Malisons, malisons mair than ten,
> That harry the Ladye of Heaven's hen!

The curious title here given to the bird is similar to that in a widespread rhyme which links the Robin and Wren as a sacred pair:—

> The Robin an' the Wren
> God Almighty's cock and hen.

Another curious and interesting superstition saves the Stone-chat, or Stane-chacker as it is called in Scotland, from the rude hands of nest harriers; does not the bird itself, its poor song being interpreted, call down maledictions

---

[1] Delle Whitney Norton. [2] John Greenleaf Whittier.
[3] Lark. [4] Linnet. [5] If.

upon wanton ruthlessness? The clever Galloway imitation and interpretation of the song runs

> Stane-chack!
> Deevil tak!
> They wha harry my nest
> Will never rest,
> Will meet the pest!
> De'il brack their lang back
> Wha my eggs wad tak, tak!

Such creatures as are fortunate enough to be regarded as omens of good fortune, shelter as a whole under the protection of man. As the Stork is welcomed to the chimney tops of Holland, the Swallow and House Martin are welcomed to our eaves; for ever since the days when the swallow kind was under the direct guardianship of the household gods, the destruction of a Swallow's nest has brought ill-fortune, as assuredly as the undisturbed settling of the Swallow has brought good luck to the inhabitants of the house it selects:

> Bid the sacred swallow haunt his eaves
> To guard his roof from lightening and from thieves.

Spiders have been spared because the slaying of them was said to bring rain on the succeeding day; and an insect superstition of the Highlands shows how specific may be the protection or condemnation of a tradition. The boys of the Highlands, says J. G. Campbell in his *Superstitions of the Highlands and Islands of Scotland*, when they see a *Ceardalan* or "Dung-beetle" spare it, while they mercilessly kill the *Daolag* or "Clock" beetle. And the reason, they will tell you, is that when the former was asked by the man who went to seize the Saviour, how long it was since He had passed, the Dung-beetle answered "twenty years ago yesterday," but the latter said "only yesterday." Hence when boys hammer life out of a "clock" they chant:

> Remember yesterday, yesterday,
> Remember yesterday, wretch,
> Remember yesterday, yesterday,
> That let not the Son of God pass.

I must not be taken as suggesting that Dung-beetles are more plentiful than "Clocks" in the Highlands owing to the prevalence of such a superstition, for the efficiency of protection or destruction depends on its intensity relative to the numbers of the species concerned, and it would take much fortuitous slaughter to affect the insect hordes. But nevertheless superstition and the side winds of religious cults have had a real effect in protecting, throughout the world, certain creatures to which legend and tradition have added new associations of piety or reverence, of awe or dread.

CHAPTER V

THE DELIBERATE INTRODUCTION OF NEW ANIMALS

*Nature's road must ever be preferr'd;
Reason is here no guide, but still a guard.*
                                            POPE.

MAN pays little heed to that balance of Nature which has arisen out of centuries of struggle and competition, and which makes the fauna and flora of an old but uncivilized country established and stable. When the immigrant reaches the new country of his hopes, across the ocean, he refuses to take the experiments of Nature for granted, and sometimes forgetting that new soils, new climates, and new associations of living things demand a new outlook and require new treatment, he proceeds to transform the new found land to the pattern of the old. This transformation, which may proceed to such lengths as radically to alter the general aspect of a fauna or a flora, progresses at first mainly by the introduction of plants and animals from the homeland.

No sooner does man enter upon a new heritage than he endeavours to keep alive the memories of home by surrounding himself with the familiar flowers, beasts, birds and fishes of the old country, irrespective of their fitness to survive in fresh conditions, or of the fitness of the aboriginal fauna to assimilate them without ill effects. Why else, do you think, did the early immigrants to New Zealand set free in the bush the British Robin Redbreast, the Bullfinch, the Greenfinch, the Turtle Dove, the Lapwing and half a dozen other old favourites? Further, the immigrant's amusement must be catered for by the establishment of animals whose sporting qualities he knows. So the Trout of Loch Leven has been placed in rivers half the world over, Rabbits and Hares, Pheasants and Partridges, have been transported to Australia and New Zealand, Blackgame to Newfoundland

and British Columbia, and in New Zealand even the Red Deer and the Moose have been set at liberty. Lastly, he needs must be ministered to by the domestic creatures and cultivated crops which stood him in good stead in his earlier home. The habit of transportation of animals is older than the Neolithic invasions of Britain, and to-day transoceanic trade has become a constant factor in the home breeding of stock.

Often the introductions are successful and turn out to be of the greatest value. How much poorer Europe would be did it lack the domestic animals of the East and the American potato and tomato; how much poorer America, with its domestic stock limited to its native llamas, and without the wheat, rye and oats, the pears and apples, the hemp and flax which reached it from Europe after Columbus had pointed the way. Still more is the prosperity of the "new countries" bound up with man's power of trying afresh the successful experiments of the old countries. The sheep and cattle, the sugar-cane and wheat of Australia; the wheat and wool of New Zealand; Canada's grain crops, her flocks and her herds; all of them are foreign to the lands they have made prosperous, one and all man has brought them to the new soil and tended them to new fruitfulness.

Not all the experiments of introduction have been so successful as those just mentioned. There have been many failures. Often conditions of soil and climate, of food, or of relationship to the original inhabitants, prove too trying to the enforced immigrants and they or their few successors wither and die out. So it has been in Scotland, to cite only two examples, with the Reindeer and the Beaver, which have been allowed to run free in the hope that they would again make a permanent home in a country they once frequented; and so it is with several of the foreign species of Trout with which time and again attempts have been made to stock our rivers and lochs. So it has been also in New Zealand with the Bullfinch, the Turtle Dove, the Robin Redbreast, the Grey Linnet, the Lapwing, and the Partridge.

On the other hand, sometimes the results of such an experiment outrun the expectations or wishes of its originator, so that its success becomes a measure of its harmfulness and an index to the rashness of man in endeavouring to

DELIBERATE INTRODUCTION OF NEW ANIMALS

improve upon the establishments of nature. Of many instances, one need only recall the notorious case of the introduction of one animal after another to Australia, to the annoyance of the farmer and the detriment of the country. Brought from Somerset by some well-meaning individual, four pairs of Rabbits were set free in the neighbourhood of Geelong in 1858. They found conditions so favourable, and multiplied so rapidly, that already in 1875 an Act was passed in New South Wales to encourage their extermination. Recently as many as 20,000 have been destroyed in one or two days at a poison trap; it was estimated that in 1917 the export of rabbit flesh to the United Kingdom from New South Wales would amount to 1,250,000 crates containing 30,000,000 individuals; the country is overrun, and the destruction of grass is so great that the average stock of sheep in New South Wales has fallen off by many millions during the last quarter of a century. To keep the Rabbits in check, Foxes were introduced and encouraged. They too have so increased in numbers and are so destructive to lambs that the Vermin Act of 1914 imposes on every landholder the duty of killing them. In the year 1915–16, 679 were slain. As a further check upon the Rabbits, Dogs, which also had been introduced, were largely employed, but they, too, having become wild, and with increased numbers having developed a taste for domestic flesh, have fallen under the ban of legislation, so that under the Vermin Acts of 1912 and 1914, rewards have been paid in New South Wales alone for the slaughter of 75,000 wild Dogs.

GENERAL RESULTS OF INTRODUCTION OF NEW ANIMALS

The case of the Rabbit in Australia is only one of many similar cases, and it does no more than represent in extreme the result of any successful experiment in foisting an alien beneficiary upon a native fauna, which has settled its differences and has become established as regards the inter-relations of its own members and the food supply of its country.

Should alien stock be introduced with success the animal life of a country alters appreciably, in so far as the import is conspicuous or increases greatly in numbers or comes into direct relationship with man. To-day we could not exist without our introduced domestic creatures, which in number far exceed the original population of large wild animals they have replaced. We who are not farmers would regret the disappearance of the Rabbit from the hill-sides, or of the Pheasant from the coverts; we who are not foresters would miss the gambollings of the Squirrel, or the solid presence of the Capercaillie in the woodlands. Yet even these mild exotic pleasures are bought at a price. After all, the food supply of a country, so far as most wild creatures are concerned, is almost a fixed quantity, and the total amount of animal life in a country depends upon the quantity of vegetation available throughout the year, for even carnivorous creatures ultimately depend upon the vegetation which nourishes the herbivores. If then additions are made to the native animal life, by so much as the imports consume is the food of the native stock diminished. Thus arises a struggle between the aboriginal population and the newcomers, a struggle which spreads from food to living spaces and breeding sites, and which is none the less real because in its earlier stages it is almost imperceptible.

The results of the contest are clear enough in most countries. It may be that the introduction obtains a precarious foothold and lives, as it were on the crumbs that fall from the rich native's table, without seriously affecting the creatures into whose environment it has dropped, or that it taps a source of food supply hitherto almost unused, as the Squirrel did in feeding upon the growing shoots of Scottish pine trees. But an unobtrusive existence is the exception, and either the newcomer finds the struggle for food too severe and declines in health and numbers, until it disappears from the fauna it invaded; or, as often happens, it becomes firmly established, and increasing the difficulties of those old races of residenters upon whose food supply it trenches, causes the weaker of them to languish and to forsake the invaded districts, and finally, driven from one stronghold to another, even to succumb to the new and unaccustomed competition.

Frequently the introduction of strange animals to a new land has a more immediate and more noticeable effect on the native fauna. For man in transporting creatures for his own purposes often seems to forget that their nature, and not his desires, will remain their guiding law. So the Stoats and Weasels taken from Great Britain to New Zealand to check the spread of the Rabbit, having surfeited on rabbit diet, promptly turned to the flightless Kiwis and have brought them and other small natives to the brink of extermination—an ignoble work in which they have been strongly abetted by the wild descendants of the Pigs introduced by Captain Cook on two occasions in 1773, and by the Cats, Dogs and Rats of later importation. The creatures set loose in Australia to police the rabbit warrens have themselves become poachers, and the unique native ground fauna of Australia, as well as the shepherd's flocks, has suffered severely. The Macaque Monkeys introduced to Mauritius by the Portuguese have all but exterminated Mayer's Pigeon (*Nesœnas mayeri*), whose nests, eggs and young they destroy.

Scotland furnishes many illustrations of these varying results of the introduction of new animals; but since, even in countries colonized in comparatively recent years, there is already insuperable difficulty in tracing the steps of the spread or decline of an introduction, it is natural to suppose that the existing records of Scottish experiments, some of them carried out hundreds of years ago, would furnish only hints for a story rather than a detailed history.

For the sake of imposing some arrangement upon what are after all results of a medley of human whims, I shall consider the introductions in three groups in sequence of their human importance; the first comprising those animals which man brought, thinking of his own welfare, the second those which he brought for the ends of sport, the last those which he set free for a variety of reasons or for no reason at all, but mainly for his amusement or his pleasure.

V. 1

ANIMALS INTRODUCED FOR THE SAKE OF UTILITY

THE utilities to which man has paid regard in acclimatizing new animals cover a wide range, but they may be said broadly to depend either on the value of the animal in itself, that is to say on the products it yields when alive or after its death, or on the work it can do, whether that be the artificial but efficient labour resulting from years of man's guidance and training, or the instinctive habit of the creature, which in following its own law of life benefits also the race of man.

Chief amongst the useful animals in these groups must rank the races of domestic stock, for human existence as it is would be impossible without the flesh of flocks and herds and the warm wool of sheep and goats, and intolerable without the services of the ox, the horse and the dog. The introduction to Scotland of the more interesting domestic animals has been sufficiently touched upon in the chapter dealing with their domestication. It is enough to add that the acclimatization of domestic stock is a human operation of wide significance. Scarcely a country is known, but man has planted there his ox and his ass, his swine and his sheep, always to the checking and restriction of the native fauna, and sometimes even to the extermination of its weaker elements. Indeed, in several of the island faunas, such as that of New Zealand or of the Canary Islands, the introduced domestic stock far outnumbers in variety, and in absolute numbers entirely swamps the native mammalian species of the land. With the advance of knowledge as to the stamina of different breeds and crosses, the work of acclimatizing old animals to new localities progresses with more lively assurance than ever. The climatic conditions of Alaska have proved so trying to the introduced old-fashioned domestic stock that a new breed of cattle—a cross between the

Thibetan Yak and Galloway Cattle—is being created at the experimental station to supply the need for unusual hardiness, for already crosses between the Yak and the domestic cattle of the East have proved of great service in Turkestan and other parts of Asia.

APPEARANCE AND SPREAD OF THE RABBIT

Apart from the introduced domestic animals, the most important creature to have been planted in Scotland from foreign parts simply for the sake of its yield, is the Rabbit. Its universal presence belies its alien blood, and one would scarcely associate with its homely and unassuming aspect the interest of its history. It is generally held by naturalists that the Common Rabbit (*Oryctolagus (Lepus) cuniculus*), as we know it to-day, spread from the south-western portions of Europe bordering upon the Mediterranean Sea, an area to which it had been relegated by the severities of the Ice Age. From these regions, partly by natural roving, much more by the deliberate influence of man, the Rabbit has spread over the western and central countries of Europe, and from there it has been transported to the uttermost ends of the earth.

At what period the Rabbit was re-established in Britain after its extermination during the Glacial Period, is a matter of great uncertainty. In Scotland its bones have been found amongst the debris of the kitchen-middens of Neolithic and later times, as well as in the upper layers of cave deposits; but little weight need be attached to these occurrences, for the Rabbit is a burrower and a vandal which makes short cuts through the neat layers and classifications of the excavator, so that a contemporary of our own might rest its bones by the side of the long extinct Mammoth to the confounding of interpreters of the past.

As a matter of fact, testimony points strongly to the absence of the Rabbit in prehistoric Britain. In Scotland its remains are absent from the bone deposits which have been found deep below the surface, beyond the reach of modern burrows. Julius Caesar mentions the occurrence in Britain of the Hare, the Hen and the Goose, but omits the Rabbit. Nor was it introduced by the Romans as is commonly believed.

It is not represented in the remains of the Roman settlements at Newstead on the Tweed, nor at Inveresk near Musselburgh, and the single jaw of a young individual which I examined from Traprain Law in Haddingtonshire, was almost certainly recent, for burrows penetrated the kitchen-midden in all directions.

There are no pre-Norman allusions to the Rabbit in Britain; it has no native name in the English or Celtic languages; and warrens are not mentioned in Domesday Book. We are therefore justified in assuming that it did not occur, at any rate as a wild animal, in Britain even in the eleventh century.

The earliest name by which it is known in English historical records is "cony" or "coney," a name clearly derived from a Norman-French word, the plural of which was "coniz" or "conis," becoming in English "conys" or "conies," and in the singular "cony" and "conie." It seems probable then that we are indebted for the introduction of the Rabbit to the Norman Conquest—a supposition strengthened by the frequent and steadily increasing references to the animal after the twelfth century.

The sole object of its introduction was utility, since at no time has the Rabbit been considered worthy the lance of the true sportsman; for, said the irreproachable author of *The Master of Game* (MS. Bodl. 546) in the fifteenth century, "Of conies I do not speak, for no man hunteth them unless it be bish hunters [fur hunters] and they hunt them with ferrets and with long small hayes [i.e. nets]." Notwithstanding *The Master of Game*, Bishop Leslie tells us that in sixteenth century Scotland Rabbits and Hares were hunted by special dogs. But if their sporting qualities were limited, the value of their flesh for food and their skins for fur was apparent, and these first led to their encouragement and rapid dispersal. As Reyce put it in 1618 in *The Breviary of Suffolk*:

Of the harmlesse Conies, which do delight naturally to make their aboad here,...their great increase, with rich profitt for all good housekeepers, hath made every one of any reckoning to prepare fitt harbour for them, with great welcome and entertainment.

An excellent summary of the history of the Rabbit in England, will be found in Barrett-Hamilton and Hinton's

ANIMALS INTRODUCED FOR SAKE OF UTILITY 249

History of British Mammals, from which the above quotations are taken. Here I shall confine myself to a few extracts from the old laws and descriptions of Scotland indicating the progress made by the Rabbit in its conquest north of the Tweed. Whether it was here introduced direct from the Continent or made its way through England is difficult to decide, but the earliest records indicate that in the thirteenth century it was as common in North as in South Britain, and suggest, therefore, that its introduction must have been almost contemporaneous in both countries, and may have been independent in each. In this connection it must be remembered, as Prof. Hume Brown has pointed out, that even in the reign of David I, from 1124 to 1153, the Norman element, which had already played an important part in Scottish affairs, became predominant at the court and in the councils of the Scottish monarch. It is not at all unlikely that this friendship may have led to the beginning of a secondary Norman conquest of Scotland by the introduction of the Norman-bred and Norman-titled " cony."

Be that as it may, the earliest unmistakeable references in historical records to the occurrence of Rabbits in England seems to be that noted by Prof. Rogers, recording the payment of 2½*d.* at Waleton in 1272 for the taking of Conies and Partridges with Hawk, Dog and Ferret ; and the report to Edward I in 1274 regarding the produce of Lundy Island, where the "taking of rabbits is estimated at 2000, £5 10s." and the estimate is at " 5s. 6*d.* each hundred skins, because the flesh is not sold."

But before this time the value of rabbit-warrens had become well known in Scotland, and to others than the proprietors, for even in the reign of Alexander II (1214–1249), it was found necessary to protect the royal warrens by statute, the penalty for trespass being death and confiscation of property. In the succeeding reign, of Alexander III (1249–1286), the keeper of the warren at Crail in Fifeshire was paid a salary of 16s. 8*d.* from the royal purse for his services during the year 1264. The King's Chamberlain's accounts of 1329 show that he paid a sum of 8s. to four men for crossing to the Isle of May, at the mouth of the Firth of Forth, to catch Rabbits; and David II, who reigned from 1329 to 1371, granted the keeper of the warrens in Fife, William Herwart, a charter in life-rent of the office of Keeper of the King's Muir in Crail and of its "cuningare" or warren, his salary being 40s. a year.

It should be explained that in early times the word "cony" (or some modification of it), alone was in use, and that even when the diminutive term "rabbit" or "rabette" was introduced in the fourteenth century, it was applied only to the young. "The Conie...beareth hyr Rabettes xxx dayes, and then kindeleth" (*The Noble Arte of Venerie or Hunting*, 1575).

Already in the fifteenth century, the Rabbit had made good its introduction and become a recognized article of commerce in Scotland. In 1424 a law was passed exacting a custom of 12d. on every 100 "cuning" skins exported, and in 1457 a further degree of protection was afforded the animal, the destroying of "cunings" in time of snow being made a point of dittay or criminal prosecution. It is quite clear that at this early date, the Rabbit was established throughout a wide extent of the country. The Rental Book of Cupar Abbey shows that in 1474 the Abbey possessed a "warander of kunynyare" or keeper of the warren, and in the following year, Gilbert Ra or Rae undertook to keep for the Abbey the "conyngar [warren] fra all scaith and peryl, and promoofe and put that to all profit at [h]is powar." Even in the smaller islands of Orkney it flourished, and as we learn from a Rental Book of the Earldom dated 1497–1503, formed part of the rental in kind paid to the Earl—the links of "Dernes, Burra, North Sandwick, Pappa prope Westray, and Sanday," combining to supply annually "114 cunnings," and "1274 cunningis skinnis."

The sixteenth century marked a notable increase in the interest taken in the Rabbit both by the public and the law. The value of a warren contributed not a little to the "common good" of a township, so that the introduction of Rabbits to a suitable piece of waste ground was to be desired wherever they could be "gudly had." Already the common warren of Aberdeen, "cunicularium de Abirdene," was in full activity on the links to the south of Don-mouth, near the "Gallowhills," as we learn from a casual reference to its site in a charter of James IV to the Provost and Baillies of the Burgh in 1583.

The value of the warrens gave rise to a series of laws dealing with offences against their "breakirs." In 1503 it was ordained that stealing of "cunings" was to be a point of "dittay," or criminal prosecution, the "unlaw" being £10; and in 1551, in consequence of "the great and exhorbitant dearth," young Rabbits were given protection for three years, except from noblemen with Hawks, the law demanding "That na maner of persoun tak upone hand to slay ony Lapronis," *lapron* or *laprinn* being a common Old Scots term for a young Rabbit. At this time, the price of a Rabbit was fixed by a long forgotten food controller: "*Item.* the cunning 2 shillings unto the feast of Fastens evin [Shrove Tuesday] next to cum, and fra thine furth twelve pennies"—equivalent to the price of a brace of Blackcock,—while the "best Lapron" fetched only 2*d.*

Nevertheless the success of the Rabbit's introduction proved a burden even in the early days of the century; witness the grievance of "Schir Robert Egew, Chaiplan to My Lord Sinclair" who complains in 1511, "Ther wilbe our [over] mony *cunningis* [:] with[in] twa yeir thai have riddillit all the erdis of the Linkis richt weille."

Perhaps nothing illustrates the progress of the Rabbit in the sixteenth century so strikingly as its wide distribution among the islands of Scotland. In 1529 "Jo. Ben." in his slovenly Latin, describes its abundance in the Orkney Islands: in the parish of Sandwick, in the uninhabited isle of Lambholme, where many Rabbits were slain by men of other islands, and in Sanday, where in winter the Rabbits became so tame, owing most likely to overstocking and consequent lack of food, that they were caught in the houses of the people. Monro found similar evidence of great numbers in the Hebrides: on Mull, on "Inche Kenzie...full of cunings about the shores of it," and on "Sigrain-moir-Magoinein, that is to say the Cuninges ile, quherin ther are many cuninges," as well as in the Orkneys on "ane little iyle, with a chapel in it, callit Cavay." And Von Wedel, a German nobleman and traveller, remarked upon the many Rabbits of the Bass Rock when he visited the Lowlands of Scotland in 1584.

It is unnecessary to enter into further detail regarding the spread of the Rabbit within our borders, for with the increase of travelling and of written descriptions of the country in the seventeenth and eighteenth centuries, the evidence becomes unwieldy, and each item in itself of less significance in the general movement. The trend of the evidence is none the less clear—that the Rabbit continued to spread more and more rapidly, partly through natural migration and still more by deliberate introduction or transportation to new areas. There are records of the planting of many fresh colonies in the latter half of the eighteenth century, while some counties, such as Kincardineshire, did not know the Rabbit until its deposition there in the first quarter of the nineteenth century.

Two factors made for its success and rapid increase in the later centuries: first, the great strides which have taken place in the improvement of agriculture and in the consequent increase of the yield of the soil, for the increase of the farmer's crops is an increase of the Rabbit's food; and second, the growing attention paid to the preservation of sporting animals, and the consequent destruction of the very creatures which kept the Rabbit (as well as the Pheasant and the Grouse) in check—the Fox, the Polecat, Stoat and Weasel, the Eagles and Hawks, and in more distant days, the Wolf.

So it is that even at the end of the eighteenth century we find it established throughout most of the mainland and in the islands—from the Lowland counties of Berwick and Roxburgh, Dumfries and Ayrshire, through the midlands of Fife, Clackmannan, Kinross and Stirling to the wilds of Perth and Argyll, Easter Ross, Sutherland and Caithness. Many of the writers in the Old Statistical Account (1792–8) describe it as "rare" or as a newcomer, but in some places it was very common. About this time, at the warren of Dowally near Dunkeld, the tacksman killed a yearly average of 125 dozen; in one year at Stromness 36,000 rabbit skins were sold, at 8*d.* each; and in the last century as many as 200,000 skins have been sold in a winter at Dumfries (see also under "Destruction," p. 166).

Need more be said to prove the success of the introduction of the Rabbit from the point of view of acclimatization? Perhaps only that its firm establishment and increase in number have made it so destructive to crops that in 1917 a

Government "Rabbits Order" was found necessary, giving Local Agricultural Committees power to authorize *any* person to kill rabbits wherever they have become a nuisance.

EFFECTS OF INTRODUCTION OF RABBIT

It must not be supposed that a far-reaching introduction, such as that of the Rabbit, stands by itself. Like a stone cast in a placid pool, it gives rise to ripples of influence of first, second, third—one cannot tell how distant—degree, spreading one cannot tell to what depth and extent. Thus numbers of Rabbits alter to a remarkable degree the vegetation of the districts they infest, changing the nature as well as the amount of plant life, as Mr E. P. Farrow clearly shows in a well illustrated paper in the *Journal of Ecology* for March 1917. This change in vegetation in turn affects the animals, mammals and birds, insects and other invertebrates. Some disappear and their places are filled by newcomers, while others of the old residenters flourish and multiply under the new conditions.

In yet another way, Rabbits influence their co-habitants in a region. There is, for example, a definite connection between over-abundance of Rabbits and scarcity of Hares, or between the appearance of Rabbits and the disappearance of Hares. It is quite likely that the general decrease of the Common Hare in Scotland, a decrease which has been noticed for many years, may be due in part to the increase of the Rabbit in recent times. Local Scottish instances of this relationship are known; but an Australian example offers more simple and direct evidence. A writer in *The Field* for 26th May 1917 states that Hares, introduced into Australia, were at one time so abundant in the district of Goulburn in New South Wales that drives had to be arranged to keep their numbers in check—as many as 800 on occasion being killed in a single day. At that time Rabbits had not yet reached Goulburn from the region of Geelong, where four couples had been liberated by settlers about 1858. As the Rabbits spread over the intervening 400 miles, the Hares began to disappear, until after the extraordinary increase of the Rabbit, few were left. Yet so soon as rabbit-fencing was put up, and the enclosed areas, sometimes covering several square miles of country, were cleared of the pest, Hares made their way within the fences and there became numerous again. The apparent antipathy between the two rodents has a simple explanation, for it is due to the fouling by many Rabbits of grass and herbage, which thus become quite unpalatable to Hares. I have no doubt that careful observation of the smaller inhabitants of a rabbit-infested region would show that besides these, many faunal changes have been caused by the influence of this hardy importation.

FAILURE TO ACCLIMATIZE FOREIGN DEER

Fortunately for man's welfare, not all the animals which he introduces, take possession of the land of their adoption as the Rabbit has done. What dreams of swelling numbers and profitable herds accompanied the reintroduction of the Reindeer to Scotland, I do not know, but it is certain that the dreams, such as they were, have come to naught. Why this should be is difficult to say, especially as native Reindeer survived for long in the northern counties, and as suitable food seems to be sufficiently common on many moors. Be this as it may, no fewer than fourteen Reindeer were brought to Dunkeld and released by the Duke of Athole on the hills of Athole at different times in the eighteenth century, and of these only one survived for two years. Similar attempts to establish them in the Forest of Mar in Aberdeenshire and in the Orkney Isles have been equally unsuccessful. I do not of course refer to the preservation of such examples as are to be seen in the Park of the Zoological Society at Corstorphine near Edinburgh (Fig. 58, p. 339), where young Reindeer have been born and successfully reared, for there the animals are strictly tended and dieted, and cannot be regarded as surviving on their own merits, or as forming an addition to our fauna.

Similar failure met Sir Arthur Grant's attempts to acclimatize the American Wapiti Deer at Monymusk in Aberdeenshire, where in the nineties of last century they could be distinguished by their large size and fine antlers. Although they seemed to thrive well and crossed freely with the native Red Deer, the stock finally died out. A like fate has befallen the Virginian Deer introduced into Arran about

ANIMALS INTRODUCED FOR SAKE OF UTILITY

1832 (see p. 287), but their importation and that of the Wapiti were measures of amenity rather than of utility.

DESTRUCTIVENESS OF INTRODUCTIONS

It is strange with what persistence Nature has mocked man's efforts to introduce new creatures in any country for the sake of their usefulness, domestic animals apart. I have already alluded to the unfortunate results of the introduction

Fig. 48. Little Owl—an introduction to Britain which has become a nuisance. ⅓ nat. size.

of Hares and Rabbits to Australia, and the equally disastrous effects which followed upon the setting free of Foxes and the escape of Dogs. In New Zealand a somewhat similar series of disasters met man's efforts. Introduced Rabbits multiplied and became a pest which cleared the ground of cultivated crops. Ferrets, Stoats and Weasels were released to destroy them, and themselves attacked the native fauna and the flocks of sheep. In New Zealand, as in Australia and America, the common House Sparrow has been set free to destroy the hosts of insect pests brought into existence by cultivation; but in all these countries, the Sparrow has multiplied at an alarming rate, and having shown a strong and increasing preference for grain, has become a vagabond and outlaw, whose death is sought at a price by agricultural councils, municipal authorities and governments. The Common Starling, introduced to Australia and New Zealand, has fallen into similar evil ways, for the settler, hoping to make an end of his insect pests, little thought that his protégé would make a beginning with fruit.

Has not Nature also mocked us in this country in the case of the introduction of the Little Owl (*Athene noctua*) (Fig. 48, p. 255)? Brought by Lord Lilford from Holland in the eighties of last century, and set free in Northamptonshire to rid country belfries of sparrows and bats, and fields of mice, and by Mr E. G. B. Meade-Waldo to Kent in 1874 and later, the Little Owl in a few years has spread into all the neighbouring counties and to some far away. It has even reached Scotland, where in 1912 one was shot at East Grange in Fifeshire. And everywhere it has betrayed its trust and, hawking by day, has destroyed Warblers, Finches, and Thrushes, and has been convicted of stealing the young of Pheasants and Partridges from the coverts, and chickens from the poultry yard.

BALANCE OF NATURE UPSET

Why should these unlooked for and destructive results so uniformly greet the efforts to introduce, for the sake of their utility, new elements into a fauna? Such misfortunes do not dog the introduction of domestic animals or of animals introduced for purposes of sport. The reason is connected with food supply, and seems in part to be this: When animals are introduced for utility's sake, they are required in large numbers, and are encouraged to breed. It comes about, therefore, that if climate and conditions are suitable the foreigners multiply with rapidity, and the result is either that the food supply relied upon to support them falls short of requirements, or that increasing competition drives the aliens to try a new food which in the end, usually to the dismay of the farmer, they come to prefer. Domestic animals are in as great or greater numbers than the introduced pests, but then careful provision is made by cultivation for their food supply, and their tastes are given no opportunity to wander. Sporting animals on the other hand, are not encouraged to multiply to the same extent, and the very reason of their introduction almost certainly insures that the breeding stock will be periodically reduced within reasonable limits.

SUCCESS DEPENDS ON AN EVEN BALANCE

All cases of utility importations, however, are not failures. Where a balance is struck between natural food-supply and numbers, an easy-going adaptation results. In 1911, forty Reindeer were introduced into the Pribilof Islands off Alaska. These thrived and by 1914 had increased in number to 150; but soon the increase must cease for the supply of lichens is limited, and more Reindeer might mean the loss of all. The Honey Bee was imported into New Zealand in 1842 and has flourished; for the white clover previously introduced, which had never seeded till the Bee was brought to fertilize it, alone would have supplied abundance of food. So too with the Humble Bee: for long the imported red clover bloomed in New Zealand fields, but the flowers were sterile, and the settlers had to bring their red clover seed from overseas. Then the Humble Bee was introduced, and Humble Bee and red clover struck a working balance—the red clover, fertilized by the visits of the Humble Bee, now ripens its seeds, and the Humble Bee thrives on the red clover's stores of honey.

Of equal interest are the extraordinarily successful results which have followed upon the introduction to California of another fertilizer—the Fig Insect (*Blastophaga grossorum*), a small Hymenopteron belonging to the family of the Chalcids. For ten years a Californian orchard-owner grew figs with uniform ill success. Doubting the quality of his stocks, he then imported cuttings of Smyrna figs from their native home in Asia Minor; but this too proved an utter failure, and for fourteen more years his sixty-acre orchard never yielded any financial return. It was known that wild figs containing the Fig Insect were regularly placed in the Smyrna orchards. Mr Roeding therefore imported in quantities from Smyrna wild Capri figs containing Blastophaga alive, and at the time of the blossoming of the figs, hung the insect-laden wild figs amongst his Smyrna trees, in order that the female insects might enter the curious hollow flower-cluster of the figs, and in seeking a suitable place for the deposition of their eggs, might dust the seed flowers with pollen, and so cause the "fruit" to set and swell. The results fell as had been hoped: the tiny Fig Insects, imported from Asia Minor, visited and fertilized the hitherto sterile flowers; the trees, unfruitful for fourteen years, blossomed and were fruitful; and in 1900 a first crop of Smyrna figs, 60 tons in weight, was obtained from a Californian orchard. Here no natural balance has been disturbed, for the introduced insects confine their attention solely to the figs introduced before them, and find in these their natural food, shelter and breeding places.

SOME INTRODUCED FISHES

In our own country, sport rather than utility has regulated the introduction of freshwater fishes (see p. 276), but in other countries they have been introduced towards the end of creating a profitable supply. California has been remarkably successful in its endeavours to modify its fauna for commercial advantage. Between 1871 and 1880, Mr W. H. Shebley tells us in *California Fish and Game* for 1917, as many as 619,000 Shad fry were imported from a hatchery in New York and turned down in the Sacramento River. As a result, the Shad (*Alosa alabamae*), a native of the Atlantic coastal waters of America, is now one of the commonest fishes in Californian waters. The introduction of Carp (*Cyprinus carpio*), a native of the rivers of China, has been equally successful, and this fish "will probably become one of the State's most valuable food fishes." It has, moreover, prepared the way for the Black Bass (*Micropterus dolomieu*), which feeds upon it, and the successful introduction of which from the rivers east of the Alleghany mountains is regarded as "one of the greatest feats of acclimatisation of new species of fish in the history of fish culture." Yet even in successful California, there is a fly in the ointment, for the success of the alien Carp has involved the destruction of the Californian Perch, and the increase of the foreigner has been so phenomenal in the Chatauqua Lake that the water has become fouled and unfit for use. Even the savage Pikes and

Muskelunges brought to the lake to exterminate the Carp have themselves fallen victims and have disappeared, and the Carp still flourish.

For utilitarian purposes the gigantic Australian "Murray Cod" (*Oligorus macquariensis*) and, about 1883, the Canadian Black Bass have been introduced into English rivers, but the experiments have met with no success.

THE MEDICINAL LEECH (*Hirudo medicinalis*)

The universal use of the Leech for blood-letting in the old days led to its abundant appearance in our own country. Sir J. Dalzell illustrated it in his *Powers of the Creator* (1853) from a Loch Leven specimen, and there are other records of its presence, for it is probable that many a doctor dropped his Leeches into a convenient pool, where he could readily find them again. But the Medicinal Leech has failed to gain a firm foothold. Its gradual elimination from use in medicine led to its disappearance, for the stock is no longer replenished and the pools where once it was common know it no more.

COUNTER-PESTS

I cannot close this section without mentioning a new development of scientific research, which promises on account of their utility to add many insects to old faunas. The typical results of this fresh method, however, are to be seen in America and on the continent of Europe rather than in this country. I refer to the introduction of insect counter-pests—insects which at one stage or other feed upon some pest of cultivated crops. America has found special need for such destroyers. Records show that 407 introductions of foreign plants were made to the United States of America during the first three months of 1913, and the importations of field and garden plants from other countries exceed a thousand a year. With this great annual inflow of plants come their native pests, and these, unwittingly admitted, have spread, in some cases with a rapidity beyond belief, since the new land may harbour no destroyer—bird or insect—such as kept the pest in check in its old country. This factor in the success of many accidentally introduced pests was soon diagnosed, and so long ago as 1860, the use of insect counter-pests had been tried, for the United States Census Report of that year states that the New York Agricultural Society " has introduced into this country from abroad certain parasites which Providence has created to counteract the destructive power of some of these depredators."

In recent years the introduction and use of counter-pests has been carried out with great skill and deliberation. Insects known to attack certain pests have been brought from their native land and reared, particularly in America, in such numbers that they could be distributed wholesale to the affected areas. One or two examples will be sufficient to make clear the significance of such introductions at the hand of man.

In 1868, in the early days of the Californian fruit-growing industry, some young orange trees were brought from Australia to the neighbourhood of San Francisco. The trees were infested by an Australian Hemipterous insect—the Fluted or Cottony Cushion Scale (*Icerya purchasi*), a small inert-looking creature which subsists on the juices of the tree.

Fig. 49. Cottony or White Scale being attacked by the Cardinal Lady-bird imported to America from Australia to combat it. (Twice natural size.)

For many years the pest worked unnoticed, and became so successfully acclimatized that by 1880 it had spread all over the State, causing such terrible devastation in the orange groves of Southern California that in a single year the orange crop was reduced from 8000 to 600 waggon loads. Such havoc was caused during the next eight years that Mr A. Koebele, an expert entomologist, was sent by the United States Department of Agriculture to Australia to find out by what means Nature there kept the Fluted Scale in check. Koebele found that a brilliant red and black Lady-bird Beetle (*Vedalia cardinalis*) preyed extensively upon the Scale, and some 500 specimens of the Lady-bird, carefully packed, were sent alive across the seas to California, where they were fed and tended. After a few generations, it was found that they had increased so greatly in numbers that some could be distributed and set free in the hope that they would become naturalized and seek to feed, as they did in their own country, upon the Scale-insects. The result was a triumph for scientific investigation, for the Australian Lady-birds settled comfortably down to their labours and multiplied so satisfactorily under natural conditions in the orange groves of California, that in a few years the increase of the Scale was checked, and ever since it has been held in subjection. Solely on account of its utility, this fine Lady-bird has been added, not only to the fauna of America, but, since its value there was tested and proved, to the faunas of South Africa and Egypt, where it has achieved similar excellent results.

Other Lady-birds have been introduced into the United States to fight various Scale-insects, one of which, the San José Scale (*Aspidiotus perniciosus*), first imported from China, caused terrible destruction in American orchards in the early nineties of last century. Careful observation showed that a Lady-bird (*Chilocorus similis*), common to China and Japan, was the most efficient counter to the Scale in its native haunts. A larval beetle has been seen to eat six Scales a minute, and, since it never seemed to weary of its repast, about 8000 a day. The Lady-birds were transported from the far East to the United States where they were bred, distributed and set free in great numbers, to the discomfiture of the Pernicious Scale. This introduction, however, finally died out in its new country.

A last and more recent illustration of the influence of the counter-pest method in adding to a country's inhabitants is that concerning the Mulberry Scale of Japan (*Aulacaspis pentagona*). This Hemipterous insect, like other Scales, attacks the outer surface of trees, which it pierces in order to abstract the plant juices within. It is almost an omnivorous pest, for in a limited area it has been found on as many as 50 different kinds of trees, but its greatest damage is worked on peach-trees, plum-trees, and in southern Europe on mulberry trees. From Japan the Mulberry Scale has been carried on young trees to all parts of the world, and in most countries it has become naturalized and has recommenced its evil work: in Asia it has spread from Japan to Ceylon; it has reached Australia and the islands of the Pacific; in America it has made its home from Canada, the United States and isles of the West Indies to Brazil and Argentina. In Europe it is mainly confined to the southern countries, especially France and Italy, though in 1898 it made an appearance in England, where it was speedily suppressed owing to the energetic action of Professor R. Newstead. In infected countries many methods have been tried to check the plague, but the most efficient is undoubtedly that of a counter-pest, parasitic upon the Scale, which has been recently bred by A. Berlese from material in Italy. In the efficiency of this tiny Hymenopterous insect, belonging to the family of the small active Chalcids, and known after its discoverer as *Prospaltella berlesei*, great faith has been placed, for in Italy large trees, entirely covered with Mulberry Scales within the bodies of which the larvae of *Prospaltella* live and feed, have been completely cleared in eighteen months. The efficiency of the parasite is increased by its fertility, for *Prospaltella* is said to have in a year four or five generations of females, each of which lays about 100 eggs—so that the potential progeny of one individual at the end of a season lies between one hundred millions and ten thousand millions. The Chalcid is distributed in infected districts on mulberry twigs bearing Scales parasitized by the counter-pest, and these twigs are attached to the branches of scaly trees.

It was little wonder, then, that Uruguay, troubled with the Mulberry Scale, should apply in 1913 to Italy for a supply of her Chalcids. These were sent; they survived and thrived, spread and colonized the land. Thence in 1914 some of the Chalcids were transported to the Argentine, where Mulberry Scale had made its appearance some ten years before. The extent of man's influence in adding new species to a fauna for utilitarian ends may be judged from the statistics of the National Commission nominated by the Argentine Ministry of Agriculture for the express purpose of establishing *Prospaltella berlesei*. From June to mid-September 1916, 4650 fruit-growers in Argentina applied for the parasitized twigs, and received in all over 530,000 twigs, each of which bore many individuals of the

Chalcid parasite buried in the noxious Scales. In three years a total of 3,000,000 twigs have been distributed in Argentina alone, and the Commission considers that *Prospaltella berlesei* has now been so satisfactorily and plentifully established as to check the invasion of the Mulberry Scale.

So successful have been the results of the introduction of counter-pests in many foreign lands, that sooner or later conservative Britain, when it has awakened to the value of scientific methods, may be led to forsake its easy policy of *laissez faire*, and to add a few benefactors to the host of pernicious insects it has already allowed to be added to the native fauna. In this hope the above illustrations of the value of counter-pests have been inserted here.

V. 2

ANIMALS INTRODUCED FOR THE SAKE OF SPORT

A familiar tradition regards sport as one of the chief ends of the Briton's existence. It is not surprising therefore to find that attempts have been made to establish many creatures in order that their subsequent pursuit may afford the primitive pleasures of the chase.

In most parts of Scotland the sly Fox is destroyed by fair means or foul, for too well is it known that

> There's a tod aye blinkin' when the nicht comes doon.

But in several districts of South Britain, Foxes and Badgers also, are actually imported from the Continent and from Scotland, and are set free to establish themselves in the hope that the hunt may benefit thereby.

The large Red Deer of Germany have been imported and released in western Ross-shire to mend the breed of their degenerate Scottish relatives, but it is doubtful if their influence will long survive those harder conditions of life, which have adversely affected the size and quality of the Scottish race.

THE PHEASANT

Birds, however, supply the most striking examples of the influence on our fauna of sporting introductions. Take the Common Pheasant. It has long since spread over the whole countryside; its long tail may be seen gliding through almost any covert, and its harsh crow is familiar to every dweller in the country. It is a great insect destroyer, yet it has turned to the crops of field and garden, and in some places has caused such destruction that it has been threatened with capital punishment by the law, for permission has been given to tenants to shoot at sight game damaging their crops.

Who would imagine that this moderately common bird, at home in our woods and thickets, eating the wild and cultivated produce of the land with the zest of a native, could be a stranger of strangers? Yet so it seems to be, for tradition says that the Pheasant's home is in far Colchis, on the banks of the river Phasis in Asia Minor. Nevertheless tradition may be wrong, for the fact that fossil remains of several species of Pheasant have been found in prehistoric deposits in Europe, suggests that the Common Pheasant (*Phasianus colchicus*) may have originated nearer home. Its remains, however, have not been found in any early deposit in this country, and the oldest record, from the station at Silchester, points to its connection with the Romans.

Other evidence makes it probable that the Pheasant, perhaps in company with the Peacock, was introduced into Britain to grace the villas of the Romans. A manuscript in the British Museum shows that in 1059, during the reign of King Harold, "unus phasianus" was regarded as equal to two Partridges in the bill of fare of the canons of Waltham Abbey. The importance of this evidence lies in that it rules out of count the Normans, the importers of the Rabbit, for the Pheasant was naturalized in England before their Conquest, and it is regarded as unlikely that the Anglo-Saxons, who are not known to have introduced any animal, could have brought it to this country. In 1100, in the reign of King Henry the First, the Abbot of Malmesbury obtained a licence to kill Pheasants and Hares; in 1299 the price of a Pheasant was fourpence, as against three halfpence for a Mallard; about 1512 the Pheasant's value had risen to "xiid" and a Mallard to "iid a pece." Thereafter there are many references to the presence of Pheasants in England and to their value in the dietary; they had become one of the common food birds.

I can find, however, no evidence that Pheasants were known in Scotland until long after they had been established in England. No remains have been found amongst the organic accumulations of the Roman station at Newstead, or in any Scottish deposit of the Romano-British period. In that wonderful palace, to which I have already referred (p. 210), wherein the Earl of Athole feasted King James the Fifth and "ane ambassadour of the Paipes" on the braes of Athole in 1529, everything that could be obtained was brought that the king and his guests should be "verrie weill entertained." One of the manuscripts of Lindsay of Pitscottie's account of this feast enumerates the viands served to the royal guests; and although some authorities regard the list as an interpolation of later date, this but increases its significance from our point of view. The birds set before the king were "capon, cran [crane or perhaps heron], swan, pairtrick [partridge], plover, duik, drake, brissel-cock [turkey-cock or perhaps guinea-fowl], and paunies [cf. Lat. *pavones*—peacocks], black-cock, and muirfoull, capercailles." Here in a wonderfully detailed list, in spite of the fact that a rare bird like the Peacock is referred to, there is no mention of the Pheasant. There is some reason, therefore, for concluding that the Pheasant was not available.

Other evidence supports this conclusion. In 1551, in the reign of Queen Mary, an attempt was made to regulate the price of wild fowl and game, a standard price being fixed for each variety; and though wild and tame Geese, Blackcocks, Plovers, Curlew, Moor-fowl, Partridges, Woodcock, Snipe, Quails, and even Larks, are specified (see p. 140) there is no allusion to the Pheasant. Four years later a law was passed by the Scottish Parliament, enacting that Pouts [young Partridges], Moorfowl, Ducks, Drakes, Teal, and "Goldings," were not to be killed before Michaelmas under pain of £10, but Pheasants are not mentioned. There is here presumptive evidence that in the middle of the sixteenth century, Pheasants were either absent from Scotland or, if present, were very rare, being, it may be, kept for ornament in the gardens of the great. Indeed, even a quarter of a century later, Bishop Leslie remarks upon their rarity in so many words, which read in the quaint translation of Father Dalrymple:

> Farther because nathing is althrouch [throughout] fortunat and happie, quhat ane way abundes with ws, another way inlakes [is lacking] with vs, and is indigent: for the foul called the storke, the fasiane, the turtle-dwe, the feldifare, the nichtingale, with vthiris natiounis ar frequent, bot skairs with us are fund.

If the scarcity of the Pheasant was to be compared with that of the Stork and the Nightingale, it must have been very scarce indeed.

It is interesting that Leslie should have mentioned the abundance of the Pheasant with "vthiris natiounis" (France perhaps, whence he came in the train of Queen Mary, or maybe England), for this abundance makes its scarcity in Scotland at so late a date all the more striking. Yet before the sixteenth century had ended, the Pheasant had been naturalized in Scotland, and had found a protector in the ready arm of the Scots law. In June 1594 an Act was passed by which it was

ordained that quhatsumever person or persones at ony time hereafter sall happen to slay deir, harts, phesants, foulls, partricks, or uther wyld foule quhatsumever, ather with gun, croce bow, dogges, halks, or girnes, or be uther ingine quhatsumever or that beis found schutting with ony gun therein

should pay a penalty of "ane hundreth [100] punds."

I think the Scottish laws make clear, however, that even yet the Pheasant was a bird of great rarity, for many subsequent enactments omit reference to it. Take, for example, the very comprehensive Statute of 1600 already quoted (p. 203), the Statute of 1621, or even that of 1707, none of which specifies Pheasants though many other game birds are mentioned by name. John Taylor in his *Pennyles Pilgrimage* recounts with relish the variety of fare he received in the year 1618 at the hands of "my good Lord Erskine" and mentions of birds "pidgeons, hens, capons, chickens, partridge, moorecocks, heathcocks, caperkellies, and termagants [ptarmigan]" but no Pheasant.

The establishment of the Common Pheasant as a game bird in Scotland may, therefore, be said to date from the close of the sixteenth century, when it was probably brought across the borders from England. I have no doubt that its late appearance in Scotland was due to the poverty of the country as a whole, and of the barons in particular. For even under favourable conditions of shelter and cultivation the Pheasant will not thrive, and probably could not exist, without some protection and attention. No one can allege that the bleak Scotland of Queen Mary's time was a sheltered or well cultivated land, and the barons, ready enough though they were to hawk or hunt the native birds, which multiplied of their own free will, could bear neither the expense nor the trouble of rearing and tending Pheasants in a bare land.

The establishment of the Pheasant in our coverts has led to an interesting secondary introduction or transportation and establishment. Ants and their larvae are well known to form a favourite food of the Pheasant, the larvae being the main support of young birds. To meet this demand large species of ants have been set free in coverts, where they have sometimes become established, building their nests and hatching their young in ever multiplying numbers—increasing at one and the same time the food of the Pheasant and the variety of the local fauna.

In recent years many importations of wonderfully beautiful eastern varieties of the Pheasant have been made, the best known being the Chinese or Ring-necked Pheasant (*Phasianus colchicus torquatus*), the "common" pheasant of central or south China; the handsome Mongolian Pheasant (*P. c. mongolicus*), from the mountains of eastern Asia; the Japanese Pheasant (*P. c. versicolor*), distinguished by the dark green of its under parts, and the gorgeous Reeve's Pheasant (*P. reevesii*), from the highlands of northern and western China. Several of these varieties and their crosses may now be seen at large in one part or another of the countryside.

INTRODUCTION OF THE CAPERCAILLIE (*Tetrao urogallus*)

If man has added the most beautiful of game birds to our coverts, he also has placed in the woods the most handsome of their modern inhabitants—the "capercaillie," "capercalye," "auercalye" or "horse of the woods," as our old writers variously termed him. This fine bird (Fig. 50), the large size and glossy black plumage of which would make it remarkable in any association of bird life, was well known to early historians of Scotland. Boece (Bellenden's translation), says of it in 1527,

Mony uthir fowlis ar in Scotland, quhilkis ar sene in na uthir partis of the warld, as capercailye, ane foule mair [in size] than ane ravin, quhilk leiffis allanerlie [lives entirely] of barkis of treis.

But the gradual disappearance of the pine woods, upon the

ANIMALS INTRODUCED FOR SAKE OF SPORT 269

shoots of which it lived in the winter time, brought about the extermination of the "old man of the woods,"—the "caber-coille" as he is affectionately named in Gaelic. Fuller reference to this unfortunate extinction will be made in discussing the effects of the destruction of the forests (p. 354); it is sufficient here to say that after a long period of dwindling numbers the last of the native Capercaillies seem to have occurred in Scotland about the seventies of the

Fig. 50. Capercaillies (Cock and Hen)—reinstated in Scotland after extermination. $\frac{1}{8}$ nat. size.

eighteenth century, one having been seen about 1762 in the woods of Strathglass, while Pennant says the rare bird was still to be met with in 1769 in Glenmoriston to the west of Inverness.

The Capercaillie, however, continued to survive in the great forests of continental Europe and Asia, from Norway and Sweden in the west to Kamschatka in the east, and from Siberia in the north to the pine-forests of Germany, Austria and the Balkans. A first and fruitless attempt to re-establish these birds in Scotland was made by the Earl of Fife at Mar Lodge in Aberdeenshire in 1827 and 1829, when Capercaillies were brought from a Swedish forest. But the numbers which arrived were too few: in 1827 only a male reached the end of his journey alive, and in 1829, when a single pair was imported, notwithstanding that they reared apparently healthy broods, parents and progeny soon died out.

Astonishing success, however, met the efforts made to establish the Capercaillie in the vicinity of Lord Breadalbane's castle of Taymouth. Through the instrumentality of Sir Fowell Buxton, a Norfolk squire, at least 13 cocks and 19 hens (some accounts say 48 birds), arrived at Taymouth in 1837 and 1838 and were set free in the woods. They at once settled down in their new surroundings, formed nests and hatched and reared healthy broods, with such success that twenty-five years later the Marquis of Breadalbane estimated the number on his estate at 1000 birds, although his head-keeper believed that 2000 was nearer the mark.

From Taymouth adult birds were transported to various localities, as to Arran in 1843, where, reinforced by fresh importations from Sweden in 1846, they became established. Other deliberate endeavours to found new colonies, as at Dunkeld and in the counties of Ayr and Argyll, were unsuccessful, although some, as the introduction of eggs at Tulliallan in 1864 and ten years later of birds at Lathirsk in Fife, met with better fortune. From the naturalist's point of view, however, the most interesting result of the Taymouth introduction has been the extraordinary way in which natural processes of increase and migration have led the birds from this limited station to colonize the greater part of Scotland. For the details of this gradual conquest of new areas, the reader must turn to Dr J. A. Harvie-Brown's *Capercaillie in Scotland* where the movements up to 1879 are carefully recorded. Here I can give only the main features up to the present day of the distribution from the centre at Taymouth Castle in central Perthshire.

It is natural to suppose that since the Capercaillie depends on pine woods both for shelter and food, its dispersal would be regulated by the position and suitability of such woods, together with its own rate of increase and the consequent necessity for the discovery of new feeding and breeding grounds. This is exactly what was found to take place. Fortunately the years that intervened between its extinction about 1770 and its introduction in 1837, were years fruitful in planting, and in the interval, woods largely of coniferous trees had sprung up on the hillsides and along the valleys of many of our rivers. Dr Harvie-Brown was of opinion that the Capercaillie viewed prospective sites from its old establishments, and this very probable selection by sight, together with the fact that most of the woodland lay along the water-courses, would determine the Capercaillie's dispersal along the valleys. Indeed, judging from the dates of the advent or establishment of birds in new areas, the valley systems ranked second only to the presence of fir-woods in determining the course of the migrating Capercaillies. It is a curious feature of these migrations, that, as a rule, hen birds alone prospect new dwelling-sites, and commonly settle in a new area two, three or four years before any cock bird arrives. During the interval before the coming of their true, even if polygamous, partners, the hens frequently mate with the closely related Blackcock (*Tetrao tetrao*), and even with the Pheasant, the result being that the extending margin of a Capercaillie country is marked by the presence of hybrid birds.

The Tay and the Forth valley systems offer highly instructive evidence of the influence of such routes on the dispersal of a species. Follow for a moment, with the aid of the accompanying map (Map I), the Capercaillie's wanderings from the centre at Taymouth Castle, at the east end of Loch Tay, where it was established in 1837. I give as a rule the earliest recorded appearance at any particular place, but have ignored sporadic occurrences clearly out of touch with the general movement of the birds. From its dispersal-centre the "Caper" followed the valleys radiating east and west from the northern end of Loch Tay. It reached the meeting place of Tummel and Tay in 1844, and passing north and south along the valleys, reached Blair Athole on the Garry in the following year, sending off, by the way, a side branch which followed the Tummel itself to Loch Rannoch.

The main migration down the Tay was well defined. It reached the junction of the Isla shortly after 1847, and splitting into two bodies as the two valleys presented themselves, passed onwards up the Isla and down the Tay. A glance at the map will show how the former body, keeping to the main drift of the valley, gradually colonized the tributary valleys, settling in the upper parts of Strathardle in 1860 and of Glenshee in 1865. The Glen Isla settlements not only supplied the tributary valleys, but overflowed into Forfar by the low land at the base of the Grampian foothills, and reached Glamis Castle near Forfar in 1863, and the neighbourhood of Brechin in 1870. This northward migration was reinforced by an introduction at Cortachy in 1862. Thence progress along the easy flats of the South and North Esk was peculiarly slow, for the Capercaillie made no general appearance in North Kincardineshire till 1878, when it had spread up the wooded valley of the Dee as far as Banchory and Inchmarlo. Widely spread colonies were in course of time established in Aberdeenshire and Banffshire. Col. Sir Arthur Grant, Bart., informs me that on the estate of Monymusk on the Don near Alford, Caper were first seen in 1889, when a hen in a very exhausted condition was picked up during a snow storm. Here the birds were first shot in December 1891, when six were killed. Since that time, the increase in this district has been extraordinarily rapid (see p. 274). About 1896 I learned that birds had been seen in the woods at the "back o' Bennachie." In Glass parish on the Deveron they appeared in 1897, and in the woods at Methlick a pair had a brood in 1911. This great extension of range, of over one hundred miles as the crow flies, was the direct outcome of the break-away that followed the Isla valley, while the main movement of Capercaillies kept close to the line of the Tay.

To return to this main movement:—From the junction of the Isla (1847) the Caper reached the mouth of the Tay about 1852, having sent a branch up Glen Almond in passing. This meeting-place of the midland vales near Perth became an important centre of distribution, whence the midlands and southern counties were invaded. The Earn valley was conquered in progressive stages, Loch Earn being finally reached in 1876. The Ochil Hills were turned on their north-eastern flank; easy entrance being thereby gained to the plains of Fife and Kinross, while

INTRODUCTION & SPREAD OF THE CAPERCAILLIE IN SCOTLAND.

○ Introduction at Taymouth—the main colonizing centre.   ○ Other effective introductions.
□ Introductions which have failed to become established.   → Direction of spread; '27 etc. years of arrival or establishment at localities indicated (from 1827 onwards); isolated dates indicate casual appearances.

progress along the fertile flats of Strathmore gave access to Clackmannan and the Forth Valley. Some fifteen years served to cover the valley system of the Forth, the earliest records of breeding being those at Blair Drummond in 1860 and at Dunmore at the mouth of the river in 1863. Up the Teith Valley, the Caper settled near Callander in 1872. From the southern end of Strathmore the movements southwards were continued by two channels, one leading by the eastern shoulders of the Campsie Fells to the Lothians and the south-eastern counties, the other by the western shoulders to Dumbarton, Renfrew and the south-western counties.

Westwards from the main centre at Taymouth, the wanderers passed up Glen Lyon, and along the sides of Loch Tay, whence, following the valley of the Dochart (1865), they debouched upon Glen Falloch. From this they reached the shores of Loch Lomond in 1874, and crossed to Argyllshire where they had penetrated almost to the line of the Crinan Canal in 1910, and whence they crossed to the island of Bute about 1913.

The records of the natural dispersal of the Capercaillie in the Lowlands are less easy to interpret, owing to the independent introduction of the bird at Glenapp in Ayrshire, where it survived for several years after 1841, and at Tulliallan near Kincardine-on-Forth, where a strong establishment was made in 1864.

Neither of these supplementary introductions, however, has so much contributed to the spread of the Capercaillie as the persistent and successful efforts begun in Strathnairn in 1894, when a commencement was made by turning down 31 birds brought from Norway and Austria, to which stock fresh birds were added each succeeding year till 1900. I have no doubt that the Strathnairn centre is responsible for the appearance and increase of the Capercaillie to the west and north, at Inverness (1912), Beauly (1912) and in Ross-shire (1912), as well as in Nairn and Moray (1907) to the east.

There are many features of interest in the recolonization of Scotland by the Capercaillie, but sufficient has been said to show the widespread significance of the introduction of this fine bird, which through man's intervention has spread from the western slopes of Argyllshire to the plains of eastern Fife, and from southern Wigtownshire to the hillsides of

Ross-shire. The accompanying map (p. 271), which interprets in a somewhat dogmatic way the dispersal of the Capercaillie in Scotland, should be compared with one showing the distribution of woodland in modern Scotland. Only thus can the importance of the influence exercised by valleys and forests on the movements of the birds be fully appreciated. I would add in closing this account, that the effectiveness of the establishment of the Capercaillie in our woods is demonstrated by the numbers that have been killed in a day's shoot on widely separated estates. Thus, on Sir Arthur Grant's estate of Monymusk in Aberdeenshire, 84 Capercaillie were killed in a day in 1910, and 73 in 1911; Mr J. G. Millais records that in a single day 107 were killed at Fotheringham in Forfarshire in 1894; and on Blackhall in Kincardineshire 150 were got on one day in 1908, the record day's shoot for Scotland.

It may seem a simple and natural thing that a bird, which like the Capercaillie had flourished in the country at a recent date, should on being reinstated, again obtain firm hold in its own land. But the problem is not so simple nor the result so dependable. The disappearance of the Capercaillie was due mainly to the gradual destruction of woodland; the success of its reinstatement depended upon the presence and suitability of the new woodlands which had sprung up during the years of its absence. It is possible that the new destruction of woodlands which the War has entailed may again restrict the distribution of the Capercaillie, or at any rate check its dispersal by removing the forest stepping-stones which offered it gentle stages for progression from one area to another.

THE GREAT BUSTARD

At first glance, the case of the Great Bustard (*Otis tarda*) (Fig. 62, p. 366) appears to be exactly comparable to that of the Capercaillie. At one time, this huge game bird was common on the plains of Scotland, and the pen of Boece testifies to its breeding in the district of the Merse in Berwickshire in the sixteenth century. To this record I refer in greater detail in discussing the influence of cultivation on animal life (*loc. cit.*). The Great Bustard died out in Scotland as it did throughout the United Kingdom. Several attempts have since been made to re-establish it in Britain, but with uniform failure; the once native bird persistently refuses to adopt again its former home. The secret of the failure lies in this: that while times and conditions have changed, they have in no sense changed for the better from those which drove the British Bustard to extinction. It is a bird of the plains which nests in the open and trusts to its keen sight to warn it of danger still afar off, and to its speed of limb to carry it to safety. But the new cultivation and the growth of sheltering plantations, coverts and hedgerows, afford possible shelter to lurking foes, and here the Bustard cannot dwell and thrive.

RED-LEGGED PARTRIDGE

Another introduction which has met with little success in Scotland, though it is common over great parts of England and Wales, and has actually defied attempts to exterminate it in East Anglia, is the pretty Red-legged Partridge (*Caccabis rufa*), a native of south-western Europe, first brought to England in 1770. It is a bird fond of sand-dunes, but somewhat less so of highly cultivated areas, and this may account for the contrast between the success of the English experiments and the comparative failure of the Scottish, although in recent years it appears to have become established in Fifeshire.

INCIDENTAL GAME-BIRDS

From Norway the Willow Grouse or Ryper (*Lagopus lagopus*), near relative of our own Red Grouse, has been introduced into Argyllshire, but so far it has taken no important place in the native fauna. The same is true also of such importations to England as the Virginian Quail or Bob White (*Colinus virginianus*), the Button-Quail (*Turnix sylvatica*) of Southern Europe, and the Barbary Partridge (*Caccabis petrosa*), but the persistence with which Hungarian Partridges have been turned down in Sussex, where in six successive seasons numbers varying from 50 to 175 brace have been released, may enable this stranger to obtain a firm foothold.

SOME SPORTING FISHES

Reference has already been made to freshwater fishes introduced for the value of their flesh as food, but another consideration—that of sport—has been a contributory factor in bringing to our rivers a number of denizens of foreign lands. The most venerable of such sporting fishes, from the point of view of its establishment and supposed associations, is undoubtedly the Grayling (*Thymallus thymallus*), so long a recognized inhabitant of our rivers that the date of its coming is forgotten. Its introduction, like that of several other fishes, has, for lack of a better suggestion, been attributed to the monks; but it has been pointed out that while many of its local fisheries are in neighbourhoods where monasteries once stood, yet in English counties where there were many monasteries there are no Grayling, and further, that so sensitive a fish could scarcely have been conveyed alive from the Continent with the means at the disposal of the monks. However that may be, the Grayling, a native of the continental countries from Italy and Hungary to Lapland, is now firmly established as an angling fish in many British rivers. In Scotland, it is common in the upper reaches of the Clyde where 10,000 eggs from the Derwent were planted in 1857, as well as in the Rivers Ayr, Lugar and Greenburn in Ayrshire, and in the Gryffe Water in Renfrewshire. In Dumfriesshire it was introduced into the Nith in 1857 or 1858, and a few years later into the Annan. Fewer than a dozen were set free in the Tay some twenty years ago, and already in 1905 they were so well established that "it was not astonishing to catch one anywhere between Perth and Kenmore." It occurs also in the Tweed and its tributaries, amongst them the Teviot, where the first example was caught in 1855, having escaped from a pond at Monteviot where it had been introduced by the Marquis of Lothian about that time. So abundant is it now in Tweed that netting for coarse fish at the instance of the Tweed Commissioners resulted in the capture of 5791 Grayling in 1913, and 7178 in 1914. Low (1813) stated that it was common in Orkney.

If the Grayling first made itself at home in our rivers,

the Carp came as a good second, for although we may suspect the accuracy of the old doggerel lines

> Turkeys, Carps, Hops, Piccarel and Beer
> Came into ENGLAND all in one year—

they at least faithfully indicate that the Carp was introduced to this country and that at an early date. The Carp (*Cyprinus carpio*) is a native of the rivers of China, but so long ago as 1496 it had been planted in the streams of Britain, for Dame Juliana Berner's *Boke of St Albans* says that it is a "dayntous fysshe, but there ben but fewe in Englonde and therefore I wryte the lesse of hym." The "fewe in Englonde" of the fifteenth century have increased to many in our day, for the Carp thrives in lakes, rivers, and ponds, natural and artificial, from Northumberland to Cornwall. In Scotland it is less common than in England, yet it occurs in many lochs and ponds throughout the country and, according to Stoddart, exists even in the lakes of the Outer Hebrides, in Lewis, Harris, the Uists, and Barra. In other countries, such as America, valuable and extensive fisheries have been created by the introduction of this native of the Far East (see p. 258).

The Trout of Lake Geneva (*Salmo lemanus*) was placed many years ago in several lochs and tarns in Mull, as well as in Loch Lomond, where it has now become very rare. A somewhat similar tale of half-hearted success has to be told of the efforts to introduce into Scotland the American Brook Trout (*Salvelinus fontinalis*) often misnamed in this country the "Rainbow Trout." This Char, related to the red-bellied Char which inhabits some of our deeper lakes, has been set free in many lakes and rivers, but though it survives for a time, it appears seldom to thrive and become firmly established. Mr H. A. Woodburn says of its occurrence in the Clyde drainage area, that it has been placed in many small lakes throughout the district and also in Loch Lomond, where it still maintains its identity but has not thriven. It has also been distributed throughout Renfrewshire and Ayrshire, and is thriving in the Rivers Ayr and Irvine, and in the Waters of Borland, Kilmarnock, Cessnock, Carmel and Alnwick. It is impossible to follow the details of the introduction of the American Brook Trout in other areas, for it has been widely distributed throughout Scotland, even in islands such as Mull, and there is the less need for so tedious an enumeration, since the citation of the Clyde area alone sufficiently indicates to what extent man may encourage an alien fauna to replace that of nature's assembling. Introductions of the true Rainbow Trout (*Salmo irideus*) of California have been made on several occasions and in several places, as in Loch Uisg and the Lochbuie lochs of Mull. In 1898 over a thousand " Rainbow Trout" were set free in the river Buchat, a tributary of the Don in Aberdeenshire[1].

TRANSPORTATIONS

Apart from the deliberate introduction of new creatures from a strange land, there is a minor form of introduction, which, while it involves no fresh addition to the fauna as a whole, has yet some influence upon the numbers and distribution of its members—I mean the transference of an animal from an old to a new area. There would seem to be little scope for such transportations in a country so small as Scotland, yet many have taken place, and as they relate in the main to sporting animals, a few typical illustrations may be given here. Perhaps the most noted of all such transportations concern the short-headed Trout of Loch Leven (*Salmo levenensis*), famed for the red colour of its flesh and its peculiar delicacy of flavour. There is scarcely a Scottish lake or slow-flowing river much frequented by anglers, but there Loch Leven Trout have been released, as many as 150,000 fry having been set free in one lake, Loch Awe, in the course of the three years preceding 1890. England, too, has shared in the spoil, and amazing success has followed the introduction of Loch Leven Trout to the still and running waters of New Zealand. The enthusiasm of anglers, again, has stocked with the common Brown Trout of the brook (*Salmo trutta*), many a Highland loch or lochan, which till that time knew no fish except perhaps the migratory Eel, whose elvers take to land when they encounter a waterfall such as would completely check the passage of Trout or Salmon. Trout from below the great Smoo Cave, near Durness, were placed many years ago in the river above the Cave and beyond the impassable cascade which bursts through its roof, and, isolated in their new habitation, they

[1] I suspect that these "Rainbow Trout," recorded by Mr G. Sim, may have belonged to the species referred to above—the American Brook Trout.

have developed on their sides a peculiarity of their own—great red splashes of colour. Many troutless lochs in the wilds of Sutherland, such as Loch Bealach na Uidhe, Loch na Ganvich and Loch Unapool, now contain a bounteous stock of trout, thanks to the energy of anglers such as Dr Harvie-Brown.

The habit of the sluggish, mud-loving Tench (*Tinca tinca*) has led to its introduction into many slow-flowing rivers, and natural and artificial ponds and reservoirs, such as Pressmennan Lake formed in 1819 in Haddingtonshire, or Hirsel Lough, in 1876, in Berwickshire, the Pitfour Ponds in Aberdeenshire and the Tay. In such areas the "Fisher's Physician," as he was long ago dubbed, lives and breeds in security.

Perch (*Perca fluviatilis*) have also been spread abroad in our rivers by the agency of man, for though naturally they are rare north of the Forth, individuals have been successfully introduced during the past hundred years to the Loch of Spynie near Elgin and the river Deveron, to the river Don and Strathbeg Loch in Aberdeenshire, and to many lochs and streams in the Tay basin and other valleys beyond their natural range.

Even the voracious Pike (*Esox lucius*) has been granted new areas in which to work its will on its lesser cousins, examples of the success of its transportation being shown in Gartmorin Dam on the Forth and Migdale Loch near Kyle of Sutherland; while the Char (*Salvelinus alpinus*), interesting relic of the cold torrents which rushed in our valleys as the ice-sheets of the late Glacial Period melted, has been transferred from its cool, deep glacial lakes to many it would not have chosen of his own accord.

The gentle sport of angling is also responsible for the increasing range of distribution of the Minnow (*Phoxinus phoxinus*), for this tiny but active denizen of the river shallows does not extend under natural conditions further north in Britain than the waters of the Deveron, or at furthest the Lossie. On account of its use as a bait for larger fish, the Minnow has been set free in the Spey (although it may have existed there before its artificial introduction), as well as in the streams of Argyll and Arran, and of several other localities whence it was originally absent.

Even birds and mammals have not escaped this artificial means of dispersal. Brand (1701) mentions that in his day "Moorfowls" or Grouse, were absent from the Shetland Isles, but that a few had been brought over from Orkney. These, however, soon died out, as also did a few imported to Weisdale Voe in 1858, though the descendants of the latter survived for 14 or 15 years. In 1882 an endeavour to acclimatize Grouse on the heather-clad slopes of the island of Yell met with greater success, for owing perhaps to the more suitable nature of the ground, these established a slender colony which existed at any rate till comparatively recent years.

In Orkney on the other hand, where the Red Grouse was plentiful, Partridges were unknown; "There are here no Partridges," wrote Brand in 1701, "but plenty of Muirfowls." Nor have any of the many attempts which have since been made to plant Partridges on the Mainland, on Hoy and Walls, on Sanday, on Shapinsay (1883) and Rousay (1883), met with success, although for several years, Rousay boasted the presence of four or five individuals.

Of mammals, Roe Deer, which are natives of few of the Scottish islands, have been established on Bute, Islay, Mull and Jura; but the Hares best illustrate the results of deliberate transportation. There is no satisfactory evidence that the Common Brown Hare (*Lepus europaeus*) is indigenous to the islands of Scotland, yet now it is almost impossible to name an island of any significance from which it is absent. In the Firth of Clyde, the "bawtie" has been set free and thrives upon Arran, Bute and Cumbrae; in the Inner Hebrides it was introduced on Islay (before 1816) and on Jura, on Coll (about 1787) and on Tiree, on Mull (1814–15), and on Skye, as well as on other lesser isles; amongst the Outer Hebrides, Professor Macgillivray tells us in 1830, that it

has been naturalised in the neighbourhood of Stornoway and in Barvas, in the district of Lewis and in one of the Barray islands; but it does not appear that it ever occurred indigenously in any part of the range.

The Blue, Alpine, or Variable Hare (*Lepus timidus*) has shared in wanderings like its cousin, for since the disappearance of the last traces of the Ice Age, its natural range in the British Isles has become limited to the Highlands of Scotland (see Map II). It too has been planted on the islands, so that wherever it occurs to-day in the Outer or Inner

ANIMALS INTRODUCED FOR SAKE OF SPORT 281

Fig. 51. Blue or Mountain Hares (male, female and leverets)—established in many new areas in Scotland. $\frac{1}{9}$ nat. size.

Hebrides (Lewis and Harris, Skye and Raasay, Mull, Islay and Arran) it has been introduced, for if it were ever native in these parts, the original breed, as on Hoy in the Orkneys, has disappeared. Recent introductions to Hoy and Gairsey seem to have failed.

On the mainland the artificial colonization of the Alpine Hare has been more extensive and successful. A century ago it was unknown south of the Forth, but from its Highland home it has been planted on the uplands of the southern counties. In Manor Parish in Peeblesshire, Alpine Hares were set free in 1834, and further colonies were planted in the same county in 1846 and 1847; in the Pentlands they were set free in 1867 or 1868, and in 1861 or 1862 on Cairntable on the borders of Ayrshire and Lanarkshire. On this mountain, Dr B. N. Peach tells me, they were so abundant when he made a geological survey of the district in the years 1867 and 1868, that they formed the staple food supplied to the geologists by the hill shepherd. So congenial has the Alpine Hare found these Lowland haunts that not only has it held its own and increased vastly in numbers in the areas in which it was placed, but it has spread into the surrounding country, overrunning the counties of Renfrew, Ayr and Lanark, in the first of which 300 have been killed in a season on the Misty Law Hills. Spreading eastwards from their Peeblesshire stations by way of the Moorfoot Hills the Mountain Hare reached the Lammermoor Hills about 1860, the Lauderdale moors in Berwickshire some four years later, and the extension of the Lammermoors in east Haddingtonshire about 1880. A southward movement from Cairntable colonized the Lowther Hills and Queensberry Hill in 1865, and advanced until between 1878 and 1880 the southern march was checked by the Solway, which was reached in the neighbourhood of Kirkgunzeon and Criffel Moors to the south-west of Dumfries.

Now there is scarcely a moderately high hill within these areas of the south country but has its colony of this interesting relic of glacial climes. Not only has the Scottish Mountain Hare (*Lepus timidus scoticus*) succeeded in colonizing new districts in its own country, it has also been successfully transferred to many English counties and to the mountains of Wales.

DISTRIBUTION OF THE MOUNTAIN HARE IN SCOTLAND.

▓▓▓ Areas to which Mountain Hare is indigenous. ▓▓▓ Areas colonized by transported individuals.

Dates surrounded by lines indicate times and localities of introduction by man. Italic dates, underlined, are approximate times of arrival in new areas from the centres of introduction.

V. 3

ANIMALS INTRODUCED FOR AMENITY

SEVERAL creatures of interest and a few of some importance in the composition of our fauna owe their presence to little more than the whim of the lover of Nature or the surplus energy of the naturalist. The reason for some of these introductions stands in plain view, for others it is indiscernible.

THE PEACOCK

No one can doubt, for example, that the feeling for colour led to the transfer of the "beauteous Peacock" far from its native home in India and Ceylon. We, in Britain, probably owe its presence to the Romans. In mediaeval days it seems to have been moderately common in Scotland, more common than the Pheasant, for it graced the table of the Earl of Athole at the great feast in the "wilderness" of Perthshire in 1529 (see p. 266), when its lesser relative was absent; and in the old Scots laws it was ranked as wild-fowl, under the plea that "the nature of the Peacock is wild, although they are wont always to return to the same place." (Frag. Coll. I. 750 b). Nevertheless, Juno's bird, despite the decrees of law and the long ages of its habitation, is still no more than an exotic flower in our dull clime, a bird the "painted plumes" of which add to the pleasure of park or shrubbery, but which remains outwith the native fauna.

BIRDS OF BRIGHT PLUMAGE

Other strange birds have been set free on account of their beauty, though most of them have rapidly disappeared from the fauna they invaded. Red-winged Starlings (*Agelæus phœniceus*) from America, were released by the late Duke of Argyll in Argyllshire, but soon died out, and in recent years "Pekin Robins" (*Liothrix luteus*) and many exotic Doves— Bronze Wings, Turtles, and Crested Pigeons—have been released in London with little success. On account of their interest, Lord Carmichael of Skirling transported Nuthatches from England to southern Scotland, but these importations also left no trace.

GOLD AND SILVER FISHES

Nor is there room for doubt that the love of colour has been the secret of the introduction of the Gold and Silver Fishes of the East, not only as tiny captives of the indoor aquarium, but as inhabitants of ornamental and other ponds throughout the country. It can hardly be said that these have become part of the fauna of Britain, yet Goldfishes (*Carassius auratus*), natives of China and Japan, have been established in so many artificial lakes and warm engine-ponds of mills, and multiply so successfully in these sheltered havens, that, like the Peacock, they may be regarded as aliens, which, though they are still clearly foreigners, have nevertheless come to stay.

On the other hand, there are animals, first brought to this country as objects of interest or amusement, which have so easily settled in the land of their adoption that they have become part and parcel of the native fauna.

FALLOW DEER

Take the case of the Fallow Deer (*Cervus dama*). Fallow Deer, closely related to, if not identical with our present species, inhabited Europe and England in the warmer interglacial periods of the Ice Age. Thereafter they seem to have disappeared so completely, as, according to the usual statement, to have left no trace in the later deposits of the British Isles. It is possible, however, that, like other harassed wild things, they survived longer than is suspected in the security of the outer islands, for Barry in his *History of the Orkney Islands* (1805) states, I know not on what authority, that the superficial deposits of Orkney have yielded animals strange to the present day fauna, "such as hares, and several sorts of red and fallow deer, the horns of which have been often dug from the earth in various parts of the country." Nevertheless it is a strange but generally accepted fact that while Red Deer and Roes continued to survive in our land, the Fallow

Deer entirely disappeared, leaving no trace in the peat or marl of the Scottish lakes, nor in the shell-mounds, kitchen-middens, or dwelling places of the Scottish peoples, even to the time of the brochs, which extends almost to the ninth century of our era.

The Fallow Deer owes its subsequent introduction to the grace and interest which its lithe body and handsome antlers contribute to the pleasure parks of the great. There is no record to tell when it first returned under the protection of man, but we know that shortly after the Norman Conquest it was regularly hunted in England by the king and his barons, and there can be little doubt that it was imported from one of those districts of southern Europe, from Portugal to Greece, in which it retained a hold through the changes of post-glacial times. To Scotland it was in all likelihood transported a short time after its establishment in England. It appeared at first as a guarded and pampered tenant of the pleasure park, for the earliest historical reference to the species is that mentioned by Professor Cosmo Innes, recording how in 1283 an allowance was earmarked in the accounts of the King's Chamberlain for mowing and carrying hay and litter for the use in winter of the Fallow Deer which lived in the park at Stirling. Now as this "new" park was formed and surrounded with a palisade in 1263 for King Alexander III, it is possible that the Fallow Deer were imported at that time specially to grace the new royal pleasure ground.

Three centuries later, in 1564, Queen Mary watched from the Braes of Athole the great gathering of Deer arranged by the Earl of Athole for her entertainment. In Dr William Barclay's account of this notable hunting, as translated from the Latin by Pennant, occurs a passage describing how the Highlanders of the "tainchel," forming a ring round the area to be raided, "went up and down so nimbly, that, in less than two months' time they had brought together two thousand red deer, besides roes, and fallow deer." If both translation and original account are accurate, the presence of Fallow Deer in the wilds of Glen Tilt would indicate that they had already escaped from the narrow limits of the parks, and become established in the highlands of Perthshire.

Standing by itself this evidence seems little to be depended on, but it is supported by more trustworthy hints, which indicate that, more than a century before Queen Mary's gathering on Athole Braes, Fallow Deer were well established in the country. Already in 1424, they were afforded the protection of the law, being grouped with Red and Roe Deer in an Act which provided that "stalkers that slayis deare, that is to say, harte, hynde, dae and rae," should incur a penalty of "fourtie shillings" and their employers a fine of £10. "Dae" is clearly a Scottish form of "doe," the ordinary and only appellation of the female Fallow Deer, just as "rae" is the female Roe in contradistinction to Roe-buck (witness the use of the terms in an Act of 1503 which includes both "raes" and "rae-bucks"). The reference to Fallow Deer in the Act of 1424 is made still more positive by the fact that the other deer of the country are carefully specified—the "harte" and "hynde" of the Red Deer, and the Roe.

A perusal of the early Scottish laws, however, makes quite clear that the Fallow Deer stood in a category distinct from the Red and the Roe Deer; for these being native and far more common are frequently mentioned in laws which ignore the presence of the Fallow Deer. Yet in 1587 we find the Fallow or Dae mentioned with several of the most familiar of wild animals: "slayers and schutters of hart, hinde, dae, rae, haires, cunninges, and utheris beasts...sall be lyke cryme to their committers as the stealers of horse or oxen." And since, using caution as with men, we may judge deer by the company they keep, it is apparent that the Fallow Deer were well established and at least moderately common in the latter part of the sixteenth century.

The law, however, gives us no ground for supposing that these deer were free or wild; for the specific mention in a statute of 1503 of "parked deare, raes, or rae-bucks," is sufficient indication that the Fallow Deer were still the deer of the parks. In the same century, King James VI himself, when he returned with his consort, the Danish Princess Anne, is said to have brought from Denmark the first examples of the black variety of Fallow Deer—and these he doubtless deposited in his deer park at Dalkeith Palace.

Nevertheless the experience of more recent times would lead us to suppose that if the "parked deer" escaped, as time and again they were almost sure to do, they would have had no difficulty in establishing themselves in the wilder

ANIMALS INTRODUCED FOR AMENITY

districts. Indeed, so suitable do conditions in Scotland appear to be that, once established in freedom, Fallow Deer are not easily kept in check. A single authentic example will be sufficient to illustrate this adaptability. About 1780 a dozen Fallow Deer were brought to Raehills in Dumfriesshire from Hopetoun, where they have existed since at any rate 1700. They were carefully tended for some time, but finally broke out of their enclosure and became wild. So rapidly did they increase in numbers, and so annoying did their depredations on growing crops become, that serious attempts were made to exterminate them. Expert marksmen were employed, and liberty was given to one and all to shoot the runaways. In a single week, 50 of the deer were killed, and yet the efforts of the deer-slayers could not keep pace with the increase of the deer, and in 1845 they had become exceedingly wild and were supposed to number upwards of two hundred.

Indeed it is clear from other evidence that Fallow Deer had become established in the forests of mid-Scotland before the middle of the seventeenth century, for the Wardlaw Chronicles record how in 1642 a "gallant, noble convoy, well appointed and envyed be many." went a-hunting with the master and his lady, and in the Forest of Killin in mid-Perthshire "got fallow-deer hunting to their mind, and such princely sport as might alleviat the dullest spirit."

In our day, the Fallow Deer, introduced to Scotland at various times by man's agency, and holding its place by virtue of its own adaptability, is to be found in a wild state from Drumlanrig in Dumfriesshire in the south, through the wilds of central Argyllshire, the woods of the Tay Valley from Stanley to Blair Athole, in the Islands of Mull and Islay, to their northern outpost in the Dornoch woods of Sutherland. And, whereas it was introduced to add to the amenities of the pleasure park, it has become, on account of its numbers, an object of sport as well as the source of a useful supply of venison.

VIRGINIAN AND JAPANESE DEER

The elegant and graceful Virginian deer (*Cariacus virginianus*), a native of North America, was set at liberty about 1832 in Arran, where it succeeded in establishing itself, and Mr E. R. Alston recorded that it still occurred but in dwindling numbers forty years afterwards. So far as I can discover, the Virginian Deer have now entirely disappeared, and are reported to have died out not many years after Alston referred to them in 1872.

It has been found, too, that the Japanese Deer (*Cervus sika*), native though it be of far distant China, Manchuria and Japan, takes kindly to our woodlands. In 1887 Sir Arthur Bignold turned out at Lochrosque, in Ross, one buck and four does, and these have so thriven and multiplied that 30 years after, they have colonized far beyond the bounds of their first settlement, and have straggled at one time or another over the greater part of Ross. At the present day the naturalized descendants of the original five aliens are believed to form a herd of from 150 to 200 in number.

THE AMERICAN GREY SQUIRREL

If one race of animals more than another has benefited by territorial acquisitions in this country through the whims of man, it is the tribe of the Squirrels. The Grey Squirrel (*Sciurus carolinensis*) belongs by rights to the American continent from southern Canada to southern Mexico and Guatemala, but its silver-grey coat and lively habits have made it a favourite pet in this as well as in its native land. The escape or release of such pets has already set up many potential centres of distribution in England: in Regent's Park, populated from the Zoological Garden, in Richmond Park, whence it has wandered in force into the open country of Surrey, in Hampstead, in Buckinghamshire, in Bedfordshire, colonized from the Duke of Bedford's collection at Woburn, and in Rougemont Gardens, Exeter, from which the first west of England specimens have recently been reported.

In Scotland the Grey Squirrel has made good an establishment on the west coast and is spreading there. A single pair was released at Finnart on Loch Long about 1890, and mainly from Mr J. Paterson's records of their dispersal we learn that the Squirrels had spread northwards to Arrochar and Tarbert in 1903; eastwards over moderately

ANIMALS INTRODUCED FOR AMENITY 289

high ground to Luss in 1904, and to Inverbeg on the banks of Loch Lomond in 1906; south-westwards to Garelochhead in 1907, and along the western shores of Gare Loch to Roseneath in 1915; southwards, till in 1912, individuals were obtained at Row, Helensburgh, Alexandria and furthest south at Culdross. From this area an eastern

Fig. 52. American Grey Squirrel—acclimatized in many parts of Britain. ¼ nat. size.

extension carried the migrants to the eastern bounds of Dumbartonshire at Drymen about 1915. These records make it apparent that in a quarter of a century the Grey Squirrel has taken possession of a strip of country roughly twenty miles long by fifteen miles broad. There is every likelihood that the Zoological Park of the Scottish Zoological Society, situated at Corstorphine on the outskirts of Edinburgh, will become a strong colonizing centre in the east of Scotland, for already Grey Squirrels have taken to the woods outside the Park's boundaries, and have been found in the policies of Dalmeny. The pets of confinement turn out too often to be the pests of the open, and already many plaints have been raised regarding the mischief wrought by these aliens, which, Sir Frederick Treves protested in 1917, drive out our Red Squirrel, "eat everything that can be eaten, and destroy twenty times more than they eat." So that the spread of the Grey Squirrel threatens us with a plague as grievous as that which has rewarded the well-meant efforts of the enthusiasts who set the Common Red Squirrel free in our woods, that his interesting presence might add to the delights of Nature lovers.

THE SPREAD OF THE RED SQUIRREL

The Common Red Squirrel (*Sciurus vulgaris*) is so familiar and so much at home in our woodlands that one seldom thinks of him as an alien brought to our shores and encouraged to settle—to the loss of our small native birds and of the woods themselves. And, indeed, there is some ground for regarding the Red Squirrel as of native stock; for, as I shall show in more detail in discussing the effects of the destruction of our forests (see p. 351), it was at one time a familiar denizen of the country north of the line of the Forth and Clyde—

> I saw the Hurcheon and the Hare,
> The Con, the Cuning and the Cat,

sang the writer of *The Cherrie and the Slae*—*Cuning* being the Rabbit, and *Con* the Squirrel.

The demolition of forest, however, banished it entirely from the Lowlands, and gradually drove its diminishing numbers from one stronghold to another, till, by the end of the eighteenth century, it seems to have everywhere disappeared, except, perhaps, in the recesses of wilder and more remote forests, such as the native woods of Rothiemurchus at the base of the Inverness-shire Grampians. The present-day numbers and distribution of the Squirrel in Scotland are undoubtedly due in the main to the introduction of new animals from outside the country's bounds, and to their establishment in certain definite areas which have become centres of colonization for the surrounding districts.

The spread of the Squirrel, as also the steps of its disappearance, have been minutely worked out by Dr Harvie-Brown in a series of three papers published in 1880 and 1881 in the *Proceedings* of the Royal Physical Society of Edinburgh, and to these I would refer readers desirous of knowing the details of the early movements, contenting myself here with a general survey, carried down to the present date, of the results of the various introductions.

Fig. 53. Common Red Squirrel—a former native of Scotland, reintroduced. ⅛ nat. size.

It may be premised that the spread of the Squirrel was not equally rapid in all the areas, that it was regulated in direction as well as in speed by the physical characters of each district as well as by the state of its woodland; bleak mountains, and to some extent, rivers, obstructed or checked the onward movements, while close-set woodlands acted as stepping-stones, guiding and hastening the progress into new districts. Reference to Map III will aid in the comprehension of the influence of the different centres, and of the direction and speed of the various migrations.

In the Midlands

While the Red Squirrel was on the verge of extinction in even the wildest districts of Scotland, the Duchess of Buccleuch, about the year 1772, added to her menagerie at Dalkeith, a few Squirrels brought from England. These escaped to the woods, became established, and so increased in numbers that they were forced to seek fresh outlets. They spread east, populating most of the woods in Midlothian and East Lothian by 1802; they spread south, entering Peeblesshire about 1801, but staying their course on the slopes of the Moorfoot and Lammermoor Hills. Their western and northern advance from Dalkeith was extraordinary for its rapidity, as also for its extent. The Squirrels made an easy journey along the Lothian plain to Stirlingshire, which, with the eastern extension of Dumbartonshire, was reached about 1810. At Campsie they appeared in 1827, and in 1830 had reached the west of Stirlingshire and western Dumbarton at Luss and Killearn, while a branch penetrated into north Lanarkshire. On a northerly track from Stirlingshire they appeared in Perthshire (Kincardine-on-Forth) about 1821, having probably crossed the Forth by the Bridge of Frew, arrived in Clackmannan about 1837, and overrun Fifeshire between 1834 and 1859.

In the Lowlands

The remainder of the south country was peopled from three distinct centres of dispersal: (1) The south-eastern area, lying about the valley of the Tweed, was stocked from an introduction to Minto in Roxburghshire about 1824. Hence the Squirrels reached northwards to Duns Castle in 1830, and were so abundant at Wolfelee near Hawick in 1835 that regular attempts were made to destroy them. Westwards they pushed into Selkirkshire, Peeblesshire and southern Lanarkshire entering that county at Lamington Parish about 1841, while southwards they were checked only by the Cheviots on the northern bounds of Dumfriesshire.

(2) Dumfriesshire and the southern counties seem to have been colonized from a source outwith the Scottish borders—Houghton House near Carlisle, where Squirrels were abundant at an early date. In 1837 they entered

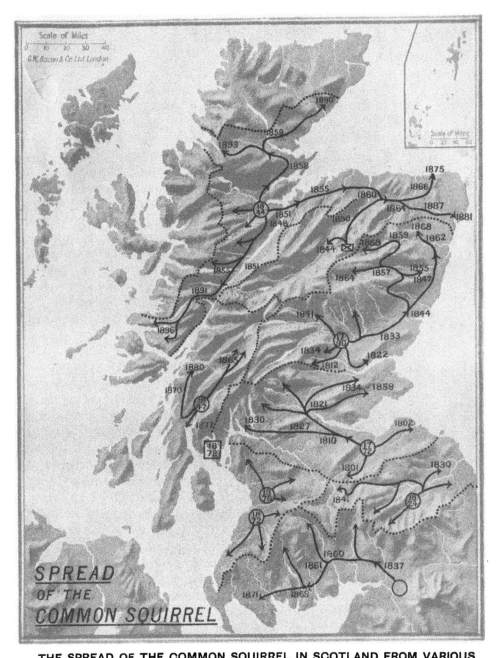

THE SPREAD OF THE COMMON SQUIRREL IN SCOTLAND FROM VARIOUS CENTRES OF INTRODUCTION, Showing the influence of valleys on dispersal.
◯ Centre of introduction and subsequent dispersal.   ⊠ Centre of supposed natural resuscitation.
☐ Introduced but subsequently died out.   .... Approximate boundaries of areas colonized from each centre of introduction. The approximate dates of introduction and arrival at different localities are indicated.

Canobie parish, which abuts against Cumberland, and crossing Dumfriesshire in 23 years, appeared in the neighbourhood of the town of Dumfries in 1860. In that or the following year they had crossed the march into Kirkcudbrightshire, and in 1869 had penetrated to the south of that county. Four years later they crossed the Cree into Wigtownshire, and there in 1892 were recorded by Sir Herbert Maxwell as "now becoming plentiful."

(3) Three separate introductions stocked Ayrshire—one about 1866 at Mauchline, the second a somewhat mythical importation "by navvies" on the Water of Ayr about 1870, and the third two years later by the butler of the Marquis of Ailsa. The influence of the Ayrshire centres, however, was limited by the western extensions of the Lowther Hills, as well as by contact with the armies advancing westwards from the eastern centres of Houghton, Minto and Dalkeith.

In 1872 Squirrels were set free on Bute but they eventually died out.

In the Central Highlands

Comparable to Dalkeith in the extent of the area it has affected is the centre of Dunkeld, in Perthshire, whither prior to 1793, Squirrels were imported by the 4th Duke of Athole probably from Scandinavia. Hence their southward course lay most open, and with speed they flowed down the valley of the Tay, Methven being reached in 1812, and the Carse of Gowrie ten years later. Westwards they made slow progress, owing, one must infer, to the absence of plantations suitable for shelter and food. Woods were planted at Glenalmond only after 1825 and in 1834 Squirrels appeared here. Nevertheless they formed colonies along the upper valleys of the Tay and its tributaries till the mountain barrier of the southern Grampians checked their march.

Their northward advance, also, seems to have been checked by the Grampians. Not till 1841 were they established at Blair Athole, while following the plain of Strathmore, as did the Capercaillie a few years later, they appeared at Glamis in 1833, and reached Brechin in 1844. Thence they passed across the North Esk, and traversed Kincardine probably on two routes, one leading by the coast (Dunottar woods, 1847) towards the lower Dee, the other by the valley of the Feugh to Banchory, where they appeared about 1855. From this station they seem to have spread up the Dee, reaching Glen Tanar near Aboyne in 1857, and becoming distributed throughout most of the woods of Upper Dee by 1864. Two routes apparently led them to the fertile valley of the Don, one from Aboyne up the Tarland Burn and down the Corse and Leochel Burns to Vale of Alford, where one was killed by a dog in 1859. Thence an easy journey took them to Kildrummy about 1860 and Glenkindie some three years later. A second route lay by the haughs between the lower waters of Dee and Don, for Squirrels were seen on the estate of Thainstone near Port Elphinstone in 1862 and at Manar, three or four miles farther up the river in 1868.

Whether the waves of this invasion from the Central Highlands now found themselves spent, or whether they spread farther across the less wooded portions of Aberdeenshire is difficult to decide. Since I am of opinion that the former was the case, and that the consecutive dates of arrival in these areas indicate invasion from another centre, I shall discuss the colonization of northern Aberdeenshire in the section dealing with the Northern Highland dispersal.

In the Western Highlands

An isolated introduction took place in Argyllshire about 1847, when, along with the Rabbit, the Squirrel was brought from England and set free at Minard House on the western shore of Loch Fyne. This establishment formed the centre of a small and isolated colony. The strongest army of migrants moved northwards, reaching Inverary, twelve miles from the starting point, in 1855, and Glen Orchy ten years later. The western advance rounded the south end of Loch Awe, and speedily spread along the western coast, so that by 1880 its outposts had reached the environs of Oban. As in most other areas, the southern movements lagged seriously behind the others, for not till 1877 and 1878 were Squirrels seen even in the neighbourhood of Lochgilphead, only eight miles from the starting point.

In the Northern Highlands

Beaufort Castle, on the Beauly River in Inverness-shire, the last to be considered of the main centres of squirrel dispersal in Scotland, was stocked by Lady Lovat in 1844. Northwards the vagrants moved, settling and multiplying as they went, so that in 1848 they were common in northern Inverness-shire, and ten years later had invaded Ross and Cromarty as far as Kilmuir Castle and Tarbat House on the northern shore of the Cromarty Firth. In 1859 they appeared in Sutherland, and were plentiful in the east of the county in the 'seventies, an increase attributed by Dr Harvie-Brown to the advantage taken by the Squirrels of the Highland Railway Bridge built at Invershin in 1869. It is probable that Squirrels would make use of this bridge in moving across the river, but it seems to me that the argument makes too much of the possibility of a stream acting as a barrier. Squirrels are excellent and deliberate swimmers. They have frequently been observed swimming from one point to another across the fjords of Norway, and in the *Field* of January 27, 1917, the distances of some of their journeys are given—the most striking example being a direct swim of some 500 metres, about one-third of a mile, which was made without break, notwithstanding that there were two places on the way where the tiny navigator could have rested had it wished. We can, therefore, hardly regard the River Oykell as a serious barrier: the less so when we take into account that it had already been crossed for many years by Bonar Bridge, three miles down the river.

So common and destructive did Squirrels become in east Sutherland that, in the seven years between 1873 and 1880, 942 were killed at Dunrobin. In recent years the increase of Squirrels in this area has been phenomenal, and some account of the extraordinary numbers slain is given in the discussion of the Squirrel as a pest (p. 181).

The heights of Morven and the ranges of hills which form the boundary between Sutherland and southern Caithness, as well as the treelessness of the latter county, seem to have stayed their northward progress, for as late as 1887, Squirrels were unknown in Caithness, although they have since become common in the Berriedale district. It is not surprising to learn, however, that they have found an easy and natural route from the North Sea to the Atlantic coast by way of Strath Oykell, Glen Einig, and, on the western watershed, Glen Achallt, so that in 1893 they appeared on the shores of Loch Broom.

Great as was the progress of the northern advance from Beaufort, the southern and eastern movements were as rapid and as extensive. In 1848 Squirrels had reached the banks of Loch Ness near Dochfour, and from Glenurquhart proceeded to Glenmoriston where they became common in 1864. Still holding along the northern bank of Loch Ness they reached Fort Augustus about 1851 and Glen Garry some three or four years later. Thence they penetrated to northern Argyllshire, where they were seen in 1891 on the shores of Loch Eil, and pushing westwards they reached the Atlantic coast in Ardnamurchan in 1896.

To the east from Beaufort, the fertile and wooded levels bordering the Moray Firth allowed of easy progress. The town of Inverness was surrounded in 1851, and by regular advances the Squirrels reached Nairnshire (Cawdor) in 1855, Morayshire (Glen of Rothes) in 1860, and then, in company with wanderers from the forests of Speyside, crossed the narrow waist of Banffshire to Aberdeenshire, where they reached Drumblade near Huntly, about 1864, Turriff in 1865 or 1866, Fyvie in 1867, and Troup Head in 1875 or 1876. In steady stream they seem to have trooped down the valley of the Ythan from Fyvie to the woods about Haddo House, and thence to the neighbourhood of Ellon, where Dr R. M. Wilson noticed them in the early 'eighties. They are now also common in the isolated woods of Pitfour House near Old Deer. So the north-eastern corner of Aberdeenshire is no longer the untrodden ground which Dr Harvie-Brown found it to be close on forty years ago.

In the Spey Valley

While artificial restorations of the Squirrel were being made at these scattered localities, it is possible that the old native race itself was making new efforts to retrieve the ground it had lost, stimulated by the growth of new woodlands in places which before had been barren. We can hardly believe that, so long as some of the great native forests still spread over miles of territory, a few pairs of Squirrels did not survive in their depths, awaiting the dawn of a new era to multiply and spread.

Dr Harvie-Brown was of opinion that such a resuscitation took place in the Forest of Rothiemurchus on the northern flanks of the Inverness-shire Grampians. Squirrels were seen in 1844 in the woods of Upper Strathspey and in 1856 had reached Grantown-on-Spey. From this point onwards their movements merged with those of the eastern army from Beaufort, with which they apparently combined in the over-running of the southern shores of the Moray Firth and northern Aberdeenshire.

So the Squirrels, like the Capercaillies, followed the wooded valleys, checked here and there, it is true, by bare stretches of mountain or treeless slopes, but persevering until they had replaced in Scotland a race of Squirrels it had all but lost. Now they range the country from Wigtownshire to Sutherland and from the Atlantic Ocean to the North Sea. The country of their adoption has favoured them; they have multiplied so enormously that they have come to be regarded as one of the prime pests of the forester, for they destroy the young shoots of pine trees, remove the bark, devour the seeds, and commit these enormities in such overpowering numbers that, in the woods of Glen Tanar, Aberdeenshire, 1000 trees valued at £500 were destroyed in the first fifteen years after the Squirrel's appearance there. Thereafter an average of about 200 Squirrels was shot in these woods every year. In the Cawdor plantations in Nairnshire, where a small reward of a few pence was given for each Squirrel killed, 14,123 Squirrels were presented in the course of sixteen years, for the slaughter of which the sum of £213 13s. 9d. was paid. And the rate of increase north of the Caledonian canal has been still more remarkable (see p. 181).

A success from the acclimatization point of view, the introduction of the Squirrel, like so many introductions, has proved a failure, if not a disaster, from the economic standpoint.

THE BEAVER

Passing reference may be made to an interesting attempt of the late Marquis of Bute to acclimatize the Beaver (*Castor canadensis*) in Scotland, although strictly speaking the experiment has little more bearing on the Scottish fauna than the placing of animals in the Scottish Zoological Park, since the creatures were kept in an enclosure. In the three or four acres of this enclosure, situated in the woods west of Kilchattan Bay in Bute, the Beavers lived happily and prospered so that the four imported (from Canada) in 1874 had increased to sixteen in 1878. They felled trees, built dams across the stream which ran through their enclosure, erected a lodge and seemed to have thoroughly settled down. Yet in 1890, less than twenty years after their establishment, the Beaver had died out in Bute.

A similar result ended a similar experiment begun about 1868 near Wangford in Suffolk; but there the Beavers were cast derelict by man and had they not been trapped or killed, seemed likely to thrive, for they bred and colonized new localities.

TRANSPORTED MOLLUSCA

To what motive do we owe the spread of the tiny Mollusca that have been placed in artificial ponds and canals? —not to the love of beauty, for they are insignificant in appearance, but probably merely to the desire of the collector to have in his neighbourhood a creature in which he took an interest. In any case his activities in this direction have not had much significance. Some Pond Snails, *Limnæa stagnalis* and *L. glabra*, have been added to the fauna of Scotland, the former occurring, for example, at Possil Marsh, and in a dam at Greenock where it was introduced in 1882, the latter in Frankfield Loch, where it was placed before 1876. The little *Sphærium lacustre*, once absent from Scotland, now occurs through man's agency in the Forth and Clyde Canal and in Possil Marsh near Glasgow.

"ESCAPES"

Occasionally it happens that animals deliberately brought into the country to be kept as pets regain their freedom and for a time rove the country as though they were of its denizens. In a few cases, hardy creatures such as Red and Grey Squirrels have succeeded in establishing themselves, but as a rule, "escapes" leave no trace upon the native fauna. The Smooth or Ringed Snake (*Tropidonotus natrix*) has been found at large many times, in the suburbs of Edinburgh, in the neighbourhood of Paisley, in the woods of Carluke, Lanarkshire, but there is no reason to believe that the apparently native specimens were other than escapes, or that, in spite of its opportunities, the Ringed Snake has ever become established with us.

So also it has been with the favourites of the bird-fancier, the Baltimore Oriole, the Canary, the White-throated Sparrow, all of which having at one time or another escaped from confinement have spent a short life, if not a merry one, in freedom in Scottish localities. The escape from ornamental ponds of Ducks, such as the Muscovy and Summer, of Geese, as the Spurwinged, Chinese and Canada, of Demoiselle and Crowned Cranes, and many others, also frequently occurs, but, except for the Canada Goose, established at Loch Leven, Loch Lomond and elsewhere, the venturers have never succeeded in making good a footing.

A NEW MOTIVE

Man has caused many an unpremeditated injury through his rash and thoughtless introduction of new animals, but it has been left to a Falconer of Shetland to furnish the solitary example of deliberate ill-intent.

"There are no Weasels in all the northern Isles of Zetland, as I am informed," wrote Brand in his careless style, "tho' numerous in the mainland, which they report thus came to pass: The Falconer having a Power given him to get a Hen out of every House, once in the Year; but one Year they refusing or not being so willing to give, The Falconer out of Revenge, brought the next year two Weasels with him, which did generate and spread, so that now they are become very destructive to several goods of the Inhabitants, whereof a Gentleman our Informer told us he had killed several half an Ell long."

This importation of the seventeenth century, the Stoat and not the Weasel as Brand alleges, is said still to be represented in the Shetlands, although in the Orkneys where it was likewise dumped, it has disappeared.

Whether the story be true or not, the deliberate result is on a par with the results of many thoughtless introductions of

300 DELIBERATE INTRODUCTION OF NEW ANIMALS

more significance—the Rabbit and Squirrel in Scotland, the Rabbit, Weasel and Fox in Australia and New Zealand, the Sparrow and Starling in many parts of the world—pointing a warning finger at the naturalist and reformer who, by introducing animals would revise Nature's order, and by short cuts and unimaginative experiments tends to make a wilderness where he had looked for a paradise.

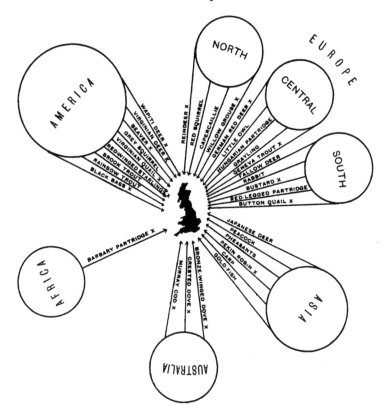

Fig. 54. Some Deliberate Introductions to Great Britain; shown diagrammatically associated with the country of their origin. Introductions which have died out or have failed are indicated by an ×.

PART II

MAN'S INDIRECT INTERFERENCE WITH ANIMAL LIFE

PART II

MAN'S INDIRECT INTERFERENCE WITH ANIMAL LIFE

> Nothing is foreign; parts relate to whole;
> One all-extending, all-preserving soul
> Connects each being, greatest with the least;
> Made beast in aid of Man, and Man of beast;
> All serv'd, all serving: nothing stands alone;
> The chain holds on, and where it ends unknown.
>
> POPE.

FAR beyond the bounds of his original intention man's influence upon Nature has extended, till scarcely a "Beast, bird, fish, insect, what no eye can see, No glass can reach" but has felt the touch of his power. Indeed, balancing the effects of his direct influence against the indirect or secondary results of his interference, one is forced to the conclusion that the latter have been more far-reaching and ultimately more effective in altering the aspect of the countries and of the faunas he has invaded. There is no action in nature but has its equal and opposite reaction.

To some of these indirect results passing reference has already been made. The introduction of new animals, and especially of domestic stock, has entirely altered the face of Scotland, for it has led to the disappearance of forests, as well as to the lowering of the upper limit of forest growth and to a transformation in ground vegetation. Many native animals have been reduced in numbers or banished from the country, and others have been encouraged to multiply and spread.

The following chapters, making closer scrutiny of this aspect of man's influence, endeavour to trace the tendencies and extent of the three types of interference with Nature, which, rebounding upon animal life, have made the greatest impress upon the composition and distribution of the fauna of to-day—the destruction of the forests, the cultivation of the soil and amenities of civilization, and the development of commerce with foreign lands. A few additional illustrations of the unforeseen effects of the coming of man will be found in the concluding chapter of this volume.

CHAPTER VI

THE DESTRUCTION OF THE FOREST

"The broad woodland parcelled into farms";
There
"Lies the hawk's cast, the mole has made his run,
The hedgehog underneath the plantain bores,
The rabbit fondles his own harmless face,
The slow-worm creeps and the thin weasel there
Follows the mouse, and all is open field."
                                              TENNYSON.

ONE inevitable result has followed in the wake of civilization wherever man, the transformer, has penetrated—the disappearance of woodland. It was not so in the days of primitive cultures. The great forest tracts of America, which covered the whole country except in prairie regions, survived without significant change during the undisturbed occupation of the simple Indian races. It is true that the Indian hunter made his clearings in the forest, and tended his narrow fields, but so soon as he moved to new regions in pursuit of game, Nature resumed control, and in a generation the old clearing was obliterated by fresh growths of scrub and timber. "In the husbandry of Nature" says Marsh, "there are no fallows."

It must have been thus in the days of the early occupation of Scotland, when Nature ruled the wilds; but, as his power increased, man gathered the reins from Nature's hands and drove his chariot of progress through the heart of the forest, leaving scars which time could no longer heal. In America, under the rule of white men, the native forests are fast disappearing; in Scotland they have all but gone.

VI. 1

SCOTTISH FORESTS IN TIMES PAST

Long before man made his appearance in the country, Scottish forests had their vicissitudes. Even after the settlements of the early Neolithic peoples had been made, Nature still retained a firm grip of her own, and her strong hand can be traced in the successive rise and disappearance of great forest tracts, such as we can scarcely imagine at the present day. A glance at these changes will supply a perspective from which later happenings can be the better gauged.

THE LOWER FOREST OF THE PEAT

No records remain to tell of the strange fluctuations of Scottish climate and vegetation, except the scant stores of insignificant debris hidden in the bottoms of ancient lakes or in the depths of peat-bogs; but to the investigations of a small band of workers, James Bennie, Clement Reid, and especially Francis J. Lewis, these hidden treasures have revealed a marvellous tale of change.

After the great period of the Ice Age had gone, Arctic plants spread from the hills to the level of the sea, but in the western and northern islands these glacial immigrants from the continent had scarcely obtained a foothold before a warmer and drier climate ushered in the Lower Forest of the Peat. Trees of Birch, Hazel and Alder found foothold upon the valley floors, and gradually climbing the hillsides, spread in a vast forest, here more dense, there more sparse, extending from the lowland valleys to an altitude approximating to that of our modern woods, close on 2000 feet. Even to the Hebrides and the Shetland Isles, where now no trees can grow, the Lower Forest spread. In the times of the Lower Forest of the Peat, man had not yet penetrated to northern Britain.

THE UPPER FOREST OF THE PEAT

On the heels of the warm and dry period, during which the woods of Birch, Hazel and Alder flourished, came a second period of moist cold when glaciers again appeared in the highland valleys and Arctic plants crept downwards into the

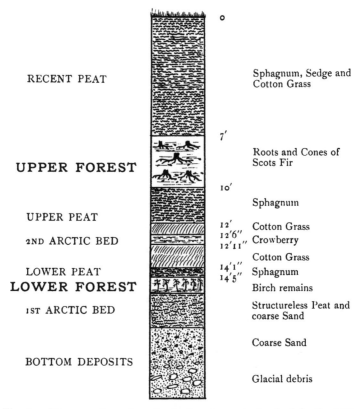

Fig. 55. Diagrammatic section of upland peat-moss in Lowland Scotland (Merrick and Kells District) showing relationship of former plant deposits and of the Lower and Upper Forests, the latter contemporaneous with man. (Founded on investigations of Mr Francis J. Lewis.)

lowland plains. In many areas, the Lower Forest, hampered and swamped by the development of boggy pools and sphagnum moss about its roots, decayed and fell, and was buried by great stretches of morass and peat moss. It is probable that man arrived in Scotland during the inhospitable conditions of these times.

SCOTTISH FORESTS IN TIMES PAST

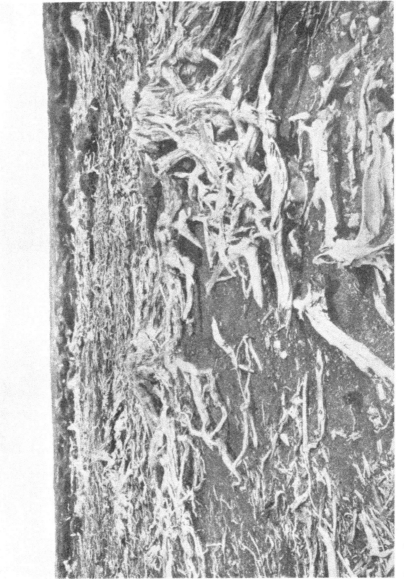

Fig. 56. Remains of the Upper Forest of the Peat—roots of Fir trees, laid bare by the wastage of the peat in which they had been preserved. 1 mile S.E. of Daless, Findhorn Valley, Nairn. Photo by Geological Survey.

But such conditions too eventually gave way, and for a second time a drier climate ushered in a new forest growth, the greatest forest that Scotland, as we know it, has ever seen. Yet the new forest was different from the old. It spread from the lowlands of Wigtownshire to the north of Sutherland, but nowhere does it appear to have extended to the islands of the Hebrides or to Shetland. On the mainland, however, it reached far beyond the limits of the woodland of to-day, clothing the mountains with its verdure to a height of over 3000 feet, a good thousand feet above our highest present day birch woods. Of equal interest is the fact that the Upper Forest of the Peat was in the main composed of a tree which has come to be particularly associated with Scotland, the Scots Fir (*Pinus sylvestris*), which now for the first time invaded northern Britain.

It is no simple matter to compare forests of some eight thousand years ago with those of to-day, but a rough calculation, checked by reference to the conditions of a land so little influenced by cultivation as Scandinavia, indicates that the Upper Forest covered an area at least ten times as great as that of our modern woods, that where they cover a meagre twenty-fifth to twentieth of the land surface, it overspread half the country (see Fig. 86, p. 484).

In this great forest, the settlers of the New Stone Age hunted the Reindeer and the Elk, the Red Deer, the Roe, and the great Urus; in it they defended their flocks from the ravages of the Brown Bear, the Wolf and the Northern Lynx, and captured for the sake of their flesh and skins the Badger, the Otter and the Beaver.

NATURE OR MAN

Great as has been the destruction wrought by man in our woodlands, the series of natural changes revealed in the layers of our peat mosses should warn us lest we lay too much to the charge of human interference. Man was well established in Scotland at the period of the greatest development of forest growth, and from that time till now has been in continuous occupation, yet it would be a mistake to imagine that the reduction of wooded area was entirely due to his efforts. The peat mosses have another revelation to make.

The swing of the climatic pendulum brought again a period of wetter conditions during which the main masses of "recent" peat were formed. Physical conditions favoured the extension of peat and moorland areas at the expense of forest, for the humic acid of the peat bogs acts as a destroyer of trees, and winds favour low growing moorland vegetation rather than saplings exposed to the full fury of the blast. As peat and moorland spread, woodland decayed, and the destructive influence of man was for long aided and abetted by the processes of Nature.

At the present day the moist conditions of the time of the "recent" peat formations have somewhat ameliorated, for in many places, peat, instead of forming, is itself decaying and being washed away by the weather. So that had not man stepped in to interfere, forest areas would probably now have been undergoing a natural extension.

EARLY HISTORICAL FORESTS

Between the records of the peat-bogs and the annals of history there is a chasm difficult to bridge. At what period of human development the great Upper Forest of the Peat began to decay under the excessive moisture of later times, we do not know. The forest had certainly been replaced to some extent by swamp before the opening of the Christian era, and the Roman historians are at one in describing Caledonia as a land of clouds and rain, of bogs and morasses. Yet they are equally emphatic as regards the great extent of wood and the variety of trees which clothed the land. Other evidence, such as that deduced from the distribution of primitive iron-smelting furnaces, points to the occurrence of great tracts of forest in late prehistoric times and in places whence woodland has almost or altogether disappeared.

The earliest references to Scottish woodland are, however, of comparatively late date, and show that in the twelfth century the necessity had already arisen of conserving forest growth by laws and penalties. The *Leges Forestarum*, generally ascribed to the reign of William the Lion (1165–1214), though by many held to be of later date, prohibit the taking of fire or of domestic animals into the woods, as well as the cutting of oak trees. Some of the penalties are curious: if fire, a horse, or a dog were brought to the wood, eight cows were to be forfeited, while if goats were found there at large, one was to be hung up by the horns in a tree. Yet in the twelfth century Caithness still possessed the woods it now lacks, for the Orkneyinga Saga relates how at that time the Yarls Harald and Rögnvald were accustomed nearly every summer to fare to Caithness, and there to go up into the woods and wastes to hunt.

From this time onwards Scots laws and charters make frequent reference to forests throughout the country, but it is necessary to bear in mind in interpreting the significance of these references that their "forests" were not our forests. The old "forest" was an area given over to hunting, and is perhaps best defined by the negative statement that it contained no arable land. It bore some resemblance to the "deer forests" of the present day: it did indeed contain woodland and covert for the shelter of its wild inhabitants, but it also contained open areas of browsing pasture or "vert," as it was termed. The "forest" of Scots law must not be regarded as an area entirely covered with dense woods. Thus when we read of a vast forest in southern Scotland extending from Chillingham to Hamilton, a distance of about eighty miles as the crow flies, and including the famous Ettrick Forest and many others, we may imagine no more than a stretch of wild and desolate country given over to wild beasts, and containing many extensive woods, but containing also stretches of barren moor and lush meadow.

An idea of the extent of woodland in the early days of the consolidation of Scotland may be gained by reference to two widely separated areas selected almost at random.

In Dumfriesshire

The names of many places in Dumfriesshire—Mouswald, Thortorwald, Tinwald—clearly indicate the former presence of wald or wood, where none now exists. In a Charter of Alexander II (1214–1249) Richard de Bancori quits claims to his lord, Robert de Brus, of the woods of Musfaud or Mouswald; and substantial tradition asserts that a great oak forest spread from that place, by Thortorwald to Tinwald, so dense that a man could have traversed the distance from tree to tree without putting foot to ground. The extent of this wood probably exceeded twenty square miles.

IN ABERDEENSHIRE

Or take the case of Aberdeenshire. Here there were no fewer than eight forests, Dyce, Drumoak, Birse, Mar, Stocket, Kintore, Buchan and Bennachie. The last was included in grants by the Earl of Mar to Sir Robert de Erskine (1358)—"cum pastura in foresta de Benechkey [Bennachie]." In 1324 King Robert the Bruce granted lands in Buchan "cum nova foresta" and the forest of Kintore to Robert Keith, Marischal of the Kingdom. But the most interesting of the Aberdeenshire forests, in view of its history and early disappearance, was that of Stocket, wherein, in 1010, Malcolm II came near to losing his life by the attack of a wolf. In 1319 this forest was granted by Robert the Bruce to the burgh of Aberdeen, for an annual *reddenda* of £213. 6s. 8d. the King reserving the right of "vert" and hunting therein. The Provost of Aberdeen for the time being, was Keeper of the Forest of Stocket, which extended to some thirty square miles and was ultimately held by Aberdeen in free Burgage of the Crown at a nominal rent. Under the town's auspices this ancient forest, which lay just outside the bounds of modern Aberdeen, was soon "improved and alienated from its original purpose."

IN THE FIFTEENTH CENTURY AND AFTER

After the period referred to in these charters it is evident that the woods of Scotland suffered rapid destruction at the hands of man, for law after law was passed aiming at the preservation of the old timber and the planting of new. In 1424

it is ordained, that...them that be night steelis greene woodde, or pealis the bark off trees, destroiand wooddes...sall paie fourtie shillings to the King for the unlaw and assyith the partie skaithed;

and in the middle of the century, freeholders, temporal and spiritual, were commanded to order their tenants to plant woods and trees, to make hedges and sow broom in convenient places and according to the extent of their holdings. These regulations seem to have done little to stay the destruction. Parliament after Parliament passed similar enactments, putting the old acts "in sharp execution," increasing the obligation of landholders on the plea that the woodes of Scotland being utterlie destroyed....everlik Lord and Laird make them to...plant at the least, ane aiker of Woode, quhair there is na great Wooddes nor Forrestes;

and increasing, also, the penalties for disobedience, until in 1579

quhat-sumever person stealis, pealis, and destroyis green-wood, pullis or cuttis haned Broome...in case the offendours be not responsall in gudes,... for the first fault be put in the stokkes, prison or irones auct [eight] dayes on bread and water; And for the second fault, fifteene dayes....And for the third fault hanging to the death.

Yet in this sixteenth century, in spite of the decadence the law records, Scotland was by no means a bare country. Bishop Leslie, who wrote from first hand knowledge, refers again and again to fair woodland;

heir agane sall ye se...a dry knowe, or a thin forrest, thair a thick wodd, all meruellouse delectable to the eye, throuch the varietie baith of thair situatione, and of the thing selfe that thair growis;...Paslay quhilke is situat amang cnowis, grene woodis, schawis[1] and forrest fair onn the River of Carronn;...Vuir[2] Clydisdale or Cludisdale...as lykwyse nathir[3] Cludisdale, amang fair forrests and schawis schene; with thicker woodes sum are decored[4].

And again

From thir cuntreyes that wyde and ample forrest, called the Tor Wod [*Caledonia Silva*], hes the beginning; quhais boundis war sa large, that frome the Callendar and Caldir Wod evin to Lochquhaber war extendet.

Leslie's testimony notwithstanding, visitors to Scotland like Aeneas Sylvius and Fynes Moryson were struck by the bareness of the country. In Fife, says the latter, "trees are so scarce that I remember not to have seen one wood," and "one of the senseless gibes of that splenetic southron," as Professor Hume Brown dubbed Sir Anthony Weldon, was that (in 1617) Judas could not have found a tree in Scotland whereon to hang himself.

The truth seems to lie in this, that in the populous and cultivated areas to which the foreign visitors naturally paid most attention, long usage had seriously reduced the woodland, but that in the ruder regions forests were still to be found in plenty. Surely there is transparent ignorance of the north country in Andrew Boorde's letter of 1536 to Thomas Cromwell, where he says of Scotland:

---

[1] Thickets. [2] Over or Upper. [3] Lower. [4] Adorned.

The part next England is the hart and best of the realme; theirin is plenty of fyshe and flesh, and snell ale, except Leth [Leith] ale,...the other parte of Scotland is a baryn and a waste country full of mores.

And it seems to me that the other travellers were equally in error in generalizing from impressions gained in the centres of population and industry. In proof of the great ignorance of the Highlands which prevailed even amongst the elect of Scotland I need do no more than cite the joy of the Parliament sitting in Edinburgh in 1609, at "*the discovery* of woods in the Highlands which, by reason of the savageness of the inhabitants, had been unknown, or at least unused."

It is fair to assume, therefore, that the plaints of Parliament "how Woods, Parks, and all sort of Planting and haning decayes within this Realme" (1607), refer mainly to the familiar country, roughly to the south of the line of the Forth and Clyde estuaries.

Here certainly the nation was in straits for timber. Ettrick Forest, which actually gave the name "Forrestshire" to a great stretch of country "south from Tweeddale," was "dissolved from the Crown to be set in few¹" in 1587; and King James VI, when he made the wise proposal to his Privy Council that, in order to conserve timber for home uses, exports of Scottish timber should be totally prohibited, had to be reminded that within the memory of man no timber had ever been exported from Scotland.

A generous reading between the lines of the law, seems to me to indicate that the drastic steps taken by the Parliaments of the fifteenth, sixteenth and seventeenth centuries were due, not so much to the absolute lack of woods, as to the great and increasing importance of timber on account of the national uses to which it was put, and to the difficulty of finding sufficient supplies to meet the constant demand. Statutes of the latter half of the seventeenth century, mention by name at least thirty different forests in Scotland; and it is a pleasure to turn to Gilpin's cheerful account of Scottish forest scenery at the end of the eighteenth century, and to his appreciation of the natural pine and birch woods of the north, of the oak mixing "its cheerful verdure with the dark green tint of the pine" in the county of Stirling, of the "great quantity of very fine timber" in the woods of

¹ Feu.

VI. 2

CAUSES OF THE DESTRUCTION OF THE FOREST

How came it that the fine forests of Scotland decayed so completely? Nature, we have seen, with changing climes and conditions played her part, but man must bear the heavier share of the guilt of destruction. A short study of the immediate causes of the destruction of the forest will show how deeply his ways of life and the advances of his civilization bit into the woodland.

THE NEEDS OF THE HOUSEHOLD—Fuel

In point of time the earliest uses to which the woods were put were those of the household, and of these one of the first was fire. Excavations of prehistoric settlements almost invariably reveal the presence of hearth-stones blackened and burnt by fire, for fire was essential even where a cave sufficed for a habitation. It is true that dried seaweeds were occasionally used for fuel, and that the dung of cattle continued to be so used even to comparatively recent times in the outlying islands, but the staple fuel was the brushwood and timber of the forest. The destruction of timber on this account must have been great, for wood continued to be the chief fuel through many thousands of years, since, although peat and turves were used to some extent in King David's reign in the twelfth century, the use of coal, first mentioned in a charter of 1291, only became general under the compulsion of lack of firewood.

Even when Continental nations still had wood in abundance, the people of Scotland were compelled to burn these curious black stones—to the amazement of Pope Pius II, who, following upon a visit to James I in the fifteenth century, wrote

In this country [of Scotland] I saw the poor, who almost in a state of nakedness begged at the church doors, depart with joy on their faces on receiving stones as alms. This stone, whether by reason of sulphurous or some fatter matter which it contains, is burned instead of wood, of which the country is destitute.

Indeed the mining of coal in Britain seems to have been almost confined to Scotland in the sixteenth century; for according to Hector Boece:

In Fyffe ar won blak stanis, quhilk hes sa intollerable heit, quhen thay ar kindillit, that they resolve and meltis irne, and ar thairfore richt proffitable for operation of smithis. This kind of blak stanis ar won in na part of Albion bot allanerlie betwix Tay and Tyne.

Housebuilding

Second only to the use of wood for fire, was the use of timber in the making of shelters for the comfort of human occupants. In one of the earliest known sites of occupation in Scotland, the Azilian settlement of Oronsay, Mr Henderson Bishop found, deeply sunk in the sand, holes which could only have contained strong upright posts for the support of a superficial structure; and the investigations of Dr J. N. Marshall and Mr Ludovic Mann at the much later vitrified fort of Dunagoil in Bute leave no room for doubt that the dwellers there inhabited well-built huts roofed and perhaps walled with wattle plastered over with clay. Mr A. O. Curle has found similar evidence of the use of wattle huts at Traprain in Haddingtonshire. As the centuries passed and men congregated in towns, the call for timber for the erection of houses increased manifold. Even in 1666, at the time of the Great Fire, London was largely built of wood, and timbered houses were characteristic of Edinburgh at a still later date. So great was the demand for building oak, that large supplies had to be imported from abroad, and when, in the reign of Queen Mary, Denmark prohibited the sale of oak to Scottish traders, the embargo, according to Professor Hume Brown, threatened to put an end to housebuilding in Scotland. The influence of timber-built towns and villages upon Scottish forests can well be imagined—one example of its actual working will suffice to point the argument. In 1792 the Rev. Thomas White, in an Account of the Parish of Liberton, wrote:

The Burrow Moor [of Edinburgh] where the scenes just mentioned [the scrimmage or "Battle" of Lousie Law in 1571] happened, is at present well cultivated, and of much value. At that time it was full of aged oaks: And it is observed, that the timber of which all the wooden houses in Edinburgh were built was taken from thence.

In gaining her towns, Scotland lost her forests.

INCIDENTS OF CONQUEST

In the old days, a necessary corollary to the conquest of a wooded country, where the country was to be held by the invading force, was the destruction of woodland, in order that the progress of the invaders should be eased and that no lurking places should remain, whence remnants of the defending army might sally forth upon the flanks of their enemy. The earliest systematic conquest of Scotland recorded in history is that of the Romans in the early centuries of our era; and tradition speaks strongly to the destruction wrought by the Roman legions in the forests of England and of southern Scotland. Nor is the tradition without some show of archaeological support. So long ago as 1701, it was noted of the buried forest of Hadfield Moss in Yorkshire that

> many of those trees of all sorts have been burnt, but especially the Pitch or Firr trees, some quite through, and some all on a side; some have been chopped and squared, some bored through, otherwise half riven with great wooden wedges and stones in them, and broken ax-heads;

and again, in Lincolnshire, under hills of sand, were discovered "roots of great Firrs or Pitch trees, with the impresses of the ax as fresh upon them as if they had been cut down a few weeks." In both cases, coins of the Roman emperors were found associated with the ruined forests. The evidence, if not demonstrative, is suggestive; and there can be little doubt that the Romans would have adopted so obvious a way of ensuring the fruits of their conquest, as the levelling of great tracts of forest, especially in the dangerous areas on the newly won frontiers of the Empire.

In much later days invasions followed a similar course. In describing John of Gaunt's invasion of Scotland in 1380, the English historian, Knighton, says that at one and the same time, it was possible to hear the sound of 80,000 axes felling the timber of the woods, and that the timber so cut was given as fodder to the fire.

Whether this be exaggeration or not, it at any rate indicates grievous destruction, typical of the progress of invading armies.

TRAVELLING AND THE MERRYMEN OF THE WOODS

Through long ages the northern parts of Scotland, that is to say the country north of the line of the Forth and Clyde, were considered by the southron to be dangerous and savage places, entered with one's life in one's hand. To facilitate travelling in these uncouth areas, roads had to be made and the woodland to be cleared. During such a process, it was the custom, based upon expediency, to destroy the trees and undergrowth for some distance on both sides of the road, in order that a clear view might be obtained and that no shelter might be left wherein a lurking highwayman could be concealed. But the desire for safe travelling led to further destruction, for bands of robbers sought shelter in the thick woods, whence they could sally upon a slow convoy, and whither they could retire again in safety with their booty. "Upon the shore of Lochebrune" wrote Monro in 1549, "lyes Ellan Ew, haffe myle in length, full of woods, guid for thieves to wait upon uther mens gaire[1]." And again, "Northwarte frae this ile lyes the ile of Graynorde, maire nore ane myle lange, full of wood, guid for fostering of thieves and rebellis." So grievous became the assaults of such outlaws upon the persons and purses of the lieges, and upon the preserved game of the forest as well, that only by the destruction of their haunts could the land be rid of them. So Hector Boece records in the sixteenth century: "The regioun [of "Fyffe"] is now bair of woddis, for the thevis war sumtime sa frequent in the samin, that thay micht na way be dantit[2], quhill[3] the woddis was bet[4] doon[5]."

THE WOLVES

The forests had more legitimate, but not less troublesome tenants in the hordes of wolves they harboured. A prayer in the old Litany of Dunkeld runs, "From caterans

---

[1] Gear, goods. [2] Daunted, defeated.
[3] Until. [4] Beaten.
[5] Cf. quotation on p. 319, each in its exaggeration contains a significant fact.

and robbers, from wolves and all wild beasts, Lord deliver us." I have already described the wolf plague and the stages of the downfall of Scottish wolves (p. 115); it is sufficient here to call to mind that their destruction involved that of many woods. In the districts of Rannoch and Blair Athole in Perthshire, of Lochaber in Inverness-shire, in the region about Loch Awe in Argyllshire, and in other places as well, local tradition or more definite record asserts that extensive forests were burned down to extirpate the wolves, by demolishing the retreats in which they found refuge.

INDUSTRIES AND WOODLAND

A heavy and long-continued drain upon the native forests was made by the demand for wood for industrial purposes. For long Scotland had held a foremost place amongst the peoples who go down to the sea in ships, and the erection of boats and ships of war was a constant tax upon well grown timber. Of one of the latter, the famous "Great Michael" built in 1511 at Newhaven near Leith, it was written that James IV

buildit the 'Michael,' ane verrie monstruous great ship, whilk tuik sae meikle timber that schee waisted all the woodis in Fyfe, except Falkland Wood, besides the timber that cam out of Norroway.

Even so the destruction due to boat-building was as nothing to that wrought by the industries in which wood was used for fuel. The manufacture of salt from sea-water was for long a national industry, practised on so extensive a scale along the Firth of Forth, that, as one visitor said, "the works are not easily to [be] numbered"; and, although at the industry's greatest development, coal was the chief fuel used, in earlier stages the use of wood fires for evaporating the brine must have made serious inroads upon the forests on various parts of the coast. But no other work of man played such havoc with the woodland as the ancient iron industry of Scotland.

From the days of the Iron Age, a thousand years and more before the opening of the Christian era, till the early years of the nineteenth century, the reduction of iron ore to a state fit for casting, was brought about by a process of fusion with charcoal. Even in the case of a single furnace, as we shall see, this entailed vast destruction of timber. But it was no single furnace that met the needs of the country. Remains of old iron workings, heaps of slag and charcoal, show that throughout Scotland, even on the wild wastes of the Highlands, the manufacture of iron was practised.

Of well defined traces of such workings, still recognizable thirty years ago, Prof. W. Ivison Macadam recorded no less than ninety-eight, scattered throughout thirteen counties of Scotland, from Sutherland to Dumfries; and certainly as many more have gone unrecorded or have disappeared with the passing of time and the disturbance of the soil by cultivation.

The number and distribution of the iron workings are of no small interest as showing the extent of the industry, and also as indicating the distribution of woodland in Scotland in past times (see Map IV). For it was an axiom of the smelters that since iron was more compact and portable than timber, it should be carried to the places where wood for its reduction grew thickest. The effect upon Scottish forests of the continuous manufacture of charcoal on a large scale, through some three thousand years, can better be imagined than described. Yet the traces left by time and the hints of the law and other records are sufficiently striking.

At Esmore, in Argyllshire, charcoal ash still covers several acres of ground. At Letterewe in Ross-shire, English miners wrought a mine, casting cannon and other implements "untill," as the Letterfearn manuscript quaintly says "the woods of it was spent and the lease of it expired" in the early years of the seventeenth century. In England in 1556 Queen Elizabeth prohibited iron-smelting in Sussex, on account of the amount of wood which was being felled for the casting of cannon. The result was to stimulate the industry of cannon-casting in other counties; so that eventually there also, the felling of trees had to be prohibited, as was the case in the Ulverston district of Lancashire in 1563, when there was passed "A Decree for the abolishing of Bloomeries [as the slag-furnaces were called] in High Furnes."

Unfortunately these restrictions caused the iron smelters to migrate to Scotland, to the ruin of the Scottish forests. The effect was apparent in an Act of the Scottish Parliament of

DISTRIBUTION OF SCOTTISH BLOOMERIES AND SLAG FURNACES.
The chief areas of distribution are shown, though many individual bloomeries have been omitted on map.

1609, "Anent the making of Yrne with Wode," in which it is stated that

being informit that some personis vpoun advantage of the present generall obedience in those partis (the heylandis) wald erect yrne milnis in the same pairtis, to the vtter waisting and consumeing of the saidis wodes....Thairfore...commandis, chairgeis, and inhibitis all and sindrie his maiesties leigis and subjectis that nane of thame presome nor tak vpoun hand to woork and mak ony irne with wod or tymmer under the pane of confiscatioun of the haill yrne...

The Scots Act notwithstanding, the smelting of "yrne with wode" and the "waisting and consumeing of the wodes" went gaily on. Even in the eighteenth century, when, after the rebellion of 1715, the military occupation of the Highlands had shown a way to fresh forests, many new slag-furnaces were erected, to which ore was brought from England.

"Thus," wrote Professor Macadam in 1887, "the following furnaces sprang into existence:—Bunawe or Taynuilt in 1730; Invergarry in the same year; Abernethy in Strathspey also in 1730; Furnace in 1750; Goatfield, Loch Fyne, in 1754; and Carron in 1760. These large works soon consumed the wood, and Invergarry and Abernethy soon ceased to be worked. Carron, having changed its fuel to coal, still exists; Goatfield and Bunawe are only a few years blown out; and there is not now a single ironwork in Scotland using charcoal as fuel, and only two remain in England."

Incidentally the distribution of these iron furnaces indicates roughly where the greatest extent of easily accessible forest still existed about the middle of the eighteenth century.

The destruction wrought by these later and larger furnaces was irreparable. In 1728, 60,000 trees were purchased for £7000 from the Strathspey forest of Sir James Grant. The trees were intended as masts to be used for the navy—an index to the effect on woodland of our "wooden walls"—but, proving too small, they were used for charring at the iron furnace at Abernethy. About 1786, the Duke of Gordon sold his Glenmore Forest to an English company for £10,000; and the Rothiemurchus forest for many years yielded large returns to its proprietor, the profit being sometimes above £20,000 in one year. That all the timber of these fine forests was used for charring, I do not imagine, but destruction on no small scale is certainly indicated by the accounts. As forests in the near neighbourhood of the bloomeries became exhausted, destruction spread in wider and wider circles, and woods more and more remote fell under the axe. At one period the Carron Iron-works Company at Falkirk in Stirlingshire, purchased for £900 a wood in Glenmoriston, on the north side of Loch Ness, in spite of the fact that it was distant eight miles of very bad road from the waterway of Loch Ness, and that the timber had to be carried thence to Carron. The Coalecken furnace on Loch Fyne turned out every year 700 tons of pig iron, and, as Macadam relates in 1887,

> the older inhabitants in the district still remember seeing the string of from 30 to 40 ponies laden with charcoal coming over the hill road from Lochawe, which is distant about 10 miles. The material was contained in bags which were placed on a large cradle saddle to protect the sides of the animal. The work ceased in 1813.

At first the furnace was taken to the forest, latterly the forest had to be taken to the furnace.

During the past century, the use of wood for smelting gradually ceased, but not until the discovery of suitable woods and the longer transport of the timber became matters of insuperable difficulty, and not before many "fair forrests and schawis schene" throughout Scotland had bowed their heads to the woodsman's axe.

ACRICULTURE AND THE FOREST

It is evident that the progress of civilization as embodied in the domestication of animals and the development of agriculture has been gained in great part at the expense of the virgin forest. The process is an old one, extending from the times when prehistoric man surrounded his settlements with clearings for his meagre fields, to the agricultural boom of the eighteen-sixties when, for example in Aberdeenshire, considerable areas of woodland were transformed into arable land. Even the pasturage of Sheep has had an enormous influence, for much of the "natural pasture" of the uplands has been gained from the original forest. There is little to show that the pastures of the southern uplands of Scotland were once covered with forest wherein the Red Deer and the Roe gave sport to kings. Yet the disappearance of the forest can be traced through the centuries keeping pace with the increase of the shepherd's flocks. Even in recent times, in the Highlands, forest has disappeared before the advance of sheep. The brothers, John Sobieski and Charles Edward Stuart, mention the occurrence of a great fire in Lord Lovat's deer forest of Glen Strath-Farar, "where twelve miles of pine, birch and oak woods, were burned in the tenantry of the late Eskedail to *improve* the sheep pasture."

In another way domestic animals have had an important effect on forest growth. Writing of Tunisia, Principal Perkins has said "In so far as young trees or shrubs are concerned, the passage of a flock of goats will do quite as much damage as a bush fire"; and in 1835 it was recorded that in the parishes of Urquhart and Glenmoriston in Inverness-shire, "goats were formerly numerous but have of late been discountenanced, as injurious to woods and plantations." In *The Origin of Species*, Darwin cites a telling example of the destructive effect of cattle-grazing upon the natural development of Scots fir at Farnham in Surrey. On our own highland moors in Inverness-shire, I have watched seedlings of birch spring up year after year, even to a hundred yards away from the parent trees, only to be destroyed by Sheep and Rabbits as regularly as they grew. In this way the natural spread of the woods is constantly kept in check, and the upper limit of our forest growth has been depressed even within the period of history.

The same process has also affected lowland woods. In primitive forests the decay and collapse of trees through old age simply make room for the growth of new seedlings; but in our own country such replacement has long been ruled out of count by the presence of Rabbits, and by the pasturing of domestic animals, which formerly roamed at large in the forest areas. By these, the seedlings as they grew were destroyed, so that the fall of each aged tree left a new and irreplaceable blank in the woods. It is conceivable that such a process, continued through the ages, may have accounted for the disappearance of not a few of our primitive woodlands. Old Scots law recognized the enmity between domestic herds and forest growth, for it ordained in 1686 that Cattle should be herded in winter and in summer for the protection of planting. Further it is on record that in recent centuries, as in the parish of Drummelzier in Tweeddale, individual woods have been destroyed by Sheep.

"ACTS OF GOD"

Nature's catastrophes, fire and wind, have shared in dooming the forest. A first thought suggests that the manifestations of elemental Nature should have no place in this account of man's influence; but a second thought will show that the greater part of their significance as destroyers, is due to his interference with the old rule. Man cuts the trees that act as buttresses upon the margin of the wood, and, laying bare to the fury of the blasts the unaccustomed growth within, opens a path for such a levelling of forest as Nature never dreamed of. As I write Scottish newspapers report the great damage done to growing timber by a gale which raged throughout the night of October 24th, 1917; and several correspondents mention that the damage was intensified by the fact that the woods had recently been cleared of their spruce and fir trees.

To man's invention of fire, many a conflagration is to be traced, but even where Nature's lightning is the spark that kindles, it is man's influence which makes of the charred tract a permanent desolation. One of the greatest forest fires on record was that of Miramichi in North America in 1825, when six thousand square miles of vegetation, chiefly woodland, were utterly consumed. Nevertheless left to herself, Nature speedily repaired the destruction; for in twenty-five years the blackened ground was covered with a natural and dense growth of trees of fair size, *except where cultivation and pasturage kept down the forest growth.* In a small area like Scotland where cultivation and pasturage are all but universal, no opportunity is left for Nature's gentle restoration.

There are evidences that both wind and fire have in former days played havoc with Scottish forests. A single example will illustrate the efficiency of each. In the parish of Coldstone in Aberdeenshire, there was discovered many years ago, buried in the peat, a pine forest which showed all the signs of having fallen before a hurricane. In describing this forest in the second volume of the *Edinburgh Journal of Geographical and Natural Science*, the Rev. J. Farquharson says that the trees were found buried ten to twelve feet deep in a moss which covered an area of about a hundred acres. Their numbers were such as to show that here a dense forest once stood. But the trees were levelled; some were broken off short across the trunk, leaving an upright stump, others were completely overthrown, roots and all. These were no accidents of the natural decay of a forest, for the trees lay uniformly in one direction, their heads towards the east. So a simple reading of the buried pine forest of Coldstone tells that a mighty west wind laid the woodland low, and at a period, as other evidences show, long after man had made his first settlements in Scotland.

As to the evidence of fire in Scottish woodland, I cannot do better than quote a comment of Sir Thomas Dick Lauder in his edition of Gilpin's *Remarks on Forest Scenery*:

The trees which are found in the Scottish mosses, and particularly the pines which are found in those of the northern parts of Scotland, all invariably exhibit marks of fire, as indeed do the stocks from which they have been severed and near to which they are always discovered. It is quite evident that these aboriginal forests of Scotland at least, have been destroyed by great conflagrations, kindled either purposely or accidentally, and perhaps in each of these ways at different times. Some of the pine logs are excavated longitudinally by the fire, so as to form spouts, such as are often supplied to the eves of the roofs of houses for catching and carrying off the rain. These appear to have been hollowed out by the fire, which had continued to burn and smoulder on the upper side, after the tree had fallen into some wettish place, the damp of which prevented its being consumed below. We have legendary accounts enough in Scotland of the burning of great tracts of forest to bear out the explanation of the appearance which these ligneous remains exhibit.

A definite example of an ancient burned wood was revealed in the neighbourhood of Tongue in Sutherland during the late Duke of Sutherland's endeavours to reclaim the district, when large trees, charred for 10 to 15 feet of their length, were discovered at a depth of 3 feet in the peat mosses. Some of the trees, Dr Harvie-Brown was informed, were 3 feet in diameter and "some of them are cored out with fire for several feet in length as if they had been burned down." Similar vast conflagrations, the brothers John Sobieski and Charles Edward Stuart relate, "afford frequent vivid similes in the old Gaelic poems."

FINAL RESULTS

These circumstantial accounts of man's inroads upon Scottish woods will, I trust, bring home more definitely than a general statement could have done, the vast influence which man has wielded upon the primeval forest. His use of timber for the building of houses and ships, for household fires and especially for fuel to carry on his great industries; his destruction of the woods to ensure his military conquests and his peaceful journeyings, to rid him of the annoyances of thieves and beasts of prey, and to make way for his smiling fields of grain; the influences of his flocks and herds and of his unwitting collaboration with the forces of nature; all these had but one effect—to demolish woodland and make a country, once rich in forest primeval as the wilds beyond the Atlantic, a land whose barrenness became the wonder and standing joke of southern travellers. The fact will stand repeating that woodland, which in the days of the Upper Forest of the Peat, clothed half the country, has been reduced to a miserable remnant scarcely covering one-twentieth of the land surface of Scotland. How far do we seem to have travelled, even since that day in the sixteenth century, when Bishop Leslie wrote of Scotland?

The woddes selfes nocht onlie proffitable to the utilitie of timber, and to that use,…are verie jocund and jellie[1] and gif we myt speik it, in a maner peirles in pleisour.

[1] French, *joli*, pretty.

VI. 3

EFFECTS OF THE DESTRUCTION OF THE FOREST UPON ANIMAL LIFE

It is a simpler matter to tabulate the changes which have taken place in animal life than to trace the actual influences which have brought these changes about. Most of the influences act indirectly and obscurely, beating against the habits of woodland creatures until the old habits are broken, and the creature reforms its ways to fit new conditions, or, since the reformation of an engrained habit is no easy thing, itself becomes broken in the process and succumbs.

PHYSICAL CHANGES WROUGHT BY DESTRUCTION OF THE FOREST

Forests keep a country moist, and tend to equalize the temperature, since they check air currents and evaporation. It is said that Canadian lumbermen can work at ease in the forest when the temperature is many degrees below zero, and when in the breezes of the open plain such a temperature would render work impossible. Professor Marsh thus summed up his study of the physical effects of the destruction of the North American forests:

When the forest is gone, the great reservoir of moisture stored up in its vegetable mould is evaporated, and returns only in deluges of rain to wash away the parched dust into which that mould has been converted. The well-wooded and humid hills are turned to ridges of dry rock, which encumbers the low grounds and chokes the watercourses with its debris, and—except in countries favored with an equable distribution of rain through the seasons, and a moderate and regular inclination of surface—the whole earth, unless rescued by human art from the physical degradation to which it tends, becomes an assemblage of bald mountains, of barren, turfless hills, and of swampy and malarious plains.

It is possible that Professor Marsh exaggerated the evils which follow disappearance of forest growth—says a critic, " Marsh found a fool in the forest, and the fool was man "—but the latest and most thorough investigation of the climatic significance of forest, a Report published by the Forestry Department of the Indian Government in 1917, still indicates that Marsh came near the truth. It has been found that in India, forest may slightly increase the rainfall, but only to an extent of not more than 5 per cent., by promoting condensation of aqueous vapour, by, as it were, cutting down the clouds, which Richard Jefferies fondly imagined lay beyond the reach of man's hands. Other important effects have been noted. The disappearance of forests in the catchment areas of some streams, in the Punjab, in Bengal and in Assam, has altered the flow of rivers, so that after the rains they now rise more rapidly and come down more torrentially. In the Punjab the exposure of the soil by the cutting of trees has caused great landslides, violent floods in the rivers, and the washing away of much of the cultivated soil.

IMMEDIATE RESULTS

Scotland has suffered in less degree. Nevertheless the more sudden and more serious flooding of the rivers after heavy rain has had accountable effect upon their inhabitants and those of the low lying valleys, drowning such creatures as Badgers often in great numbers—the two great floods of the Findhorn in 1829, say the brothers Stuart, drowned in their holes most of the Badgers in the lower banks—, washing into the stream and to destruction lesser things, fish-fry and invertebrate animals which sheltered in still water by the banks, and disturbing and dispersing the spawning beds and spawn of Trout and Salmon. Mr P. D. Malloch has stated that after a flood he has seen the sides of the Tay almost white with the eggs of Salmon, swept from the spawning beds and destroyed.

SCOTTISH FAUNA ORIGINALLY A FOREST FAUNA

But these are puny effects in relation to the fauna as a whole. The destruction of forest has told more heavily upon the inhabitants of the land, and this owing to the nature of the animal assemblage which the aboriginal woodland of Britain induced to migrate hither from the Continent, in the times succeeding the Ice Age.

At the present day the typical pine forest region, or

taiga, forms a band stretching across the temperate regions of the Old and New World, on the southern border of the barren-lands or tundra. There the creatures of the temperate forest can still be found in natural association. What animals compose the fauna which to-day selects the forest for its home? Most typical of its members are the deer—the Red Deer and the Roe, the great Moose or Elk, and the Woodland Reindeer or Caribou. These were once familiar denizens of Scottish woods. Its typical rodents are the Squirrel and Beaver; its beasts of prey the Brown Bear, the Lynx, the Wild Cat, the Wolf, the Fox; its lesser denizens, the Pine Marten, Polecat, Stoat and Weasel, and the Badger. These too inhabited Scotland. Of the birds of the pine forest, the most characteristic are the Crossbills, the Grouse and its relatives, the handsome Capercaillie and Black Game, and the Woodpeckers, Jays and Magpies; and these also we know in Scotland.

The evidences of the present day fauna, and the more significant relics of past life in Scotland, show with no uncertainty that the prehistoric fauna of North Britain was mainly a forest fauna, comparable with that which now dwells in the wild woodlands of northern Europe and America.

SOME GENERAL EFFECTS ON FAUNA

On such an assemblage of animals the destruction of the forest must have told with dire effect, ousting it from its natural habitation, limiting its range, and tending always to drive its members to extinction. On general grounds we can safely assume that the numbers of forest seed-eaters amongst birds—the Crossbill, the Bullfinch, the Siskin, the Redpoll and other finches—as well as of the insectivorous Woodpeckers and Tree Creeper, and their prey the pine-boring Beetles, must have diminished with the fall of the woods. And what of the land-shells, especially the smaller forms, of Helix, Pupa and Clausilia, which thrive particularly in wooded districts, and many of which need scarcely be looked for but in forests? They, too, must have dwindled in numbers. But of these things there is no direct evidence, so I turn to examine a few individual cases the history of which suggests more definitely the influence of disappearing woods.

SOME INDIVIDUAL EXAMPLES

It is natural that appeal should be made to the animals at present characteristic of the pine forests of temperate lands. What has the race of Deer to teach us?

It will simplify the enquiry to state, obvious though they are, the symptoms in the history of animals which betray the impress of environment. Well marked stages in animal life indicate prosperity or decline. In a progressively favourable environment an animal first increases in numbers, then spreads beyond its first bounds to new areas, and lastly may tend to develop new superfluities of structure. On the other hand in a progressively unfavourable environment an animal decreases in numbers, its range becomes more and more limited, its physique deteriorates, and finally it disappears.

How does the history of the race of Deer in Scotland respond to these criteria? There can be only one answer— it is a story of decreasing numbers, of curtailment of range, of dwindling physique, and of extinction. All the species of Deer have not shared equal disaster, but the cumulative effect of their histories is not less telling. Glance at their stories, as they have been recorded in the early deposits of the country and in history.

THE ROE DEER

The Roe Deer (*Capreolus capreolus*), the most lightsome and graceful, was once also amongst the most familiar of the denizens of Scottish woods. From north to south it roamed in freedom, not in Scotland only, but far into the woodlands of South Britain, for its bones have been found in Essex, Oxfordshire and in the southern corner of Wales. Civilization long since drove the native Roes from England.

In Scotland relics in river gravels, peat bogs and the settlements of man witness to the extent of the Roe Deer's range and the abundance of its numbers. About the time of man's first settlement in Scotland it spread to the southern borders: its remains have been found in the river-gravels of Berwickshire at Coldingham, and, in company with those of the Reindeer and Brown Bear, deep in the peat of Shaws in Dumfriesshire. In human settlements we do not expect to find traces of its presence in that extraordinary abundance which marks the Red Deer, for the Roe is a shy and superbly active creature, less likely to be caught, and its flesh is less palatable than that of its great relative. Nevertheless there are few kitchen-middens on the mainland that do not contain a bone or antler of the Roe Deer. In such deposits it has been found from the northern to the southern limits of Scotland, and the extreme cases are of special significance, since the "one or two bones" discovered at Hillswick reveal its presence even in the Shetland Isles, and its relics in the Cave of Borness in Kirkcudbrightshire, consisting of twenty-six bones, suggest that on the southern borders, it was as common as the Red Deer, represented by a like number of remains.

It occurs in the lake dwellings of Ayrshire, as at Lochlee, Kilmaurs, and Lochspouts; in the hill forts of Argyllshire, as at Loch Awe and on the Island of Luing, and of Inverness-shire, as at Craig Phaidrich; in the Roman settlement at Newstead; in the "Pictish Towers" or brochs of Thrumster and Kettleburn in Caithness, and of Cinn Trölla in Sutherlandshire; and in an underground "Pict's House" or "Eird House" at Edrom in Berwickshire. It is evident, therefore, that in prehistoric and early historic times down to about the ninth century the Roe Deer was common throughout Scotland.

When did it begin to forsake the southern country? There are many references to its protection in old Scots law, but they help us little as to its detailed distribution. It had its place amongst the list of slain in the great royal hunt in Athole in 1529 during the reign of James V; but this is beside the point for it has never deserted the fastnesses of the Perthshire Highlands. In the Lowlands it was still common in the thirteenth century, for in the reign of Alexander II (1214-1249), the Lords of Avenel, in granting their lands in Eskdale to the Abbey of Melrose, reserved the right to hunt, amongst other creatures, the Hart, the Hind and the Roe. But by the middle of the sixteenth century the disappearance of woodland, for use as timber and to make way for the pasturing of great flocks of sheep for which the Lowlands were already famous, led to the reduction of the Deer and to the narrowing of the bounds wherein they were wont to roam. Even the Statute of 1682 prohibiting the traffic in venison for seven years could not avert the evil day, and before the close of the seventeenth century Red Deer and shortly afterwards Roes were all but banished from the Lowlands of Scotland.

Henceforward, until their extension of range in recent times, they were confined to the highland fastnesses of Ross, Inverness, Argyll and Perth; for they had also been driven from their haunts in the northern counties of the mainland, and had long ceased to exist in Shetland.

Here clearly the destruction of forest resulted in reduced numbers and limitation of range. And as if to emphasize the point, the fresh impetus given to the planting of trees in the latter half of the eighteenth century, has led to a new increase in numbers and range, restoring the Roe Deer to many areas which it had long deserted. It made its appearance in the lower valley of the Tay before the close of the eighteenth century, and shortly afterwards appeared in Clackmannanshire and Stirlingshire, in Fifeshire (before 1828), and Linlithgowshire; and between 1840 and 1845, it had penetrated the Lowlands, for it then appeared on the southern slopes of the Pentland Hills near Penicuik.

It is difficult to decide whether the adverse influence due to the destruction of the woodland led to physical degeneracy in the Roe. Certain it is that the antlers recovered from the ancient deposits and the early sites of habitation are of unusually large size, but careful comparison led Mr J. G. Millais to the conclusion that the large antlers of former days could be equalled and even excelled by the best heads of the present day forests of Scotland. From our point of view the comparison is scarcely a fair one, for it is unlikely that the best heads of former days are those which chance buried in the bog and the luck of man recovered, or that primitive man caught the finest and most active animals for his meals. My own impression is that, on an average, in spite of the new life given to the Roe by the fresh development of woodland, the standard of the modern antler is less than that of prehistoric and early historic days; and the analogy of related deer in Scotland would support such a conclusion. While, therefore, degeneracy in the Roe must be regarded as not proven, it was a likely result of the adverse influences which the destruction of forest brought to bear upon its denizens.

THE RED DEER

The course of the Red Deer's story runs upon the same rough path of ill fortune as that of the Roe. But even more than the Roe, the Red has come under the influence of man, its size and the value of its venison having made it an object of more deliberate pursuit as of more strict protection.

In the days before man's arrival in Scotland the Red Deer roamed the country over. Strange to say the greater number of the prehistoric records of its presence come from an area which it has long deserted—the Lowland valleys. With these records it is impossible to deal individually, but a summary of the distribution of the Red Deer, as the natural deposits of the country have revealed it, will emphasize one significant phase of its history in Scotland.

In our oldest lake deposits—the marl clays—formed generally at a period not long after the close of the Ice Age, remains of Red Deer are abundant. Bones and antlers have been found in Roxburghshire, Peeblesshire and Selkirkshire, in Midlothian and Linlithgowshire, and as far north as Caithness. The overlying peat-mosses afford still more interesting evidence of its presence—in the Lowlands it ranged from Wigtownshire and Ayrshire eastwards and northwards through Dumfriesshire, Selkirkshire, Roxburghshire, Berwickshire, to the extreme corner of Haddingtonshire; it peopled the midland valley even at the mouths of the Earn and Tay; in the Highlands it reached the furthest limits of Inverness-shire on the west, of Sutherland on the north; and there are many evidences of its presence in the isles of Orkney and Shetland.

Neolithic man made use of its flesh for food and of its bones and antlers for tools, and there is scarcely a settlement where he or his early successors dwelt that does not contain its remains. Kitchen-middens of Bronze and Iron Age, hill forts and underground "Eird Houses," Roman settlements and "Pictish Towers" or brochs all tell of its abundance. Even in Orkney, Red Deer of large size were still common when the brochs were in full occupation; and the occupants of the cave of Borness on the shores of Kirkcudbright bordering the Solway Firth, slew many Red Deer in their Lowland valleys in times after the Roman Conquest.

From the earliest times, therefore, down to the eighth century of our era or later, the Red Deer was abundant in every corner of Scotland, not only in the uplands but in the valleys to the very margin of the sea (see Map V). From the low-lying valleys it was soon driven by the increase of man and of his rude agriculture, and by the destruction of the forest, for the Red Deer, in spite of its mountainous habitations in Scotland at the present day, is pre-eminently a forest-loving animal.

Thus we find that when Scottish history first takes up the tale of the Red Deer in the Lowlands, it had already forsaken the lower valleys, and had been relegated to forest areas then existing on the slopes of the Pentland Hills and to the great Ettrick Forest of the uplands of Selkirk and the neighbouring counties. In another connection (p. 207) I have given a short account of the main events of its history in the Lowlands as recorded in Scottish annals, until, towards the end of the seventeenth century, several factors, amongst which was the disappearance of forest in the face of an increasing culture of the soil and of a gradual extension of the pasturing of sheep, finally exterminated the Red Deer of the Scottish Lowlands[1]. At the present time, in spite of centuries of protection, the Red Deer is confined to the Highlands, where it inhabits waste moorlands and scanty woods at high altitudes, from Sutherland to Perthshire and Argyll.

That the destruction of woodlands has a very real influence upon the distribution of Red Deer has been actually observed and noted within the past century. The brothers, John Sobieski and Charles Edward Stuart, in describing in 1848 the death of the last Deer of Tarnaway Forest on the southern shores of the Moray Firth, say, from their own experience :

[1] Although the end of the seventeenth century saw the practical extermination of Lowland Red Deer, a few individuals probably still lingered in secluded places, for it is said that the last stag in the Forest of Buchan in Kircudbrightshire was shot in 1747.

DECLINE OF RED DEER—DISTRIBUTION IN PREHISTORIC TIMES, IN THE MIDDLE AGES, AND AT THE PRESENT DAY.

 Distribution at Present Day Distribution in Middle Ages Distribution in Prehistoric Times

Although the stags were never driven from the forest by hunting they went out before the cutting of timber, which left them no rest, and obstructed and marred their haunts and pastures. By degrees the hinds, calves and younger stags, ceased to return at winter, and at last the old harts were diminished to two. These, however, kept the forest till 1830.

In the reduction of the numbers and of the range of the Red Deer, we see the first signs of the decadence of its race in Scotland, and the destruction of the forest had much to do with the appearance of the signs.

Fig. 57. Antlers and portion of Skull of prehistoric Red Deer, unearthed in the Meadows, Edinburgh. $\frac{1}{12}$ nat. size.

What of the remaining criteria of decadence? True, the Scottish Red Deer has not reached the last step in the downward course, but would it not have been exterminated save for artificial fostering and protection by man? Probably it would. In any case, has it not fallen a long stage, physically, from its first fine presence in the days before history began?

The testimony of the marl-mosses and peat-bogs is unanimous as to the great size of the Red Deer of past days and the enormous development of their antlers. The chances are all against the discovery of the best heads by man during his casual excavations in the mosses, even in the unlikely event of the best heads having chanced to be entombed in the bogs. Yet what do we find?

In the days when the wild creatures of the forest wandered over the site of modern Edinburgh, a great Stag came to drink at the waters of the Loch which then covered "the Meadows" of the present day. The banks of the loch were peaty, and underneath the peat lay treacherous beds of marl in which the Deer became entrapped and, unable to free itself, died a miserable death. In 1781 the head and horns of this Stag were dug up below the roots of an old tree. It was probably no exceptional creature in its own generation, yet to-day it stands out as a giant, a seventeen pointer—a Great Hart summed of seventeen. The antlers were large, the right 3 feet in length, bearing nine points and ending in a cup-like expansion, or sur-royal, surrounded by six points; the left antler, 46 inches long, carried eight points with a fine sur-royal of five large "croches." The circumference of the beam above the second or bez-tine measured $8\frac{1}{2}$ inches (Fig. 57, p. 335).

But, as I have said, the Great Hart of the Meadows was no exceptional Stag. During the draining of Linton Loch in Roxburghshire, the entire skeleton of a Red Deer was discovered, deep in a marl bed which itself lay under 10 feet of peat, and although the bones were broken up by the workpeople the head and antlers were preserved. They are those of a "Great Hart summed of nineteen." Both antlers measured 33 inches, and their span from tip to tip, 44 inches. The skull was 20 inches in length along the profile and $7\frac{1}{2}$ inches across the forehead. Even this fine head cannot compare with another Roxburgh example found at Ashkirk, which bore two magnificent antlers carrying a total of twenty-one points.

The antlers of a very fair Royal, a twelve pointer, of the present day, would measure several inches under three feet along the *outside* of the curve, would have a spread of some $2\frac{1}{2}$ ft, and a circumference of $4\frac{1}{2}$ to 5 inches as against the $8\frac{1}{2}$ inches of the Great Hart of the Meadows.

Moreover in size of body the old Scottish Red Deer far surpassed its modern representatives. There is exhibited in the Royal Scottish Museum a complete skeleton found about

1830 in a peat moss at Smeaton-Hepburn in Haddingtonshire. It compares ill with such examples as I have just mentioned, for it is only a twelve pointer Royal, yet it must have stood 4 ft 6 inches at the shoulder, and the length of its body from nose to base of tail must have been 7 ft 10 inches. A modern Scottish Royal, a good stag, shot in the forest of Corrour and now in the Royal Scottish Museum, stands 3 ft 5 inches at the shoulder, and its body is 5 ft 7 inches long. The old-time Red Deer was a third as large again as a selected modern example. (See Frontispiece.)

It is apparent that the Red Deer of the Scottish Highlands cannot compare in physique with the Red Deer which dwelt in the Lowlands in the days before man's coming. The process of decay was a gradual one, for bones which I have examined from recent excavations in the hill-fort of Dunagoil in Bute, and in the caves of Eastern Fife,—both settlements which were occupied in the days of the early Christians—still indicate Red Deer of much greater size than those of to-day. And even in Orkney in the days of the brochs, the remains of Red Deer "often of large size" were deposited in those strange defensive structures.

What was the cause of the gradual decay of the Red Deer in Scotland? Clearly the destruction of the forest! Red Deer are pre-eminently forest animals, as their distribution in the still existing pine belts of Europe shows. But in Scotland the forest in which they thrived has disappeared, and the Red Deer have been driven to an unnatural home on waste moorlands and in bare mountain glens. As if to settle the question of the decadence of Scottish Red Deer, there are still to be found in the native forests of the Continent, in western, northern and central Europe, descendants of the original Red Deer stock—races (*Cervus elaphus typicus* and *Cervus elaphus germanicus*) the size and complexity of the antlers of which forcibly remind one of the antlers of the Scottish mosses. Where the forests have persisted the Red Deer in its original grandeur has persisted with them. It may be suggested that the shooting of the finest heads in a sporting country such as Scotland may have lowered the standard. This may be so to a slight degree, but on the Continent the Red

Deer also is stalked, yet its characters have persisted in their magnificence[1].

A broad reading of the facts shows indubitably that the destruction of the forest has impelled the Red Deer of Scotland on a downward path, limiting its numbers, decreasing its range, and whittling away its former dignity by steps of physical decadence.

THE REINDEER

As a Scottish animal the Reindeer (*Rangifer tarandus*) (Fig. 58) has long since disappeared, yet for many thousands of years this interesting animal, now confined to the northern regions of the Old World and eastern Canada, roamed upon our plains from the Solway to the Pentland Firth, and was even established in the Orkney Islands. Its antlers have been found in deposits belonging to the Ice Age at Kilmaurs in Ayrshire (Fig. 59, p. 343), where were also found the tusks of the extinct Mammoth, at Croftamie in Dumbartonshire, and at Raesgill in Lanarkshire. Clearly it entered Scotland during one of the mild periods which broke the long monotony of the age of glaciers.

But the Reindeer in Scotland survived the disappearance of the cold climate which had enticed it to enter Britain from the Continent. Cut off by the formation of the English Channel from retreat along the route of its invasion, it was compelled to make the best of changing climates and conditions. In this effort of adaptation it met with considerable

---

[1] If further proof of the implication of the destruction of the forest in the degeneracy of Scottish Red Deer were required, it is indicated by the history of Scottish Deer transported to New Zealand. In 1870 seven Red Deer from Lord Dalhousie's estate in Forfarshire were liberated in the neighbourhood of Lake Hawra, South Island. These have multiplied into the famous North Otago herd of which, in Hunter Valley alone, 1600 were destroyed in 1915–16; and the size of the Deer and quality of their antlers have responded to the abundance of food available in the heavy bush lands. Thus, although there is no strain of "park" blood in the North Otago herd, a writer in *Country Life* (9 August, 1919) recorded and illustrated two magnificent and evenly developed heads of Deer shot in 1918, one bearing sixteen points, the other with twenty points, a length of $45\frac{1}{2}$ inches and a span of 45 inches—reversions to the old standard which signify that the banishing of our Scottish Red Deer from their natural haunts to waste and barren moorlands is responsible for their physical deterioration.

EFFECTS UPON ANIMAL LIFE

Fig. 58. Reindeer—formerly natives of Scotland—in Scottish Zoological Park.

success, as is shown by its continued existence for many thousands of years.

Its distribution during the period of the formation of the peat mosses, just prior to, or even contemporaneous with the arrival of man in Scotland, shows no trace of the limitation which one associates with a decaying race. It still roamed over the length of the land, for antlers have been recovered from the peat mosses of Shaws in Dumfriesshire, from the neighbourhood of Tain in Ross-shire and from Rousay in the Orkneys. Not very far from Edinburgh its remains were found in a rock fissure at Craig Green in the Pentland Hills.

Even in later times, bones in actual association with the handiwork of the men of the New Stone Age, show that the Reindeer still ranged widely in Scotland: witness the discovery of antlers near the mouth of the Kelvin River in Glasgow, in beds of laminated clay which have yielded many dug-out canoes of prehistoric fishermen, and the more recent find by Drs Peach and Horne of Reindeer bones in the Cave of Allt nan Uamh in Sutherland, where they lay in layers containing hearth-stones burnt by the fires of Neolithic man. The men of the New Stone Age interfered little with the wild life of the country. They hunted, herded and tilled, but for their own limited needs, and their interference with woodland was of little account, as indeed is that of all primitive peoples. For so long, then, the Reindeer appears to have held its own.

The next series of records of the Reindeer, however, shows a remarkable contraction in its range. True there is a hesitating record of its presence in the south country at a comparatively late date. In the descriptions of his excavations at the lake dwelling of Lochlee in Ayrshire, which contained Bronze Age and Roman articles, Dr Robert Munro states that he found

two more or less fragmentary portions of horns which after a good deal of comparison with other reindeer horns, and with fragments of red-deer horns, I incline to set down as indicating the presence of the former animal;

but the identification is so uncertain that the record, out of keeping with the facies of the animal remains of Scottish crannogs in general, may be ignored.

To the days of the brochs, those great "Pictish Towers"

erected throughout Scotland from the Lowlands to the extreme north, "to withstand the incursions of roving pirates from overseas," we must pass for further evidence of the Reindeer's presence. But now the Reindeer is on the decline in Scotland, for its remains are confined to the northern counties. In Sutherland several pieces of antlers were found amongst the food debris of the Broch of Cinn Trölla; in Caithness, the Brochs of Yarhouse and Keiss, as well as the Harbour Mound at the latter place, yielded fragments of antlers, and at Kettleburn, horns and remains of Deer were discovered which, the excavator says "were not the Red Deer"—these too may have belonged to the Reindeer. Less uncertainty attaches to a fragment unearthed in the Broch of Burwick, in the parish of Sandwick, Orkney, where the investigator found "one piece of horn like a part of a reindeer antler, or a fallow deer's, as it is broad and palmated." It is unlikely that this fragment belonged to a Fallow Deer, of which there is no definite record in Britain in historic times till its introduction by man a few centuries ago (see p. 284). Another possibility is that it may have belonged to the Elk or Moose, but the size and palmation of an Elk's antler is on a much grander scale than that of a Fallow Deer, with which comparison is suggested, and there is no confirmatory evidence that the Elk ever existed in Orkney. We are driven, therefore, to suppose that the palmated antler was that of a Reindeer, though I cannot help regretting, in view of their interest, that this fragment and that from Burwick were not submitted to expert examination.

The brochs are supposed to have been in occupation in northern Scotland from the end of the fifth till about the ninth century, and some were occasionally inhabited till a later date. Moreover the antlers of Yarhouse, near Wick, were found not in the broch itself, but in one of the associated outbuildings which were erected at a later date than the main building, when the needs of defence were less imperative. Nevertheless a reference in the Orkneyinga Saga carries the last appearance of the Reindeer in North Britain to a much more recent time. I reproduce the literal rendering of the Scandinavian words as translated by a celebrated Icelandic scholar, Mr Eiríkr Magnusson, formerly of Cambridge. "It was the custom for the Earls nearly every summer to go over into Caithness and then up into the woods to hunt Red Deer or Reins." The period of these hunting excursions was about the middle of the twelfth century (1159 according to Jonaeus), and the Earls were Rögnvald and Harald of Orkney. The translator notes that the word *edr*, translated "or," may mean *or* (= Latin *vel*), or *and* (= Latin *sive*), and much doubt has been cast on the value of the passage on the ground that the sagaman said *red-deer or reindeer* under the impression that the two names were synonymous. This doubt seems to me to be superfluous, for who is likely to have been more familiar with these two very distinct species of Deer than the Scandinavian sagaman, who lived and wrote in a country where Red Deer and Reindeer were both common? To me the whole difficulty is one of poetic licence; did or did not the historian-poet transfer, for the sake of effect, the animals hunted in his own country to the foreign shores of Caithness? I hardly think so, for the skalds are seldom inaccurate in their descriptions of the objects of the chase. Further the statement receives some confirmation from the spirited representation of a Reindeer obviously drawn from the life, carved on a sculptured stone found near Grantown in Inverness-shire. The date of the early Christian monuments, of which this is one, is believed to range about the tenth century of our era.

The Reindeer, once abundant in Scotland throughout the length of the land, gradually became limited in range and finally disappeared. What led to its extinction? It was not lack of suitable food, for brushwood and especially "Reindeer Moss," the lichen *Cladonia rangiferina*, are still abundant in the counties where it made its last appearance. It was not wholly change of climate, for the Reindeer survived the much greater alteration from the cold of the Ice Age to the mild humidity which heralded the period of peat-formation, and back again to the drier and cooler times when the pine forest spread over Scotland. It may have been partly deliberate destruction by man, for his kitchen-middens show that he hunted the Reindeer for food; but these were days of simple weapons, the javelin and the bow and arrow, and history has seldom recorded the extermination of a fleet and wary race by a primitive people to whom the murderous slaughter of modern sport was unknown. On the

EFFECTS UPON ANIMAL LIFE

whole, I think that the destruction of the forest had more to do with the disappearance of this interesting native.

RACIAL CHARACTER OF SCOTTISH REINDEER

Hitherto I have referred to the Scottish remains as those of Reindeer in a general sense, but it may be well to enquire a little further into their special characteristics. The Reindeer of the Old and New World fall into two groups,

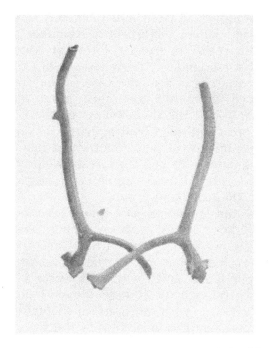

Fig. 59. Fragmentary Antlers of Scottish Reindeer of Glacial Period, found at Kilmaurs, Ayrshire. $\frac{1}{13}$ nat. size.

the "barren-ground" and "woodland" races—the former inhabiting the bleak barren lands and tundras which border the Arctic circle, the latter frequenting the belts of forest which lie to the south of the barren lands. The races are distinguished by their size and their antlers. The "woodland" Reindeer is larger than its "barren-ground" relative, whose antlers are round, slender and long, while those of the "woodland" group are heavier, flatter, thicker and more heavily palmated. In the woodland group, the brow tines are much palmated, and one is usually much more developed than the other, and the succeeding bez tines are also large and palmated, a rare feature in the barren-ground group.

Reindeer were frequently drawn by men of the Old Stone (Palaeolithic) Age who inhabited the central plains of Europe, and their drawings seem to me clearly to indicate a woodland variety. In the Aurignacian painting of two Reindeer fronting each other, on the walls of the French cave at Font-de-Gaume, in Dordogne, the brow tines of one of the Deer are strongly palmated and unequally developed. In the later Magdalenian picture of a Reindeer grazing by a pool, found engraved on a piece of bone at Kesserloch, near Thayngen, Switzerland, the brow and the bez tines are both heavily palmated, as is also the case in the running Reindeer engraved on hornblende schist, from Saint Marcel. Julius Caesar found the descendants of these Reindeer in the Black Forest and neighbouring parts of Germany at the time of his campaign in that country and in Gaul. These structural characteristics and the forest-dwelling habit of the Reindeer of Central Europe agree in suggesting that their place is with the woodland group.

What is the evidence as regards the status of the Scottish Reindeer? They probably immigrated to Scotland as part of the woodland fauna from Central Europe, for there is no evidence that any of the characteristic animals of the "steppe fauna," with which Professor H. F. Osborn groups the barren-ground Reindeer, ever found their way to North Britain. Such portions of Scottish antlers as show distinctive features also point to the woodland group. I have examined in the Hunterian Museum in Glasgow two large fragments found in 1829 in interglacial deposits of the Ice Age at Kilmaurs in Ayrshire (Fig. 59, p. 343). They represent the right and left antlers probably of one animal. The left beam is 2 ft 3 inches long, the right 2 ft 7 inches. The brow tines of both have been broken off short, but the bez tines though incomplete are well developed, 9 inches and $13\frac{1}{2}$ inches in the right and left antlers respectively; both tines broaden towards their extremities and the left shows clearly the beginnings of a well-developed palmation. The beams are moderately stout, slightly flattened, and measure, right and left, 123 and 133 millimetres in circumference above the bez tines. The right antler possessed also a back tine, a feature seldom found in the barren-ground race.

Or take the fine extremity of an antler found in the Broch of Keiss in Caithness: of this Professor Richard Owen, after careful comparison and examination, wrote

I have not seen any antlers of the Red Deer showing so much flattening or compression of the "beam" as in this specimen. I believe them to be parts of the antler of a large reindeer or variety called 'Carabou.'

While writing these words Owen seems to have had in mind the large woodland form known in Canada by that name.

There are difficulties in the way of reaching a definite conclusion regarding the exact status of the Reindeer races, for antlers show considerable variation, and the question of relationships from this point of view has not been satisfactorily solved. But I think it is justifiable to regard the Scottish Reindeer as different from the extreme barren-ground type with clean cut antlers and pronged tines, and as tending towards the woodland type, although the antlers do not attain the luxuriance of palmation of the most highly developed woodland forms of the Caribou. The modern form of Scandinavian Reindeer with palmated brow and bez tines and well developed back tines, seems to me to belong to this intermediate type tending to the woodland variety, and to bear a close resemblance to the Scottish examples of the interglacial beds and peat-bogs.

If the Scottish Reindeer was a woodland variety, as I have suggested, then the presence or absence of woods must have had a close bearing on its welfare, for although the barren-ground form may enter the domain of the woodland race during its winter migration southwards, the latter finds in the woods its permanent home. The destruction of woodland, therefore, is likely to have tended to its extermination. Neolithic man had little influence on the forest, and the Reindeer outlived him; but the Bronze and Iron Ages, with their demand for fuel for smelting, began a devastation which each succeeding age intensified, so that in the twelfth century, when Scottish laws were already. endeavouring to conserve the forest, the last reference occurs to living Reindeer in Scotland.

The evidence suggests that the destruction of the forest was an important factor in reducing the numbers of the

Reindeer, in limiting its range, and finally in driving it to extermination.

THE ELK

If there be some doubt as to the exact status of the Scottish Reindeer, there is none regarding the Elk (*Alces alces*) (Fig. 2, p. 15), a huge, ungainly member of the northern forest fauna. This giant amongst Deer, standing sometimes about six feet high at the shoulder, with antlers spanning six feet from tip to tip, is now confined to regions ranging from Scandinavia to eastern Siberia in the Old World, while a closely related form, the Moose, occurs in the New World from the New England States to British Columbia. But in the old days the Elk dwelt far south in Europe and ranged over the whole of the British Isles. In the latter, the trend of the limitation of its range has been from south northwards, as is shown in a rough way by the distribution of the remains which have been recovered. The Scottish records are far more numerous than the English, an indication of a longer period of occupation or of larger herds in the northern territories.

In Scotland the range of the Elk in time and in space was a wide one. In the period of the great marl formations, the first definite lake-deposits which accumulated after the main phase of the Ice Age had passed away, the Elk ranged from the Lowlands to Perthshire. An antler, probably belonging to this species, though described as that of the "Gigantic Irish Elk," was found in 1859 in a river gravel at Coldingham in Berwickshire. Marl beds, in most cases underlying considerable deposits of peat, have yielded remains of the Elk in Selkirkshire; at Kirkurd in Peeblesshire; in Midlothian at Craigcrook near Cramond, and in Duddingston Loch, where they were associated with many articles of bronze; in Forfarshire at one if not two localities; and in Perthshire in the parishes of Airleywight (Fig. 60), Kinloch and Muthil.

During the period of the formation of the peat, on the whole at a date later than the deposition of the great marl beds, the Elk seems to have extended its range to the northern limits of the mainland, though it still retained its hold in the southern forests. In Wigtownshire an antler has been found near the mouth of the River Cree; in Roxburghshire, Williestruther Loch yielded an almost complete skull and antlers; in Berwickshire, remains have been found at Mertoun Loch, at Whitrig Bog near Mertoun, and at Duns; in Selkirkshire, a portion of skull with antlers was discovered at Oakwood on the Ettrick; and in Sutherland, Strath-Halladale yielded a shed antler which Dr J. A. Smith regarded as wonderfully fresh in appearance.

Long after the main period of peat formation was past, the Elk still roamed at large over the greater part of Scotland, and there are sufficient records to show that it

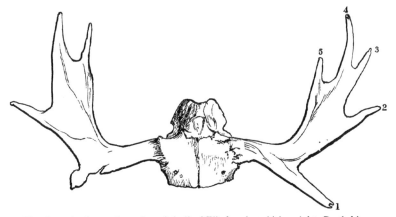

Fig. 60. Antlers and portion of skull of Elk found at Airleywight, Perthshire. $\frac{1}{10}$ nat. size.

was familiar to the inhabitants for many thousands of years. Neolithic man and his successors knew it well. I have already alluded to the suggestive discovery of horns in a bed of shell marl at the bottom of the Loch of Duddingston, which also yielded many bronze spear heads and swords, though there is no definite proof that the finds were contemporaneous. There is similar uncertainty about the relationship of "a medal of Trajan, a fibula, a patera, and a

---

[1] Such a statement of extension of range must be regarded as provisional, for it is possible that in the marl period also, the Elk had a wide northern range, though its remains have not been preserved or have not been discovered. But this is not very probable since it seems likely that while the earliest marl deposits were forming in southern Scotland, the Highlands were still buried in snow.

horn of the moose-deer" found near North Berwick. But there can be no doubt about the significance of the Elk remains recorded by Prof. J. Cossar Ewart from the Roman settlement of Newstead near Melrose.

This is the last record of the Elk in the Lowlands. It appears thereafter to have been driven to the wilder districts of the north, a fact of some interest when we recall that the earliest destruction of Scottish forests in historical times was that carried out by the Roman legionaries. The latest definite record of the Elk in Scotland is that of an antler found by Sir F. T. Barry in underground buildings attached to a broch at Keiss, a record which might carry its occupation down to about the ninth century of our era, for the accessory buildings of the brochs were made at a date considerably later than the erection of the brochs themselves.

Yet it would be a mistake to suppose that the Elk disappeared from Scotland in the ninth century. The references in Gaelic tradition to a great extinct deer, *Miol* or *Lon*— a creature spoken of as black or dark in colour, shambling in gait yet swift—so clearly point to the special characteristics of the Elk of the present day that there can be little doubt as to their significance. And when the brothers John Sobieski and Charles Edward Stuart translate a stanza of an ancient Gaelic poem, *The Aged Bard's Wish*,

> I see the ridge of hinds, the steep of the sloping glen
> The wood of cuckoos at its foot,
> The blue height of a thousand pines,
> Of wolves, and roes, and elks,

there can be little doubt that they translate a tale of things seen, for the whole poem is full of minute and accurate descriptions of nature, such as none but an onlooker could have chronicled. The reference is not an isolated one: in the old poem *Bas Dhiarmid*—the death of Dermid—a poem till lately well known in the Highlands, the following lines occur:

> Glen Shee, that glen by my side,
> Where oft is heard the voice of deer and elk.

And, as we have seen, Perthshire has actually yielded the remains of more than one individual.

I suggest that the main factor in the gradual limitation of range and final extermination of the Elk in Scotland was the destruction of the forest, which entailed the disturbance and disappearance of its haunts. Apart from the suggestiveness of its reduction in numbers and steady restriction towards the wilder and more wooded areas to the north, during periods when history tells of the woeful devastation of Scottish woods, the history of the Elk in Europe adds confirmatory evidence; for it stands in the nature of a "control" upon the events in Scotland. Julius Caesar mentions the Elk, along with the Reindeer and the Urus, as inhabiting the Hercynian forest of Germany during his campaigns in that country and in Gaul. During the third century of our era it spread over all the forest-clad parts of Germany. Here it was that the "dowghtie Siegfried" of the twelfth century *Nibelungenlied* "slowe a Wisent [Bison] and an Elk." For many hundreds of years the great woods of the Black Forest and other areas remained almost unaffected by man's destruction, affording a safe preserve for the wild creatures of the woods. *And with the survival of the forest the Elk survived;* for in Saxony an Elk was slain so recently as 1746, and in Silesia it lingered until 1776. Long before the Elk disappeared from the forests of Central Europe, a new factor, as potent for destruction as the dissolution of the woodland, had arisen—the invention of powder and the gun. It is worth remembering that although the Elk is now regarded as a northern animal, the German forests in which it lived only two and a half centuries ago, lie in a latitude far south of Scotland—an indication that climate was not a prime factor in determining its disappearance in North Britain.

In Central Europe, the forests remained and the Elk survived: in Britain the forests were laid low and the Elk, having been gradually driven to the northern confines of Scotland, disappeared.

CONCLUSIONS REGARDING THE RACE OF DEER

These summaries of the stories of such Deer as certainly survived in Scotland at the arrival of man show in different degree the stages of a general decadence. The Roe Deer became reduced in numbers and limited in range, the numbers and the range of the Red Deer became contracted and itself underwent a marked physical degeneration, the

Reindeer and the Elk were driven from one stronghold to another till both died out in the northern counties. What widespread influence was at work against these forest-loving creatures? The destruction of the forest! It may be suggested that man's deliberate destruction of Deer for food was the prime agent. But the rule of the bow and arrow could not reduce the size of the Red Deer's antlers, for the necessity of finding food meant that the youngest and most easily obtained, not the best animals (as the rifle and present day sport demand) fell to the hunter, a point abundantly proved by the large proportion of the bones of young animals in early settlements and kitchen-middens. It may be suggested that changes of climate had dire effect; but more severe changes of climate were passed through in the early Neolithic days without seriously influencing the distribution of any of the species. It may be suggested that a decrease of proper food played an adverse part in the welfare of the Deer. That is highly probable. But what caused the decrease of food, if not the increase of domesticated animals and of cultivation, and the consequent destruction of the proper feeding-grounds—the forests? I have no doubt that all the factors mentioned told against the race of Deer, but the great influence which acted with constant and ever-increasing intensity was the disappearance of the forest, and for that man was largely responsible.

SOME OTHER FOREST DWELLERS

Other inhabitants of Scotland besides the race of Deer suffered from the same cause. The story of the Elk is paralleled by that of the Great Extinct Ox or Urus (*Bos taurus primigenius*) which, common throughout Scotland in the times of the great depositions of marl and peat, was gradually driven northwards till it died out in the northern counties during the period of the brochs[1]. The Urus was a creature of the woodland, sharing with the Reindeer and Elk the forests of Central Europe in the days of Julius Caesar, and surviving in these safe retreats until, it has been stated, the beginning of the sixteenth century. Like the Elk, it survived in Europe with the forests, and with the forests, disappeared in Scotland.

---

[1] For a more detailed account of its history, see p. 49 *et seq.*

The forest-loving beasts of prey, the Bear, the Wolf, the Fox, the Wild Cat, the Pine Marten, the Polecat and their like, as well as the inoffensive Badger and the Beaver, must also have suffered in numbers and in range as their haunts were demolished. More than once I have referred to the deliberate destruction of woods for the purpose of ridding the land of Wolves and lesser vermin; but the destruction of woodland from whatever cause must have had the like effect of decreasing their numbers and driving them to resorts further afield.

THE COMMON SQUIRREL

One other creature amongst mammals, deserves mention on account of its vital connection with the forest—the Common Red Squirrel (*Sciurus vulgaris*). The refuse of the prehistoric settlements affords no evidence of the presence of the Squirrel in Scotland, nor would we expect to find the remains of so small and shy a creature in these rude accumulations. For even had they ever found their way to the kitchen-midden, the liability of the small bones to decay, and the rough and ready methods of collecting animal remains which have too often characterized the older excavations, place almost insuperable difficulties in the way of their recovery.

Yet I have no doubt that the Squirrel is an ancient native of Scotland and that this "herald of forest conditions all over the northern hemisphere," as Professor Osborn calls it, accompanied the woodland fauna which migrated to Britain from the Continent in the days succeeding the Ice Age.

During long ages it held its ground throughout the whole of Scotland. In the seventeenth century, Sir Robert Sibbald, in his *Scotia Illustrata* (1684), records its presence at one end of the country, in the woods of the southern tract of Scotland.—" In meridionalis Plagae Scotiae Sylvis reperitur"; and at other end, in the far north, Sir Robert Gordon wrote in 1630 in his *History of the Earldom of Sutherland* that

All these forrests and Schases are verie profitable for the feeding of bestiall, and delectable for hunting. They are full of reid deer and roes, woulffs, foxes, wyld catts, brocks, skuyrells, whitrets, otters, martrixes, hares, and fumarts.

And familiarity with its presence in the midlands is clearly indicated in an ancient Gaelic poem, the *Lament for MacGregor of Knaro* attributed to the first half of the seventeenth century, and referring to the region of the MacGregors in northern Perthshire:

> Tho' nimble the squirrel is,
> By patience may it be ta'en

and again

> So joyful grew my heart
> That like the squirrel could I leap.

When the Squirrel disappeared from the Lowlands, there is no record[1]. It had certainly gone long before it was mentioned as extinct, in 1841, in the New Statistical Account of Berwickshire and Roxburghshire. In Dumbartonshire, it seems to have been absent from the fauna in 1791.

In Argyllshire it was known between 1725 and 1745 when Alastair MacDonald, the Gaelic poet, likens its activity, as if he had seen it, to the nimbleness of the picked sailors of Clanranald,

> Who can climb the tight hard shrouds
> Of slender hemp,
> Nimbly as the May-time squirrel
> Up a tree trunk.

Yet here half a century later, when between 1764 and 1774 Professor Walker wrote his *Mammalia Scotica*, the Squirrel, once plentiful, had now become very rare: "In sylvis Lornae superioris, antehoc copiose, nunc rarior." In 1790, according to the Old Statistical Account, it was "now very rare, if not extinct" in Lismore and Appin, and it probably disappeared entirely about the opening of the nineteenth century. The New Statistical Account of 1842 records it as formerly abundant, "but now extinct."

About the same time it disappeared from the more northern woods. It is last mentioned in Moray in 1775, when the Rev. Lachlan Shaw, in his history of the Province, says that "there are still in this province foxes, badgers, and squirrels, weasels, etc." Two years later (1777), Pennant

---

[1] Dr Harvie-Brown contributed a long and detailed paper on the History of the Squirrel in Great Britain to vols. v. and vi. (1879 and 1881) of the *Proceedings of the Royal Physical Society of Edinburgh*, to which I owe most of the following records.

mentions it as rare in the Spey valley: "scarce in Scotland, a few in the woods of Strathspey." Still a few years later and it had gone from these regions also—"I am certain" wrote the Rev. George Gordon, "that squirrels were not known in the lower or northern part of Elginshire, or on Speyside, at least, from 1810."

It is unnecessary to follow the meagre details of the Squirrel's disappearance in other areas. Nothing could seem more fateful than this gradual dwindling and disappearance of the Squirrel, first in the Lowlands and then, almost simultaneously, throughout the rest of Scotland, until not an individual remained, except perchance a few that lurked, far from the ways of man, in the depths of such native forests as Rothiemurchus at the base of the Grampians.

What far-reaching influence was telling against the welfare of the Squirrel? It has been suggested that the Marten aided in its extermination. But the Squirrel and the Marten had lived together in Scotland for some thousands of years, without, as it were, coming to serious blows; and so far from there being evidence of any general increase of Martens, there are good grounds for believing that they also were on the down grade in Scotland. The Squirrel's disappearance has also been attributed to a series of severe winters. Yet, although this may have been a factor in temporarily reducing their numbers, it seems of itself a cause insufficient to account for their extermination over so wide an area.

The great far-reaching cause of the extermination of the Squirrel was the destruction of the forest. We have seen that at an early date the woods of southern Scotland were destroyed to make way for cultivation and sheep pasture; and at an early date the Squirrel disappeared from the Lowlands. In Argyllshire its numbers began to diminish in the middle of the eighteenth century and about the beginning of the nineteenth it had gone. But in 1730, the great charcoal-using, iron-smelting furnace of Bunawe or Taynuilt, was erected, that of Furnace in 1750 and that of Goatfield in 1754. In all, more than twenty slag-furnaces have been found in Argyllshire (see Map IV, p. 320), and these bared many districts of wood, reaching their maximum of destruction in the nineteenth century, when the countryside was so ravished of its forest that the fires of the last survivors, Taynuilt and Goatfield, had to be blown out for lack of fuel. The Squirrel was exterminated in Argyllshire through the destruction of woodland, caused by the timber-using slag-furnaces.

In the northern counties the same factor was at work. In 1730 large iron-furnaces were erected at Invergarry, and at Abernethy in Strathspey. These and their many predecessors in Nairn and Inverness-shire—the numbers recorded from these counties by Prof. Ivison Macadam are 12, 5 and 15 respectively—had ceased to work only when the forest within manageable distance had been consumed. The maximum destruction wrought by them corresponds in time with the disappearance of the Squirrel in these areas.

So it was in other regions. The Squirrel, driven from one haunt to another, sought shelter in the deepest and thickest woods, and these very woods, just on account of their depth and thickness, were those which attracted the iron-smelters of the sixteenth, seventeenth and eighteenth centuries. The terrible havoc wrought in the forests by the iron-furnaces was the main factor in exterminating the native Squirrels of Scotland. A clearer case could scarcely be found of the indirect influence upon animal life of a single industry, for the Squirrel, whose history we have traced in some detail, may be looked upon as simply a type of the forest denizen.

The introduction by man of a fresh race of Squirrels to Scotland has been described in another connection (p. 290) but this again affords corroborative evidence of the influence of forest destruction, for it was the general replanting of woodland in the latter part of the eighteenth century which gave the Squirrel a new lease of life and led again to its rapid spread throughout the country.

THE CAPERCAILLIE

The influence of forest destruction would be very partially traced were no further reference made to its effect upon bird-life; and no bird could afford a more typical test case than the forest-haunting Capercaillie, the "Cock-of-the-Woods." It was a bird well known to our predecessors. Probably at one time it inhabited the whole country, but I do not remember any instance where its bones have been found in a Scottish prehistoric settlement, although they have been found in the ancient kitchen-middens of Denmark. The furthest south record in Scotland of the native stock of Capercaillies appears to be the somewhat indeterminate remark of the Rev J. Hendrick in his *View of Arran* (1807) that they "formerly abounded" in that Island. This occurrence may represent the last survivors of a Lowland stock, for it certainly lies beyond the bounds of the general distribution of the bird in Scotland as history has recorded it.

It is true that Hector Boece's statement in 1526 may be taken to indicate a general distribution:

Mony uthir fowlis ar in Scotland, quhilkis ar sene in na uthir partis of the warld; as capercailye, ane foul mair [greater] than ane ravin, quhilk leiffis allanerlie of barkis of treis [Bellenden's Translation].

But, half a century later, Bishop Jhone Leslie, who was perhaps more familiar with the country, shows clearly that the range of the Caper was limited:

"In Rosse and Loquhaber," he writes in 1578, as translated by Father Dalrymple, "and vthiris places amang hilis & knowis, ar nocht in missing fir trie sufficient, quhair oft sittis a certane foul and verie rare called the Capercalye, to name, with the vulgar peple, the horse of the forrest, les indeid than the corbie, quhilke pleises thair mouth, quha eitis her, with a gentle taste, maist acceptable."

How Leslie could have described the bird as being less than a corbie, or having a "gentle taste," I do not understand; but probably tastes, like manners, change, for the Cock-of-the-Woods found a place as frequently as its rarity would allow on the tables of the nobility. At the royal hunt on the Braes of Athole in 1529, the Duke "maid great and gorgeous provisioun for him [King James V] in all things pertaining to ane prince" and in the long list of fowls served was the Capercaillie. In the sixteenth century, therefore, history traces its presence from the wilds of Perthshire northwards through Inverness to Ross.

The records of the succeeding century are more complete and more interesting. The Capercaillie still inhabited Perthshire: in 1617, King James VI hinted to the Earl of Tullibardine that on account of his "dutiful affection to the good of our service and your countrie's credite," he should cause to be now and then sent to us by way of present...the known commoditie yee have to provide, capercaillies and termigantis....The raritie of these foules will both make their estimation the more pretious, and confirm the good opinion conceaved of the good cheare to be had there.

In Perthshire the bird is always referred to as rare. It is no wonder, therefore, that when the "Laird of Glenorquhy" sent a Capercaillie to Prince Charles of Wales at Perth in February 1651, the King "accepted it weel as a raretie, for he had never seen any of them before." It is strange that while it was so scarce in the wild midlands the Capercaillie should have been common in the eastern counties, yet Robert Edward, the minister of Murroes, Montrose, in an account of Angus published in 1678, writes to the effect that "The mountains and heaths abound with partridge, Capercaillies, and plover, etc. etc." It is less strange that in the wilds of Aberdeenshire, Taylor, the Water Poet, should have found "Caperkellies" included in the "great variety of cheere" furnished him by the "goode Lord Erskine" during his visit to the "Brea of Marr" in 1618; or that in the dense forests of Sutherland, Sir Robert Gordon should have to record in 1630 that "there is great store of partridges, pluivers, capercalegs" and many others.

Notwithstanding a local plenty, we cannot help feeling that, even in the early years of the seventeenth century, the Capercaillie was already suffering from the disappearance of the forests, for in 1621, the Scottish Parliament prohibited the buying and selling of, amongst others, "Caperkayllies" under penalty of "ane hundreth pounds money."

But the law did little to save the Capercaillie from extinction, for the records of the eighteenth century clearly show a diminishing range with a steady drift northwards. There is now no mention of the bird in the eastern counties or even in Perthshire. It inhabited Speyside in 1745; but in 1754, Burt says it "is very seldom to be met with" in the north of Scotland, and in 1775, when Shaw described it as "become rare" in Moray, probably it was already extinct. At any rate, the last example said to have been seen in Scotland was in the woods of Strathglass in Inverness-shire in 1762, although Pennant states in an indefinite way that it was "still to be met with" in "Glen-Moriston, and east of that Straith Glas" at the time of his tour of Scotland in

1769. Its extermination is generally set down to about the year 1770.

Now the Capercaillie is a bird of the forest, without which it could not survive, for although it nests on the ground at the base of trees, it roosts in the branches and feeds upon the tender shoots of the pine. The destruction of the forest, therefore, would tell directly against its welfare, and it is not surprising that it should have gradually dwindled in numbers and that its range should have been slowly but surely curtailed during the many centuries of destruction of the woodland; until, with the climax of devastation brought about by the great slag-furnaces of the eighteenth century, it should finally have disappeared.

Just so the destruction of woodland banished the Capercaillie from Ireland.

The reintroduction of the Capercaillie from Scandinavia in 1837 and 1838, and its successful establishment and spread when new woods had arisen to give it shelter and food, have been described in another place (p. 268) and there as here I have depended for records mainly on the excellent and detailed history of the bird in Scotland published in 1879 by Dr J. A. Harvie-Brown.

THE GREAT SPOTTED WOODPECKER

It may be said that deliberate destruction by man had much to do with the disappearance of the Capercaillie. I hardly think that its slaughter for food or sport was a significant element in its extinction, but, nevertheless, it cannot fail to strengthen this account of the influence of the destruction of the forest upon animal life to trace the history of a woodland bird which is free from the animosity of man. Consider, for example, the history of the Great Spotted Woodpecker (*Picus major*), as a Scottish residenter. Its story, too, has been unravelled by Dr Harvie-Brown.

There is no reason why, in the old days, the Great Spotted Woodpecker should not have dwelt throughout the Scotland of the great forests. But also there is no reason why a bird of no interest to the hawker or the sportsman should receive the attention of the historian. So no records exist of the occurrence of the Great Spotted Woodpecker in the Lowlands, which had been denuded of wood long before the modern all-enquiring phase of natural history was born. History, and very modern history at that, records the nesting of the Great Spotted Woodpecker only in the more northern counties, in Banffshire, Aberdeenshire and Inverness-shire. It is to be remembered that autumn immigrations of the Great Spotted Woodpecker to Scotland from the Continent occur with fair regularity, so that care has to be taken lest confusion arise between mere temporary visitors and true residenters. The following records, therefore, refer only to nesting birds.

About the years 1830 to 1840, the brothers Stuart were familiar with the Great Spotted Woodpecker on the Spey, and especially in the woods of Tarnaway bordering the Moray Firth:

The Northern Woodpecker comes to breed in the spring and remains until the decline of summer. Many of the old dead firs are pierced with its holes.

Writing in 1840, Professor MacGillivray of Aberdeen University says that it is

resident in the woods [of Dee]; it occurs, but very rarely, in all parts of the district, from Banchory to Glen Lui. In Mar Forest and the Invercauld woods, it is less frequent than it was some years ago.

Already the few observers had noted that it was seriously on the decrease, and soon afterwards it had entirely disappeared. Apart from the testimony of naturalists who had found its nesting places, and of woodsmen who remembered it as a breeding species, the forests of the north country retained abundant evidence of its presence. Mr E. T. Booth in his *Rough Notes on the Birds observed, etc.*, 1881–7, says

The remains of the old timber in the Valley of the Spey, and in many other parts of Inverness and the adjoining counties, indicate that Woodpeckers were formerly numerous in those districts...On some of the largest and oldest trees, I have counted from twenty to thirty holes bored right into the centre of the stem.

Dr Harvie-Brown has also recorded his observation of widely distributed nesting sites and has concluded that while

the most noted haunts of the bird, and localities always quoted by the natives of Strathspey, were Carncruinch—once wooded to the summit with old pine—in Rothiemurchus, and the old wood of The Crannich, in Duthil; Castle Grant woods, near Grantown; Tarnaway on the Findhorn; and

EFFECTS UPON ANIMAL LIFE

Abernethy generally...it must have spread widely over all the old wooded tracts of Spey and Findhorn, as well as north of the Caledonian Canal.

The consensus of recorded observation and of opinions elicited from woodsmen by Dr Harvie-Brown set down the date of its disappearance as between 1840 and 1850.

Various suggestions have been made as to the cause

Fig. 61. Great Spotted Woodpecker—at one period exterminated as a nesting species in Scotland. About ½ nat. size.

of its gradual extermination—the enmity of Squirrels, the stealing of nesting-holes by increasing numbers of Starlings, and the destruction of the forest. The Squirrel was not a prime cause, for the Woodpeckers declined in numbers even while the Squirrel was disappearing. The Starling is an enemy to be reckoned with in the neighbourhood of human habitations, but there is no evidence that it frequented the thick forests where the Woodpeckers found a home. On the other hand, there is evidence that the progressive destruction of the forest, especially by the iron-furnaces, was accompanied by the progressive decrease of the Great Spotted Woodpecker, and Dr Harvie-Brown has shown that the years of its complete and almost sudden disappearance, 1840 to 1850, were marked by "vast and general destruction or burning of old trees," such as the bird selects for its nesting holes.

Here is tolerably clear testimony that the Great Spotted Woodpecker was banished from Scotland by the destruction of the woodland. And to-day we have corroborative evidence of the influence of forest, for the replanting of many years ago has now borne fruit in woods of such trees as have again induced the Woodpecker to remain as a nester. It has become tolerably and increasingly common in the woods of Lowland Scotland, where it has been absent at any rate since the great forests were destroyed; so that from the southern borders it now ranges to the Firth of Forth. It has taken up its abode anew in parts of Perthshire, as at Dunkeld and Pitlochry, and soon, if fresh destruction of woodland has not interfered with Nature's ways, we may hope to see it occupying the ancestral haunts in the pine forests of Speyside. With the forest it disappeared and with the forest it has come again.

CONCLUSIONS REGARDING FOREST DESTRUCTION

These typical examples of the histories of forest beasts and birds show better than any general statement could have done the vital influence which the destruction of the woods of Scotland has had upon the native fauna. It may be objected that the conclusions I have drawn are exaggerated; that had the disappearance of the forest been the prime cause of extermination, all the forest creatures would have disappeared simultaneously. But this is not so. Various creatures depend on the forest in different degree: some for temporary shelter, some for food, some for breeding-places; and the rapidity of the extermination of a forest animal is a function not only of the destruction of the forest, but also of the vital connection of the creature with the forest. Compare the cases of the Squirrel and the Great Spotted Woodpecker, both in former times tolerably free from the direct interference of man. The former disappeared, reappeared (mainly by introduction), and increased with extraordinary rapidity while the latter was steadily on the decline. The reason appears to lie in this, that while the older trees favoured by the Woodpeckers were gradually being destroyed, new planting had created a fresh supply of food which satisfied and encouraged the Squirrel. But these new plantations had no power to check the decline of the Woodpecker, to which old trees only are of value for nesting, so that, apparently in the midst of plenty, it was starved for lack of nesting sites.

While I have chosen typical forest animals to illustrate the effects of the destruction of the forest, it must not be forgotten that the influence of the destruction spread far and wide throughout the native fauna. While only creatures vitally connected with woodland suffered extinction, every animal that sought the woods at one part or another of the year, for food, for breeding, or for temporary shelter, suffered in its numbers, in its range or in its habits from the disappearance of its haunts. It is not necessary to suppose that a complete disappearance of woods was necessary to influence the forest animals, for to many of them constant disturbance of their lairs or breeding-places is almost as fatal as the total destruction of these places of rest and shelter.

Yet it must not be forgotten that if the destruction of the forest told heavily against the woodland creatures, it favoured the increase of other races, for "give and take" is one of Nature's fundamental principles. A single recent example, recorded in the *Field* of 1908 from the Game Book of a Perthshire estate, will sufficiently illustrate this truth. In the season 1820-21 the bag of Grouse on this estate numbered 259. Immediately afterwards the moor was planted, and the Grouse disappeared. In the early seventies the timber came to maturity and was cut down, and in 1873-4, where few Grouse had been seen for fifty years, 252 were killed. The destruction of woodland prepared the way for new growths of heather and for the return and increase of Grouse.

I cannot turn from this study of the relationship between forests and animals without wondering to what extent the new and wide destruction of Scottish woods, caused by the needs of the War, will affect the few truly forest animals which survive from the old fauna. That it will have its influence cannot be doubted: the Wild Cat and the Pine Marten will be driven still further into the diminishing wilds, the increase of the Squirrel may be checked, the Woodpecker may cease its colonizing. The naturalist can only hope that, without delay, new forests may arise on the ruins of the old, to save not the least interesting of our wild creatures from a fresh period of decline and death.

CHAPTER VII

INFLUENCES OF CULTIVATION AND CIVILIZATION

> No more the heath fowl there her nestling brood
> Fosters; no more the dreary plover plains;
> And when, from frozen regions of the pole,
> The wintry bittern to his wonted haunt,
> On weary wing returns, he finds the marsh
> Into a joyless stubble-ridge transformed,
> And mounts again to seek some watery wild.
>   GRAHAM'S *British Georgics.*

BETWEEN the ways of Man and the rule of wild Nature there is inevitable strife. Just as the wastes of Nature are invaded by the plough, just so the original plants and animals of a district are disturbed or dispersed; and as the economic necessity becomes more insistent and cultivation more intense, so much the deeper becomes the disturbance and the more certain the doom of the early associations. If the "lowing herd" is to wind "slowly o'er the lea," we may listen no more for the bleat of the Snipe and the call of the Curlew; if waving cornfields are to fill our valleys and climb our hill-sides, we shall no more see the Great Bustard and his companions of the open plains. The civilization of cities and villages, of mills and factories, of tramways, motor cars and railways has banished peace from the valleys, and the places where man has been most successful are those which most certainly shall know the presence of the wild no more. Yet, in spite of all the disruption, animal life has more than held its own, for cultivation and increase in numbers go hand in hand. Although many species have disappeared, the variety and the volume of life are greater.

The influence of man on animal life is most intimately bound up with the progress of civilization, and while some results of this progress have been discussed in other chapters, it is proposed to confine this to evidences of the more immediate effects of the cultivation of the soil and of the ways of civilized life. In four marked directions these influences have tended. To one set of animals, unable to accommodate themselves to the changes, they have meant reduction in numbers, restriction of range, and even extermination; to another set, less fixed in habit, the new conditions have merely afforded new opportunities of increase and expansion; in some ways they have made easy the dispersal of animals in fresh areas; and in many a creature they have actually induced the adoption of new habits of life.

VII. 1

DECREASE OF ANIMAL LIFE

In various ways, cultivation and the ramifications of civilization have tended to limit the numbers of our native animals. In the main this type of influence is due to interference with habits, destruction of breeding haunts and feeding grounds, and to a variety of side issues which have arisen through the introduction of the amenities of civilization. The causes which have led to the decrease in numbers, restriction of range or extinction of native creatures, are here discussed in three groups—the breaking in of waste land, the reclamation of swamps, and the interference of civilization.

BREAKING IN OF WASTE LAND

The appearance of Scotland when all the country was forest or moor, wild meadow, swamp or loch, can scarcely be conceived at the present day, so much has cultivation and the breaking in of the "waste" changed its aspect and nature. No change has been so far-reaching in altering the distribution of animal life. Many of the birds which to-day nest on the moorlands once spread over the whole countryside; many an animal roved in freedom upon the plains which now drags out a meagre existence in the recesses of the hills. It is not easy to trace this change in distribution from historical records, for the early historians were no naturalists, except when they had some "mervaille" to disclose, and there was nothing to appeal to their imagination in the scarcely noticeable decline of once familiar beasts and birds. Even so, in the sixteenth century Hector Boece remarked the antipathy between Nature's folk and the proximity of man:

In all boundis of Scotland, except thay partis quhair continewall habitatioun of peple makis impediment thairto, is gret plente of haris, hartis, hindis, dayis, rais, wolffis, wild hors, and toddis.

INFLUENCES OF CULTIVATION

No group of animals has suffered in this respect more than the birds of the plains. Indeed, it seems as if cultivation, which has already blotted some of them from the fauna of Scotland, may in time exterminate the whole race of these ground-nesting birds which for safety prefer speed of limb to flight—much as flightless birds have disappeared wherever they have encountered civilization.

No larger or more handsome bird ever lived and bred in Britain than the GREAT BUSTARD (*Otis tarda*). Once it

Fig. 62. Great Bustard (male and female)—formerly a native of Scotland. $\frac{1}{14}$ nat. size.

may have nested in the Lowlands of Scotland in numbers such as still are to be found in outlying districts of Spain, far from the bustle of mankind; but already in the sixteenth century the spread of cultivation had reduced it to a last remnant. Of its occurrence in the Merse in Berwickshire, Hector Boece says:

Beside thir thre uncouth kind of fowlis [capercailyie, muir-fowl, and black-game], is ane uthir kind of fowlis in the Mers, mair uncouth, namit gustardis, als mukle as ane swan; bot in colour of thair fedderis, and gust [taste] of thair flesche, thay ar litil different fra ane pertrick [partridge]. Thir last

DECREASE OF ANIMAL LIFE

fowlis ar not frequent, but in few noumer; and sa far haitis the cumpany of man, that gif thay find thair eggis aindit [handled] or twichit be men, thay leif thaim, and layis eggis in ane uthir place. Thay lay thair eggis in the bair erd.

I know of no later reference to the breeding of the Great Bustard in Scotland. It has disappeared from our country as completely as the Wolf or the Brown Bear, and others of its kind are following.

Once the QUAIL (*Coturnix coturnix*) was common in

Fig. 63. Quails—once common in Scotland, now scarce. ⅔ nat. size.

Britain, nesting often, but more frequently appearing in migrating flocks which stayed over winter into spring. Quail hunting was a favourite sport in England, and "quail-calls" for luring the quarry were articles of commerce. In Scotland too, Quails seem to have been abundant up to the middle of the sixteenth century for in 1551 their price was fixed at 2*d*. apiece—the price of a Snipe, half the price of a Woodcock. In the succeeding century, however, a decrease in their numbers, no doubt due to the advance of agriculture, caused the law to impose a heavy penalty of £100 upon any person who bought or sold Quails, and brought about in 1685 the institution of a close season from the first day of Lent till the first day of July. Thirteen years later, still dwindling numbers caused the passing of a new Act, which on the ground that Quails had become scarce, forbade anyone to make use of setting dogs with nets for taking them during the following seven years under pain of 40 merks Scots. Nowadays the Quail is a scarce bird, though its numbers vary from year to year, and as a regular nester it has all but disappeared. The intense cultivation of the valleys and the invention of close-mowing hay-cutters and reapers, which destroy birds and nests as well, has cut off the last chance of survival of the Quail in Scotland. The implication of mechanical reapers in this extermination is tolerably clear. One exception has to be made to the statement that the Quail has ceased to breed regularly in Scotland. In Shetland it nests more frequently than elsewhere in Scotland, and why? It seems reasonable to associate its persistence there with the fact that in the fields of Shetland the "reaper" is unknown. There, after the manner of the old days, the scythe or the sickle still mows the waving corn, and the Quail reaps the benefit of such peace as a primitive cultivation gives.

The CORNCRAKE (*Crex crex*) seems to be following in the trail of its relative. My own impression is that its call is now less common in the Scottish valleys I know best. Bird of the cultivated land though it be, experience has already taught many of its kind to desert the cultivated area and nest in the rough corners of the fields. It may be that the field-nesters are being gradually eliminated by the perfection of the modern reaping-machine.

Even LAPWINGS, wearied of the disturbances of cultivation, are said more and more to be seeking safe nesting-places on the wild slopes of the moors, for the Peewit, homely bird as it is, is sensitive to repeated interference, and seldom remains in a place where its eggs are destroyed time and again, either by agricultural operations or by egg-collectors.

These are simple illustrations of the restrictive effects of cultivation, but perhaps the extension of arable land or the consequent decrease of the larger members of the wild fauna have had almost as much to do with the gradual disappearance of some beasts and birds of prey, and certainly of the Badger, as the active measures which have been taken against them by man on account of their depredations. In newly settled countries cause and effect are more easily traced. Before the settlers reached California from the east, the "Jack-Rabbit" (*Lepus texianus*) swarmed in fertile valleys and plains. In the earlier settlement days it proved so serious a menace to the crops that rabbit-drives on a large scale became a feature of western life. At a single drive in 1896 in Fresno, 20,000 "Rabbits" were slain. Twenty years pass, and we learn from the *American Museum Journal* (Jan. 1917) that the custom is forgotten. Partly owing to the warfare against them, but also in great degree to the advance of cultivation, the "Rabbits" have been driven from the cultivated grounds to the rough uncultivated foothills; for several thousands of orchards and vineyards, market gardens and dairies have replaced the grain-fields of the comparatively few and large farms of a quarter of a century ago.

With the breaking in of the wild banks and braes—"the burnin' yellow's awa that was aince a-lowe, On the braes o' whin"—the nesting sites for many small birds and shelter for many small rodents and insectivores have disappeared, to the grievous reduction of their kinds.

It is not easy to lay to its charge all the evil influences cultivation has exercised upon wild animal life. Its ways of working are many and subtle. Take, for example, the case of the disappearance of Butterflies.

"Account for it as we may," says Mr William Evans in his *Fauna of the Forth*, "there is abundant evidence that butterflies were formerly more plentiful in the district [of "Forth"] than they are now; indeed we seem to have lost quite a number of species in the course of the last century. In the older lists, we find *Vanessa c-album*[1], *Pararge megæra*[2], *Pamphila linea*[3], and *Pamphila silvanus*[4], all of which must have vanished long ago, and I fear the Orange-tip, Peacock, Speckled Wood, and Ringlet—species I used to get in the Lothians when a boy but never see now—have gone also."

Cultivation is probably the culprit; the wild food plant of the caterpillar has been destroyed in the old haunts, sheltering hedgerows have been removed, the dust and grime of towns and traffic have rendered existence impossible. Even where the final stage, extermination, has not been reached,

[1] Comma Butterfly.   [2] Wall B.
[3] Small Skipper B.   [4] Large Skipper B.

the replacing of once universally distributed food-plants by pasture and arable land has led to the localization of many species of insects.

Perhaps the same vague influences have to do with the gradual disappearance of the Ring Ouzel on the Pentlands, where it nested commonly in the seventies of last century, according to Lieut.-Colonel Wedderburn's *List*, but where it is now seldom to be seen.

In odd, unthought-of ways, too, cultivation tells. Artificial manure is spread wherever arable land is known, but who takes into account its effect on the small inhabitants of the soil, except in the case of a few kinds of grubs on the farmer's black list? Yet the effect of artificial manure may be evident enough even on large creatures, for of a flock of wild geese, Pink-footed and Barnacle, which, in the spring of 1917 after the shooting season ended, moved from the marshes of the Solway to feed on the adjoining fields, no fewer than 200 were found dead; and their deaths were attributed to a new chemical manure with which the fields had been dressed. (*Field*, 5 May, 1917.)

RECLAMATION OF SWAMPS

The extraordinary extent to which the present land surface of our country is due to the disappearance of swamp can be realized only by picturing the conditions which held in Scotland about the time of the arrival of man in the country. Excessive humidity had followed upon a dry period when forest growth had made great progress. Waste waters collected; swamps formed to such an extent that the forest trees rotted, and falling, encouraged the formation of still deeper morasses, until a large part of Scotland was little better than peat-moss. Dryer climatic conditions and the growth of vegetation have done much to convert the swamp to dry land; but man also has played his part ever since the days of the Bronze Age, when the rubbish cast from the island platforms upon which his lake-villages stood contributed to the disappearance of the very lakes which gave him security.

The Romans possessed the art of draining lakes and swamps. The mansions of London rise where once the Curlew and the Bittern, Wild Ducks and Water-Hens lived and nested; Pimlico was largely built upon piles, and Sir Charles Lucas recorded in his address to the British Association in 1914, that he had heard a lady tell how her grandfather used to say that in the heart of the district where Belgrave Square now stands he had shot Snipe. The same authority records that in the last three centuries the amount of land recovered from the marshes of Lincolnshire alone must have been more than 500 sq. miles.

In Scotland a similar condition of things held. Herodian describes the Caledonians as wearing neither coat of mail nor helmet, lest they should be impeded in their marches through bogs and morasses—whence such quantity of vapour was exhaled that the air was always thick and cloudy. Even in the sixteenth century, the country was generously supplied with swamps.

"Everywhere," says Professor Hume Brown in describing Scotland in the time of Queen Mary, "there were numerous mosses, lochans, and even lochs, which have long since disappeared, and the disappearance of which has materially altered the general aspect of the country. To take but one example, in Blaeu's map of Fife, there are some twenty lochs or lochans, several of them as large as the present Loch Leven, of which there is little or no trace at the present day."

At a still later date wild-fowl frequented marshes in the heart of modern Edinburgh—in the Nor' Loch now occupied by Princes Street Gardens, and in the large Borough Loch which occupied the site of the Meadows and regarding which the proclamation was made in 1581 that "na gyrs [grass] women or utheris pas within the South Loch to sheir the gyrs thereof, hary the burdis nestis, tak away the eggs of the saming before Midsymer nixt."

How has the reclamation of such sheets of water affected the natural associations of animals? The amount of bird life which has been banished may be judged from a statement said to have been made by William of Malmesbury, who lived in the twelfth century, to the effect that the Lincolnshire Fens were so covered with Coots and Ducks and the flashes [pools] with fowl, that in moulting-time, when the birds were unable to fly, the natives could take two or three thousand at a draught with their nets. Old records show that the marshes of Scotland were not less rich in bird life. In the sixteenth century Boece said of Loch Spynie, near Elgin, now all but vanished through the influence of man,

> In Murray, is ane loch namit Spynee quhair gret plente is of swannis. The cause quhy the swannis multiplyis so fast in this loch is throw ane herbe namit Olour, quhilk burgeonis with gret fertilitie in the said loch, and the seed of it is richt nurisand and delicius to swannis. This herbe is sa brudy that quhair it is anis sawin or plantit it can nevir be destroyit; as may be provin be experience: for thought this loch be V milis lang, and wes sum tyme as the memorie of man yet beris, full of salmond and uthir gret fische, yit, fra this herbe began to burgeon in it, the watter is growin sa schauld, than ane man may waid throw the maist partis thairof; and therefore, all manir of gret fische is quit evanist out of it. [Bellenden's Translation.]

And now the Swans and cygnets, which, as other accounts tell us, were so numerous as often to darken the air in their flight, have disappeared with the draining of the loch.

Even where swamps were not completely drained the reduction in area which took place in many lochs must have greatly diminished the numbers of wild fowl. In 1810 the Loch of Forfar, a fine body of water, was so drained that its circumference was reduced from not less than four English miles to two, and its depth decreased by ten feet. The water in Loch Leven was lowered in 1830 by nine feet and the area contracted from 4312 to 3545 acres. In 1629, Lowther on his journey noted that here

> there [be] great store of almost [*i.e.* most all] kinds of wild fowl, of wild geese, there being continually seen 3000 or 4000, and swans many, the swans will not suffer any foreign swan to be with them, in stormy weather the old swans will carry the young ones on the wings off the water....They dry them in their chimneys like red herrings[1].

Though Loch Leven is still one of the most richly stocked lochs in civilized Scotland the numbers of wild animals cannot compare with those of former days. Not only have wild fowl suffered, but the greater competition for food induced by the smaller area of the feeding-ground has resulted in the disappearance of that interesting fish, the red-bellied Char, the last example of which was caught a few years before 1844, whereas in Lowther's time it was common.

Two sets of animals have been reduced in numbers or have disappeared with the draining of the lakes and marshes. The permanent inhabitants of these areas, their fishes, frogs, newts and lesser denizens have gone, and following on the disappearance of these true denizens has come that of the

---

[1] It is possible that the last sentence refers to the fishes of the Loch, though Lowther's grammatical arrangement gives no ground for that belief.

creatures which fed upon them, such as the marsh-frequenting birds, and the Otter, still a characteristic inhabitant of the Norfolk "Broads."

To take first a few actual examples of the effects upon denizens: The remains of hundreds of Common Frogs and Toads, and those possibly of the Natterjack Toad, now unknown north of the Scottish shores of the Solway Firth, have been found in deposits of the Neolithic period, in the Bone Cave of Allt nan Uamh in western Sutherland, where no marsh has existed for long ages and where now even the burn itself runs underground. It is true that, in the locality mentioned, nature and not man has been the reclaiming agent, but this does not affect the example as typifying one obvious result of disappearing marshes—the disappearance of creatures vitally associated with them.

The vanishing of two of the humbler inhabitants of the swamps is also worth recording, on account of their special interest to man. With the marshes and wet places perished the main breeding-places of the Mosquito, the larvae of which live in shallow pools; and since the Mosquito conveys to man the germ of malaria or ague, with the Mosquito disappeared that dreaded disease, which was once far more common in Scotland and England than it is at present (see p. 508). In the same way, the draining of the fields has ruined the home of a Pond Snail (*Limnæa truncatula*), the host in which one stage of the Liver-fluke (*Fasciola hepatica*) develops. But a single break in its life-history extirpates the Liver-fluke, and the ultimate result of the disappearance of the Pond Snail has been that from many districts liver-rot in sheep, which was caused by the adult Liver-flukes, has been entirely banished; so that reclamation has all but removed a fatal disease which formerly caused a loss of sheep in the United Kingdom estimated at a million a year.

Nor must we forget that the numbers of microscopically small inhabitants have been even more seriously reduced, for some of these are particularly sensitive to alterations in physical conditions. In the summer of 1897, Dr Thomas Scott found an artificial pond in the neighbourhood of Edinburgh swarming with the Water-flea, *Daphnia pulex*. Seven weeks later, as the result of a spell of dry weather, the pond was much reduced in size and not a single Water-flea could be found. If a reduction in level has so marked results, what must have been the effect on the myriads of smaller inhabitants of the waters of the draining of hundreds of square miles of swamp?

Remembering those transformations of swamp into dry land which have taken place over the whole country, we can easily understand how creatures whose sustenance depends upon the waters, such as the birds of the fens, have fallen

Fig. 64. Bittern—banished from Scotland with the marshes. ¼ nat. size.

off in numbers or have forsaken our land. Consider the cases of two marsh birds which played an important part in our predecessors' dietary,—the Bittern and the Crane.

The former abundance of the BITTERN (*Botaurus stellaris*) (Fig. 64) in the fens of England, where Bittern-fowling was long a favourite sport, itself would suggest that the "bittour" which "bumbleth in the mire" must have bred in the marshes of the northern kingdom. Fortunately there is stronger testimony of its Scottish domicile, for it is clear from a law

passed in 1600, during the reign of James VI, that the
"bittour" was a hawking quarry of the court of that period,
since it is specifically protected along with many other birds
—"sik kynde of fowles commonly used to be chased with
halkes" (see p. 203). Further in 1630 Sir Robert Gordon
wrote of the northern parts "ther is great store [in Suther-
land] of...bewters."

At the present day in Scotland only a rare straggler of

Fig. 65. Crane—a former inhabitant of Scottish marsh-lands. $\frac{1}{18}$ nat. size.

the species occurs on migration, the last I have seen being
a specimen which having lost its way, was found in Harris
in the spring of 1917, starved to death for lack of the mainland marshes where once it could have fed and nested.

There is also definite evidence of the occurrence in
Scotland of the CRANE (*Grus grus*) (Fig. 65)—a heron-
like bird distinguished by its fine large wing plumes—for
Dr Eagle Clarke identified a limb bone of this bird from the
refuse of the Roman station at Newstead near Melrose.
That it was a common bird in the Scottish marshes of

the sixteenth century is shown by Bishop Leslie's remark that in Scotland there were "Cranes anew [enough] as lykwyse herounis." Further we may justly compare the conditions here with those in Ireland, where the Crane occurred in great numbers, judging from the account of Giraldus Cambrensis who, in the twelfth century, related that in that country Cranes were often found in flocks of a hundred: "uno in grege centum, et circiter hanc numerum frequenter invenias," and a manuscript of whose work in the British Museum clearly shows a distinction between the Crane and the Heron, which are illustrated side by side. The distinction indicated by the figures is worth mentioning, because in early Scottish and other records, the name "Crane" is frequently applied to the Heron. Thus when I find that in 1541, John Soutar, in reward for acting as fowler of Cupar Abbey was to receive for a "Crane" or a Swan five shillings, or that an Act of the Scots Parliament of 1551 fixed the price of the "Crane" at five shillings, or that Martin wrote in 1703 regarding "Cranes" in Skye, that "of this latter sort I have seen sixty on the shore in a flock together," I am uncertain whether the references are to Cranes or Herons. I am inclined to believe, however, that the latter is the bird signified, for in the very complete list of prices of wild fowl fixed by the Statute of 1551, the item "the cran five shillings," clearly indicates the Heron, for it is highly improbable that the true Crane would have been mentioned and the commoner Heron omitted, as it would otherwise be, when such birds as Larks, Woodcock, and Plovers are specifically enumerated. It is still a common practice in some districts of Scotland and England to call the Heron a "Crane."

The Crane has long ceased to breed in Scotland, and now occurs only rarely even as a bird of passage. In England, as in Scotland, the reclamation of the marshes has completely banished it, for although it bred in East Anglia till 1590 and appeared as a regular migrant till a later period, there too it has been unknown as a regular visitor for many a year.

These are but typical examples of the fates of marsh-loving birds. Many more must have suffered with them. There can be little doubt that such as Herons and Curlews

DECREASE OF ANIMAL LIFE

have shared in their decline, though definite statements cannot be made in the absence of early historical references, which might give a clue to former distribution and abundance; and who knows but that the frog-haunted swamps of past ages may have afforded sustenance even to the Stork, which Fordun records as having nested upon the Cathedral of St Giles in Edinburgh in the year 1416?

The swamps of the low country are gone, for the marsh lovers "the pleasant places of the wilderness are dried up," and their tenants are exterminated, or have betaken themselves to the moors. New conditions of life have imposed new habits, and I imagine that the annual movements of Curlews and other moorland birds from seashore to uplands have been created or at any rate magnified by the removal of the lowland flats and fens, from which the seashore was but a step away. Yet less often than our forefathers did, we hear the welcome calls of the marsh birds, telling by their return that winter has passed and spring is at hand, as the old south-country rhyme has it, " Whaup, Whimbrel an' Plover, when they whustle the warst o't's over."

THE INTERFERENCE OF CIVILIZATION

Not only direct trenching on their favourite sites has influenced the wild creatures, but many of the developments of civilization have affected their numbers.

DESTRUCTION BY LIGHTHOUSES AND OTHER LIGHTS

The influence of artificial lights is not of much importance in the ultimate relationships of a fauna, yet it illustrates very effectively the destructive tendency of some of the amenities of our later civilization. Lighthouses and lightships, scattered in ever-increasing numbers round the rocky shores of Scotland to guide sailors to safety, lure every year thousands of winged migrants to their doom,—in Tennyson's words

> The beacon's blaze allures
> The bird of passage till he madly strikes
> Against it, and beats out his weary life.

The solitary beams seem to possess a fatal attraction for both birds and moths, especially on foggy nights when they are compelled to travel at low altitudes. At such times myriads of birds hover around and dash against the stout glass of the lantern with such vehemence that the sea around becomes covered with thousands of dead bodies.

"Hosts of glittering objects" wrote Dr Eagle Clarke, of migration at the Eddystone Lighthouse, "birds resplendent as it were in burnished gold, were fluttering in, or crossing at all angles, the brilliant revolving beams of light. Those which winged their way up the beams towards the lantern were innumerable, and resembled streaks of approaching light. These either struck the glass, or, recovering themselves, passed out of the ray ere the fatal focal point was reached. Those which simply crossed the rays were illuminated for a moment only, and became mere spectres on passing into the gloom beyond. Some of those which struck fell like stones from their violent contact with the glass; while others glanced off more or less injured or stunned to perish miserably in the surf below."

Even refinements of invention may make their influence felt. The *Field* records that since the lighthouse at the Skaw was fitted with a revolving in place of a fixed light, the number of birds attracted to their destruction has increased enormously. On the night of 11 October 1907 no fewer than 1000 birds of different kinds were killed by flying against the lighthouse windows. And again, the alteration of the light of the Galloper Lightship from white to red, stayed the destruction, for whereas, Dr Eagle Clarke informs me, the keepers when relieved from duty formerly took ashore clothes-baskets full of Larks, now no migrants approach the light—a simple consequence of the reduced actinic power of red rays.

Insects, as well as birds, are attracted in large numbers by the beams of lighthouses, witness the records of Mr William Evans, who obtained from a dozen Scottish lighthouses in one or two years, 7500 individuals (excluding 2000 Gnats) representatives of 241 species of Moths, Butterflies, Caddis-flies and Lacewings, Beetles and Two-winged Flies or Diptera,—while as many as 400 Moths representing 30 species have been taken on a July night at the Isle of May lighthouse. It is also on record that the Diamond-backed Moth, a highly undesirable alien, has clustered in such numbers round a lantern, that the keeper spent the whole night sweeping them off, so that his light might be visible at sea.

Probably a greater number of dusk-flying insects are yearly destroyed by the artificial lights of towns and houses.

The least observant must have noticed the hordes of Moths and Beetles which on warm autumn evenings, dash madly around the electric lights in city streets sometimes even attracting Bats to the feast; and the proverbial "singed moth" that "dreads the flame," commemorates the death of myriads of relatives which succumb to the fatal attraction of artificial light, sometimes in such numbers as to choke the oil lamp over the chimney of which they have passed in their madness.

Railways and Telegraph Wires

Most wild animals are exceedingly sensitive to sounds: a commotion amongst the Pheasants of East Norfolk and other parts of England was caused by the noises of the Jutland battle and by the great munitions explosion in the East End of London in the spring of 1917, which aroused the game-birds some 40 seconds before the sound could be detected by human ears. Can one doubt, then, that the railways which interlace with a network the busy centres of industry, and penetrate the moorland and mountain fastnesses with long lines of bustle and noise, have helped to drive some lovers of peace and solitude to seek new homes? Telegraph wires, though they supply new roosting-places for myriads of Swallows before the autumn migration, cause such mortality amongst heavy-flying game-birds that special means have to be taken to show their presence over wide stretches of moorland. Even birds of active flight frequently come to grief upon the wires. In December 1906, near Innerwick, a large flock of Golden Plover was observed to fly before a strong wind against telegraph wires, seventeen in number, with the result that thirty-one of the birds were killed.

River Barriers and Fisheries

In rivers as well as on land, the advances of civilization have told hardly upon the original inhabitants. The erection of mills for the manufacture of woollen and linen cloth, of flour, meal and paper, and the construction of dams to obtain the necessary "head" of water to drive the machinery, have given rise on almost every stream to barriers which have interfered with the free passage of the smaller organisms of the waters, and have in some cases seriously checked the movements of even the large migratory fishes. This effect was clearly shown in America where the erection of mills and their dams on the Connecticut River greatly diminished the number of migrating Salmon, though in this case a compensating increase of other fishes was the result, for the Striped Bass on which the Salmon fed, multiplied as their destroyers disappeared. But no such compensation has occurred in Scottish rivers, for our migrating Salmon seldom feed to any extent on their sojourn in fresh waters, and the erection of cruives, weirs and dykes, which impede or check migration, has definitely, and in some rivers seriously, reduced the quantity and value of the fish fauna.

Mills themselves are in some cases exceedingly destructive to fishes, as where smolts enter a lade in their descent towards the sea and so pass over the water-wheel or through the turbines. In the case of the river Don in Aberdeenshire evidence was given before the Royal Commission on Salmon Fisheries (1902) showing that during the period of smolt migration seawards, when, owing to the lowness of the river, the whole of its water passed through the mills, great numbers—"cartloads" a witness said—of these young Salmon, compelled to pass through the turbines, were cast out, bruised and dead, in the water beneath.

But this actual destruction is less common and less significant than a simple interference with the upward movements of migrating Salmon, by the erection of mill-dams and weirs. The importance of such interference lies partly in its actual reduction of numbers, but more in that it cuts off suitable spawning beds in the upper reaches from several or from all the mature immigrants from the sea, and in so doing reduces the potential stock of future years. In Scotland alone it has been estimated that as many as 50 lochs and some 360 miles of river, suitable for the existence of Salmon, are rendered useless partly by impassable waterfalls, but mainly through the interference of man in creating impediments to migration by the erection of obstructions and contamination of the waters.

The tendency of such obstacles as dam-dykes, cruives or caulds must be so clear that examples of their actual

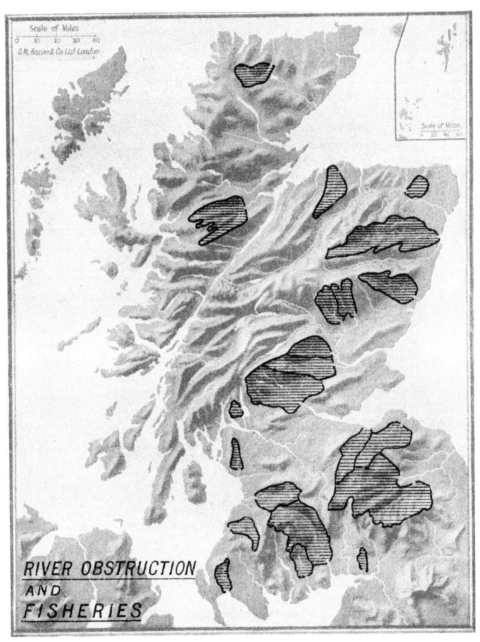

INFLUENCE OF MAN-MADE OBSTRUCTIONS (DAM-DYKES, CRUIVES, ETC.) ON SCOTTISH SALMON FISHERIES.

The shaded portions indicate river basins to which access of Salmon is interfered with by artificial obstructions
(*Founded on Report of Salmon Fisheries Commission, 1902*).

effects need scarcely be cited. Most hinder migration in some or in every state of the river, and by causing an unusual crowding of Salmon in the pools immediately below, offer easy opportunity for the spread of contagious disease or for the raids of poachers. Thus below the dam-dykes of Mugiemoss Paper Mill on the Aberdeenshire Don, as many as a thousand Salmon have been gathered in a comparatively small pool, all awaiting the flood which would give them opportunity to ascend; and it was customary for the water-bailiffs in mid-autumn to cart several hundreds of Salmon from this pool to an open stretch farther up the river. Even where a dyke does not form a complete barrier to the passage of Salmon it differentiates against the most valuable individuals: it is stated, that, before the making of a fish-pass in 1900, the Dupplin Cruive Dyke on the Earn, while allowing the passage of young and specially vigorous Salmon, formed an obstruction impassable to Salmon weighted with spawn.

Even the alteration of the river-bed, the creation of shallows above and below a dyke, may lead to the discomfiture of migrants. There appeared in the *Scotsman* of 13th March 1918 an account of a remarkable run of Salmon on a Border river, the Tweed. But a weir at Melrose interfered with the consummation of an event which ought to have stocked the upper reaches with unusual wealth of breeding fish.

"The fish ascended the cauld in large numbers," wrote the correspondent, "and in the shallow water on either side it was a matter of no difficulty to seize some of them as they made the passage. The spectacle of so many fish passing to the upper waters led to a general relaxation of ordinary conditions. On one of the days of the week-end, men, women and boys could be seen in the water, standing up to the knees, and armed with gaffs. The operations of those actively engaged were watched by large crowds on the banks. The natural instinct for capture, aided by the food stringency, became so prevalent that an unprecedented spectacle was witnessed on the Sunday. Many who had been attending the morning service found the spectacle of one particular hole, which had practically become a moving mass of fish, too much for ordinary restraint. The quantity of salmon taken at this point is understood to have been extraordinary...Two of the captured fish weighed 50 lb. and 48 lb. respectively."

The accompanying map (Map VI), where are indicated by shading the river-basins to which migration is interfered with, but not necessarily prevented by artificial obstructions, gives some idea of the extent of this influence upon the migratory fishes. It should be studied along with the succeeding map, to reveal the full measure of man's interference in our fresh waters.

Pollution of Rivers

Perhaps a more comprehensive destruction of aquatic life has followed upon the discharge into rivers of harmful by-products of manufacture. It is difficult to gauge the cumulative effects of the continuous flow of dilute solutions of poisonous by-products, and the subject has aroused much controversy; but many a river is made unsightly by their presence, and there can be little doubt that in such cases the organic life of the river is adversely affected. There can be no doubt that where deleterious by-products are allowed to enter rivers in quantity, even if intermittently, there results a destruction of fishes, and, a point as important though often forgotten, a destruction of the small creatures which are the food of fishes.

The pollution of rivers in a degree sufficient to affect the fauna is to a large extent a development of modern civilization, for although pollution follows upon the growth of industry and upon the aggregation of the people in social centres, yet its serious influence is a matter of quantity as well as of quality. As a rule it is due to the discharge from mills, mines, or sewage works of substances actually or potentially harmful, into the water of streams or rivers. Such discharges tell upon animal life in more ways than one.

Solid substances cast into rivers, even where, as in the case of sawdust, they seem to do no actual harm to adult fishes, yet cover the bed of the stream with layers of debris which ruin potential spawning beds and in others destroy fish eggs by preventing free access of oxygen. The deposition of solid matter in streams is illegal in Britain, yet here similar effects follow upon the discharge of washings from mineral works which contain fine dust in suspension. Thus the beds of streams in the neighbourhood of many coal mines are buried in layers of coal dust which settles everywhere, stifling vegetation and its animal dependents, upon which in turn the larger inhabitants of the streams subsist, destroying spawning beds, and actually choking fishes which venture within the noxious area. So also it is with the washings from writing-slate mills where, on the Seint and Llyfni, the dust is said to form in the beds of the rivers a layer as hard as cement.

Other pollutions act as direct poisons to the animal life of the stream. The waste water of bleaching works may contain chlorine; in that of paper mills have been found chlorine and sulphur compounds, as well as free sulphuric acid, to such an extent that a mixture of one part of waste to nine parts of river water killed Trout and Perch in one or two days. Distilleries poison streams with their organic acids, woollen mills with the alkalis from their wool washings, and gas-works with their carbolic acid, gaslime and cyanides. To gas-works' poisoning, fish are particularly sensitive, 1 part of waste in 200 parts of river water having been found to render an American Brook Trout moribund in 10 minutes, while other fishes survived only a few days. Waste water from iron and steel works may contain acids and may deposit a precipitate (ferric hydroxide) which adheres to the gill-filaments of fishes and checks respiration. So deadly are iron pollutions that experiment with waste water from a Canadian nail works has shown that a dilution of 1 part of the waste liquor in 1000 parts of the stream water may kill fishes such as Smelt and American Brook Trout in the course of ten minutes to half an hour.

Water discharged from old coal mines may also prove fatal to the inhabitants of a river, owing to the sulphuric acid set free by the breaking down of iron pyrites in the coal and accompanying strata. Such a pollution, accompanied by washings from a lime-kiln, recently found its way into a small loch in Scotland, with the result that the fishes were all killed off and the invertebrate life was seriously reduced, but a year afterwards Fresh-water Shrimps which had previously been scarce, reappeared and bred in great numbers in the loch.

Another frequent source of pollution inimical to animal life, arises from the discharge into rivers of organic matters such as sewage. Harmless as it may be in itself, sewage almost invariably develops noxious properties, as deadly as the poisonous disinfectants by which it is often accompanied. During its decomposition, through the action of bacteria, toxic substances may be developed, but even in their absence,

RIVER POLLUTION AND FISHERIES.
Basins of rivers in which migrations or spawning of Salmon are interfered with by pollution.
Basins of rivers in which migrations of Salmon are prevented by pollution.
(*After Royal Commission on Salmon Fisheries, 1902*).

of the lesser life can only be imagined—from the point of discharge to the sea, a distance of some 15 miles. On the Tay, the poisonous effluents of bleaching works have, according to Mr P. D. Malloch, at times left the lades quite white with dead smolts. The most general contamination, however, is probably due to the untreated sewage of towns. There are few large rivers free from this nuisance, and although in many its effects pass unobserved, in others its influence is forced upon notice: in the Forth in 1899, more than 400 dead Salmon were taken out below Stirling, and many more must have been carried away by the tide—a destruction attributed to excess of sewage during a period of low water in the river. Again on rivers such as the Clyde and the Irvine, pollution has practically put an end to the migration of Salmon.

Probably in all such cases the extermination is progressively selective, only creatures of special sensitiveness or habit succumbing in the first instance: it is on record that the influx of lead pollution to Coniston Lake destroyed the Char, although the Trout were able to survive.

The ultimate effect of pollution upon the fauna of a river depends to some extent upon the character of the river itself, especially at those parts where the noxious discharges enter. A swift river speedily dilutes the poisons added to it, and absorbs from the air fresh oxygen to replace that stolen from it by decomposing organic matter. The reverse is the case in a slow river, where a stagnant pool into which a poisonous effluent discharges may become a death trap to any fish entering it, and a barrier to migration upwards or downwards. The interference is even greater, from the point of view of fish migration, when the flow of a polluted river is held up for a long distance by the tide, for a barrier of pollution at the mouth may be sufficient to render valueless the whole course of the river, no matter how suitable its upper reaches in themselves may be. The accompanying map, following the Report of the Salmon Fisheries Commission of 1902, indicates in a general way the catchment areas of Scottish rivers and their tributaries where Salmon migration is interfered with by pollution. No information is available as to the effects of river contamination upon the invertebrate fauna, although this aspect is equally important, since upon these lesser organisms Salmon, in their tender years, and other fishes feed. But there is evidence that even larger creatures not members of the true aquatic fauna, though dependent on it, have suffered.

Even in the wider stretches of water the destructive effects of civilization upon the fauna are equally noticeable. One would scarcely imagine that a wide arm of the sea exposed to daily tidal changes, such as the Firth of Forth, would be affected by the works of man. Yet such is undoubtedly the case, for the deposition of sewage and rubbish, of coke and cinders from the increasing shipping on its waters, and from the gas works and coal pits along its shores, have played havoc with the fauna of a firth which in the sixteenth century, according to Boece, was "richt plenteus of coclis, osteris, muschellis, selch, pollock, merswine, and quhalis; with gret plente of quhit fische." The larger members of the fauna have departed, the Oyster beds are ruined, the Cockles and Mussels are not what they once were; but the most rapid change has taken place amongst the shore animals, for in many places the old fine stretches of sand and rocks whereon the people of the towns once spent happy hours, are buried beneath many inches of filthy cinders which have altered the courses of the streams and blotted out all traces of life. No naturalist would recognize in the impoverished shore fauna of the upper Firth to-day, that rich assemblage of marine things made famous, only a generation or two back, by the researches of Professor Edward Forbes, Professor Allman, Dr Strethill Wright and their fellow workers.

VII. 2

THE INCREASE OF ANIMAL LIFE

IN spite of the fact that cultivation has destroyed the breeding-places of many native animals, and has otherwise restricted their range and numbers, it cannot be denied that on the whole it has caused a marked increase of animal life. The effect is clearly seen in the Yellowstone Park in the Western United States, where birds with other animals are preserved, but where, nevertheless, they are not so abundant as in the neighbouring cultivated areas of the States. The same contrast holds good between the cultivated and wild districts of Scotland. The increase of certain species under the influence of cultivation is due mainly to the development of food supplies, and in a few cases to the increase of breeding-sites. In the present section, an endeavour will be made to trace in greater detail the causes to which the increase of animal life may be attributed.

INCREASE OF VEGETABLE FOOD

It is a truism that cultivation multiplies the plant-yield of a country, and that the more intense the cultivation the greater is the yield. It is equally evident that herein lies a vastly increased food-supply for vegetarian feeders. The result has been undoubtedly to increase their numbers, often to such an extent as to make them pests of agriculture; indeed, through cultivation, man may even be said to create the pests he denounces. Take for example the exceptionally clear case of the Colorado Beetle in America. When this Beetle, *Doryphora (Leptinotarsa) decem-lineata* (Fig. 66), was discovered amongst the Rocky Mountains in the region of the Upper Missouri in 1824, it was restricted in numbers and harmless, for it fed upon the Sand Bur of the wastes (*Solanum rostratum*), a wild species of the potato

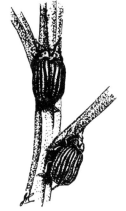

Fig. 66. Colorado Potato-Beetles. Slightly larger than nat. size.

family, peculiar to the district. Only when cultivation pushed westwards and the home of the Beetle was invaded, did it turn from the wild Sand Bur to the cultivated species, with appalling results. The abundance of the cultivated crop and its continuous distribution throughout the States gave the Beetle extraordinary opportunities for increase and spread, especially towards the east where cultivation was more intense. In 1859 it entered Nebraska. In 1861 it had reached Iowa; in 1864 and 1865 it crossed the Mississippi, entering Illinois by at least five routes. Here the distinctive influence of high cultivation was interestingly shown, for in the northern part of the State where potato-crops were more frequent than in the southern, it was noticed that "in marching through Illinois, in many separate columns, just as Shearman marched to the sea, the southern columns of the Grand Army lagged far behind the northern columns.' Wisconsin was invaded by the conqueror and possessed by the autumn of 1866. In 1867 it crossed the borders of Indiana. In 1869 it appeared in Ohio. In the following year it crossed into Canada, appearing in the province of Ontario, swarming in the Detroit River, and making passage across Lake Erie on "ships, chips, staves, boards, or any other floating object which presented itself." In 1874, it had reached the Atlantic and multiplied with such rapidity that in September 1876, Beetles were washed ashore on the coast of Connecticut in such numbers as to poison the air, and a New London vessel was boarded in such force at sea that the hatches had to be shut down. Railway trains were stopped owing to the slipping of the wheels on the rails, caused by the slaughtered thousands. A summary of the exploits of the Colorado Beetle gives a vivid notion of the influence of cultivation in increasing the range and numbers of favoured species. It travelled from its native region in Colorado to the Atlantic ocean, covering at least 1500 miles in the sixteen years from 1859 and 1874; its annual rate of travel averaged 99 miles, but so distinct was the influence of cultivation that in the wilder western States the rate of progress was less than 50 miles a year.

America shows more clearly than the old countries the actual transformations in animal life wrought by the spread of civilization, for there we can still watch the struggles of man in conquering the wilderness. Thus cultivation has increased the food and consequently the numbers of the Kangaroo Rats (*Perodipus* and *Dipodomys*) of North America, so that they have become a plague to the settlers. They were harmless until man began to cultivate the sand-hills and sage-brush country of the west, when they turned to his crops, eating and storing away in burrows for winter use the planted seeds of maize, water-melons and vegetables to such an extent that now the only alternative to destroying the animals is to give up cultivation.

In Scotland, the effects of the growth of field and garden crops on animal life, if not so striking, are still sufficiently definite. To what but the cultivation of our fields do we owe the great flocks of seed-eating birds which frequent our stubble fields and farm-yards—the Sparrows, Greenfinches, Chaffinches, Yellowhammers, Bramblings, Corn Buntings, Wood Pigeons and the rest? In the primitive woodlands of Scotland, these seed destroyers found little room, for the depths of the forest are innocent of bird life. To a limited extent they may have flourished in the open ground, but they and their like increased with the spread of cultivation and of the seed crops which supplied them with easy nourishment, and scattered until now their numbers fill the length and breadth of the land.

The increase of some of these seed-eaters, and their dependence upon cultivated crops are notorious. In spite of the constant warfare waged against it, in spite of the annual slaughter wrought by Sparrow Clubs (one such in Kent, consisting of twenty members, killed 28,000 Sparrows in a single season), the Sparrow continues to increase beyond bounds. Throughout the year 75 per cent. to 80 per cent. of the food of adult Sparrows consists of cultivated grain, and the damage done to British crops by the millions of country Sparrows and the additional millions of town cousins that move fieldwards at the first hint of ripening grain, has been estimated by a writer in the *Empire Review* (1917) at 500,000 quarters of cereals, valued at £1,500,000 a year, though another estimate of their yearly depredations places the loss at an equivalent of £8,000,000.

It is clear that the sparrow pestilence is due to the development of cultivation, abetted, as we shall see later, by the protection afforded by town and village. So it is with other seed-eating birds condemned by the farmer and gardener: in equal steps with the success of their own industry the tillers of the soil have created the plagues of the Wood Pigeon, and the Rook, of the Chaffinch, Bullfinch and their like. And many birds also which avoid the stigma of the farmer's malison owe their numbers in great part to the harvests of the fields.

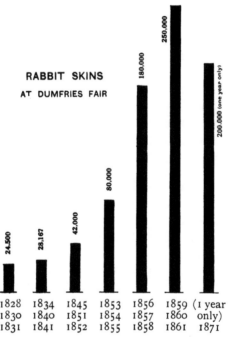

Fig. 67. RABBITS AND CULTIVATION: Numbers of Rabbit-Skins offered for sale at Dumfries Fair from 1828 to 1871. The columns represent 3-years' totals (where the years are not continuous, no record exists for the omitted years) and show a progressive and marked increase throughout the century. This increase corresponds with a period of increasing agricultural activity.

Amongst the mammals which have gained most advantage from cultivation are naturally the vegetarian rodents. The increase of Rabbits, as I have previously shown, was contemporaneous with the development of cultivation. This association is clearly shown in the statistics of the rabbit-skins offered for sale throughout the nineteenth century at the Dumfries Fur Market. The diagram above represents the numbers of skins, the records available being arranged in groups of three years for the sake of eliminating extreme annual fluctuations. It is to be seen that a steady increase is made throughout a century distinguished for its advance in agriculture; and further that the increase, which makes its first marked step shortly after the year 1840, when "notable improvements" in agricultural methods had been initiated, swells by leaps and bounds during the period following 1853, which Mr R. Prothero (Lord Ernle) has described as the "Golden Age" of British agriculture.

This close correlation between agricultural progress and numbers of Rabbits was, of course, not confined to the Lowlands nor to the statistics of the Dumfries Fair. It was probably universal throughout Scotland, and affords a simple explanation of the extraordinary increase in the annual slaughter of Rabbits on an estate in agricultural Perthshire, as revealed by quotations from the old Game Book, in the *Field* of 21 November, 1908. On this estate only 2309 Rabbits were killed in the decade 1824-1833; during 1834-1843 the bag jumped to 21,431; and in 1844-1853 it reached 51,932. Thereafter Rabbits became such a pest that special efforts were made to keep them down, with the result that the stock of Rabbits was so reduced that the continuity of the statistics was disturbed. But the significance of the history is clear—the increases in numbers were contemporaneous with notable periods of agricultural activity.

The numbers of Hares as well as of Rabbits responded to the agricultural advance of the "Golden Age," as a glance at a more detailed diagram of the statistics of the Dumfries Fur Market (p. 167) will show.

Millions of Brown Rats and of House Mice depend upon the stores of the granaries of the United Kingdom, where the damage done by the former has been estimated at upwards of fifteen million pounds sterling a year. Field Mice and Harvest Mice depend directly upon the products of the field, and had it not been for the bountiful supplies of cultivated areas, we should probably never have had experience of such a Vole plague as that which devastated many parts of the Lowlands in the early nineties of last century.

No body of animals, however, has benefited from the development of agriculture so much as Insects. Cultivation has created insect pests by the score, for there is scarcely an insect group but has many members on the black list of the farmer or gardener. Of these only a few typical examples need be mentioned. The roots of cultivated crops afford superabundant food to the Wireworm grubs of Click Beetles, the larvae of Cockchafers, and the "Leatherjacket" grubs of the Cranefly or Daddy-long-legs, which in some years, as in 1894, have so increased in numbers as to destroy hundreds of acres of pasture, so that shepherds knew not where to feed their sheep. Turnip and related crops furnish food for new hordes of Turnip Greenfly (*Rhinocola dianthi*), Turnip Moth (*Euxoa* (*Agrotis*) *segetum*), Turnip Saw-fly (*Athalia colibri*), Turnip Flea Beetle (*Phyllotreta nemorum*), and the mischievous Diamond-backed Moth (*Plutella maculipennis*), which has devastated whole fields of turnips, swedes, rape and cabbage. Upon the yield of the garden and the orchard have thriven and multiplied the Cabbage Butterflies (*Pieris brassicæ* and *rapæ*), the Gooseberry Saw-fly (*Pteronus ribesii*) and Magpie Moth (*Abraxas grossulariata*), the Red Mites of the currant bushes (*Tetranychus telarius*), the Umber and Winter Moths (*Hybernia defoliaria* and *Cheimatobia brumata*) of apple trees, which also nourish the dreaded multitudes of American Blight or Woolly Aphis (*Eriosoma lanigera*). Many more kinds of Plant Lice and Greenflies suck the juices of fruit and flower-bearing plants, and many Beetles, such as the Pea Beetles (*Bruchus pisorum*), Strawberry Weevils (*Anthonomus signatus*), Raspberry Weevils (*Otiorrhynchus picipes*), Appleblossom Weevils (*Anthonomus pomorum*), flourish upon legumes, green crops, flower blossoms and fruits, and have in one season fed themselves upon the crops of a Kentish fruit-grower to an extent reckoned in value at £500.

Within the soil as well as above ground and in the air, cultivation has increased the animal life, for not only insect larvae, but also vegetarian forms such as Millipedes and Earthworms have multiplied. Where does the angler seek for bait? Not in natural areas such as moors, nor in the leaf-mould of the forest, nor the sand-hills of the coast, but in fields, gardens and the refuse-heaps of cultivated areas. Hensen's careful examinations revealed that the highly cultivated soil of a garden yielded 53,767 Earthworms an acre, but he believed that the less intense cultivation of the cornfield would yield only about half that enormous number. Even more striking are the facts recorded by Marsh relating to the increase of Earthworms which followed upon the development of agriculture in North America.

"Forty or fifty years ago," he wrote in 1874, "they [Earthworms used by anglers for bait] were so scarce in the newer parts of New England, that the rustic fishermen of every village kept secret the few places where they were to be found in their neighbourhood as a professional mystery, but at present one can hardly turn over a shovelful of rich moist soil anywhere without unearthing several of them. A very intelligent lady, born in the woods of Northern New England, told me that in her childhood these worms were almost unknown in that region, but that they increased as the country was cleared, and at last became so numerous in some places that the water of springs and even of shallow wells, which had formerly been excellent, was rendered undrinkable by the quantity of dead worms that fell into them."

INCREASE OF ANIMAL FOOD

The increase of such animal life as depended upon vegetable food naturally led to new developments in the numbers of the creatures which subsist upon their fellows. The fresh numbers of Worms in New England, just alluded to, gave rise to an influx and increase of the small insectivorous birds which follow the trail of the settler. It was they who checked the excessive multiplication of the Worms and finally abated the nuisance.

So the augmentation of our own tiny vegetarians, earthworms and insect pests, has led to increase of carnivorous Centipedes, insect-eating Birds, and Moles. Observation of the habits, and examination of the food contained in the crops of soft-billed birds leave no doubt that they subsist largely upon the insect products of cultivated crops. Professor Newstead found that on a low average a Starling visited its young 169 times a day, and on some days 340 times, with food which included 269 injurious insects. A Great Tit was seen to make 384 visits in a day and 90 per cent. of the food carried during its visits consisted of insect pests. Professor Newstead concludes

If 20 days are occupied in rearing the young, that gives us a grand total of 7680 visits to the nest, so that the single pair of birds would be responsible for the destruction of between 8000 and 9000 insects, chiefly caterpillars.

Starlings and Lapwings have been observed to clear in four days a turnip field badly infested by the Diamond-backed Moth, and so on. Could the extraordinary increase in the numbers of Starlings which has taken place in Scotland since the forties of last century and is still in progress, have taken place, had it not been for the increasing supplies of insectivorous food afforded them by a bounteous cultivation? Their most marked colonization of Scottish areas corresponds with the "golden age" of British agriculture.

Glance at the birds which assemble in the vineyards of France, and there you will find the same factors at work. It is true that Magpies, Partridges, and Fieldfares eat the grapes and damage the bordering plants, but in 1916, M. A. Hugues cited as birds which were attracted by the insect pests of the vineyards in the neighbourhood of Nîmes, the Ortolan (*Emberiza hortulana*), Stonechat (*Pratincola rubicola*), Wheatear (*Saxicola œnanthe*), European Bunting (*Miliaria europœa*), Crested Lark (*Galerida cristata*), Short-toed Lark (*Alauda brachydactyla*), Common Linnet (*Cannabina linota*), the Warblers, the European Nightjar (*Caprimulga europœus*), and the Tomtit (*Parus major*)—a specially efficient pest-destroyer. These do not attack the grapes, but depend during a great part of the year entirely on the secondary products of cultivation.

From such examples we may safely assume that the soft-billed insect-eaters are far more numerous in Scotland than they were before man's influence began to tell; that cultivation by augmenting their food-supply has added to the numbers of Starlings, Thrushes, Blackbirds, Wrens, Robins, Tits, the Warblers and others, as well as of Rooks and Black-headed Gulls. Even the seed-eaters, such as the Finches, have shared in the benefit, for they too feed their young upon insect food.

And what of Man himself? In the pastoral and forest regions of the Old World where he depends mainly upon hunting and fishing, the population, according to a recent estimate by Mr George Philip, numbers only from 1 to 16 a square mile, but in well-developed agricultural regions it rises from 32 to 128, and in productive regions commercially developed reaches from 128 to over 516 a square mile. Granted that foreign imports contribute to the sustenance of these new commercial multitudes, could there, nevertheless, be clearer evidence of the influence of cultivation in increasing numbers?

INCREASE OF GARBAGE

In addition to the more or less natural food with which a generous culture of the fields has supplied many creatures, the increase of the refuse of civilization has tended greatly to the multiplication of the garbage feeders. In the days before man concerned himself with health matters, the self-imposed duties of scavengers undertaken by birds and beasts must have saved him from many a pestilence. At one time, Kites were common in the streets of London, and the condition of Edinburgh and Leith in the sixteenth century and earlier led to a great increase in the numbers of the crow tribe—Carrion Crows, Rooks, and probably Ravens—which, as is shown by an extract from Wedderburn's Accompt Book already quoted (p. 224), were familiar objects in the streets. It is also on record that in the sixteenth century, Ravens were encouraged to multiply in Berwick where they were actually protected in the town on account of their value as scavengers (see p. 225). As the crow tribe has benefited by the garbage of the towns, the Seagulls have flourished upon the offal of the fishing villages.

No animal has multiplied so successfully on the refuse of civilization as the Brown Rat. It lurks in the midden of the farmyard, it riddles the refuse dumps of country-towns and villages, and it thrives by thousands in the sewers of our large cities. Mr Wm Berry has recorded that the removal of a refuse dump at Tayfield in Fife, rid his district of Rats, which had appeared in such numbers that the repairing of the damage done to field dykes, through the undermining of their foundations by Rats, cost over £200.

Many garbage feeders amongst insects have developed unwonted hordes through the ways of mankind. As early as 1675 Mackaile had noticed that artificial heaps of seaweed bred large numbers of insects, probably the Two-winged Flies *Cœlopa frigida* and *Actora æstuum*, and perhaps also small amphipod crustaceans like the Shore-fleas (*Talitrus saltator* and *Orchestia gammarellus*) to the benefit of their feathered destroyers.

"The people of Orkney," he wrote, "gather the sea-ware (which is frequently and especially cast out by the sea) into heaps, which, being putrified, affordeth a very bad smell, and many insects which the sterlings do feed upon, and therefore it is ordinary to see hundreds of these birds upon each heap."

Consider again that peculiar fauna, consisting mainly of Beetles—Shakespeare's "shard-borne beetle, with his drowsy hum," and its relatives—which spends its life feeding and breeding in the dung of domestic animals. How that minute fauna, containing amongst others, Beetles of the genera *Sphæridium, Cercyon, Megasternum,* and *Cryptopleurum* in the family Hydrophilidae, and *Aphodius* amongst the Scarabeids, must have multiplied since man introduced domestic animals to Scotland, and by closer and closer cultivation increased his stock till it outnumbered many times the wild stock which it displaced! In the same category, as dependent for its present day numbers upon the excrement of domestic stock, may be mentioned a Two-winged Fly or Dipteron—the common Yellow Dung Fly, *Scatophaga stercoraria*[1], whose name sufficiently indicates its unseemly habits. Of even wider distribution, and owing its multitudes still more evidently to the presence of man, is the Common House-Fly (*Musca domestica*). In every rotting heap of garbage it finds a comfortable nursery where it lays eggs which bring forth young in countless myriads, so that even the stolid Briton has turned at last and has sworn death to the Fly, the carrier of typhoid and similar diseases.

If such are the developments of animal life due to normal conditions of advanced civilization, what can be said of the effects of the highly abnormal conditions of modern warfare? For how much has the garbage of the battlefield to answer? Beasts and birds of prey, Wolves, Vultures, Buzzards and Carrion Crows follow in the wake of the armies of Eastern Europe to-day as persistently as they have done in every war to which history bears record; but under modern conditions still new hosts have arisen. Rats, and disease-bearing insects, Fleas, Lice, the Blow-Fly, the House-Fly and many relatives, have appeared in such overwhelming numbers and have caused so much suffering to man and beast as to call for the efforts of special medical organizations.

[1] *Gr.* σκωρ, σκατος, dung, and φαγειν, to eat; and *Lat.* stercorarius, pertaining to dung.

VII. 3

DISPERSAL OF ANIMALS

DISTINCT from the potentialities of increase with which civilization has endowed animal life are the means of dispersal which man has provided in preparing for his own transport.

The opening of through waterways has been found, for example in Canada, to give to the fishes and the small inhabitants of the waters, access to areas from which they were formerly absent. In 1874 Professor Marsh stated that the Erie Canal, in spite of its length of 360 miles and its ascending and descending locks in both directions, had caused the commingling of the freshwater fishes and the native plants of the Hudson and of the Upper Lakes, so that the fauna and flora of these regions had more species in common than they had before the canal was opened. Just so it is possible that the Panama Canal may for the first time bring inhabitants of the East and West to meet; and so the great waterway of Scotland, the Caledonian Canal, by connecting the lochs of the Great Glen, and by affording a constant circulation of water owing to the passing of vessels and the inflowing and outflowing to and from the locks, has brought about a commingling of the minute but distinct members of the faunas of the lochs. The body of water which passes from Loch Oich through the Canal into Loch Lochy certainly carries with it from one loch to the other some of the Entomostraca or lower crustaceans (Water-Fleas and the like), as well as other minute organisms which people the water in their thousands.

Closely associated with the movements of the waters is the influence of the shipping which passes upon their surface, for though there is little scope for transference of animals by such means within the bounds of Scotland, yet we have seen that steamers and the flotsam and jetsam of civilization played a great part in conveying the Colorado Beetle from the United States to Canada across the Detroit River and the breadth of Lake Erie. And it is self-evident that these aliens,

Brown Rats and Rabbits, which abound on very many Scottish islands, must have been borne thither in most cases by human modes of transport, as have been those introductions from foreign lands which shall be considered in another chapter.

Roads and bridges are important factors in the dispersal of animal life. Many creatures prefer the easy means of communication man has prepared for his own comfort, and have no hesitation in turning to their own use objects planned by him for quite another purpose. In moving from one wood to another, or from one district to another, Squirrels prefer to run along roads, the open rides in forests, the top of a fence or stone dyke. They have been known to make regular use of dykes in passing between coverts. The Common Brown Hare is said (New Statistical Account) to have appeared in the mountainous districts of Lismore and Appin "not until after roads were made which opened communication with the low country."

In many cases where rivers blocked the way of progress for certain animals, man has bridged the barrier and given them easy access to new districts. Rats and Squirrels, excellent swimmers though they are, make frequent use of bridges, and Dr Harvie-Brown, from his examination of dates and places of appearance, was of opinion that the first invasion of Perthshire by Squirrels from Stirlingshire was by the Bridge of Frew over the Forth; and that they probably made their first appearance in Sutherland in 1859 over Bonar Bridge, although it was not until the railway bridge at Invershin was built in 1869 that Squirrels became plentiful in the East of Sutherland. I have already suggested that Dr Harvie-Brown laid too much stress on the necessity of the Invershin Bridge for migration, but there can be little doubt that here and elsewhere bridges have to a considerable extent aided in the dispersal of animals.

The great transport systems which cross and cross again the countries of the world have also played their part in aiding the dispersal of the beasts of the field and birds of the air as well as of the sons of men. Apart from the influence of the rolling stock of our railways in transporting creatures accidentally or otherwise, the permanent iron way and its embankments have acted as a high road along which many animals have penetrated to new districts. Several causes have led to this gradual migration. The scraps of food dropped from carriage windows, the grease used in lubricating the railway points and the axles of carriages and trucks, and the grain and other edible stuffs spilt at goods' sidings and stations have induced birds, such as Rooks and Sparrows, and rodents, like Rats and Mice, to follow the permanent way in search of food. The Sparrow has made a tardy appearance at Corrour in western Inverness-shire along the railway route, and is now fairly common at the station there, from which centre, at an altitude of 1350 feet, it is at present in process of colonizing the immediate neighbourhood. About 1910, Dr Eagle Clarke has recorded, it found its way to the Lodge and the neighbouring premises, and has become common there, although in 1917 it had not yet reached the farmhouse and its outbuildings at the head of Loch Treig. Travellers by the Highland Railway over the ridge of the Grampians may see the same hardy campfollower of domestication in numbers at Dalnaspidal, the highest station, close on 1300 feet above sea level, in the heart of the Highland moors.

Less venturesome birds than the Rook and Sparrow have made use of railways in another way. In the centres of highly cultivated areas, railway embankments and the bottoms of hedgerows are almost the only places which offer protection to wild flowers of many kinds and afford them opportunity for ripening their seeds. The railway embankments, therefore, offer special attractions not only to the seed-eating birds but to insects also, and thus to the insect-feeders amongst birds. Species which, in the first instance, have visited the embankment to feed, have remained to nest, and so pronounced has the preference become in some cases, that the embankment has come to be the predominant nesting-place of the Tree Pipits and Pied Wagtails of a district. This tendency has made for the dispersal in new areas, of embankment-nesting birds, for the birds find not only a protected nesting-place in a natural granary, but a safe roost on the telegraph wires overhead.

VII. 4

CHANGE OF HABITS IN ANIMALS INDUCED BY CULTIVATION AND CIVILIZATION

Of all the effects which the advances of cultivation and civilization have induced upon animals none are more interesting than those little modifications of habit which are clearly acquisitions of comparatively recent date. In these divergences from custom we see at its simplest the elasticity of nature which makes living creatures for the most part bundles of acquired responses, and which ensures that they do not become unresponsive automata. The modifications of habit are of very different degree, from mere adjustments of convenience, to changes which seem to have been impressed upon the actual temperament of the creatures.

In order to keep in view the more general aspect of the subject, I shall limit my remarks to changes which seem to be real modifications of habit, excluding all freak habits of isolated individuals, and including only such changes as have wide significance or have, as it were, carried with them a large proportion of the members of a species.

THE HABIT OF SELECTING A DOMICILE

There is a remarkable constancy of selection exercised by most kinds of animals as regards the nature and even the exact situation of their homes, and this constancy is by no means always dominated by considerations of safety. We do not look for a Rook's nest in a mossy bank, or a Robin's in an "immemorial elm." It is the more striking therefore that in the case of many a creature, the natural choice, engrained by one knows not how many generations of custom, has given place to a new habit at the touch of civilization.

Some animals renounce for a new habitat the type of territory in which they originally made their dwelling. The Rabbit, in its native home in the Iberian peninsula, dwelt in rocky places and on dry hill-sides, but introduced to Britain, it betook itself to low ground and arable land, to the astonishment of naturalists familiar with its continental habits. In the sixteenth century, Gesner, in recording (1550) the vast numbers of Rabbits in England, "*copia ingens cuniculorum*," drew special attention to the fact that they delighted in woods and groves, although in Spain they were confined to hilly and rocky places. Occasionally in our country, the Rabbit shows an interesting return to its old habit. Although Rabbits are fairly common on the lower ground about Corrour in western Inverness-shire, Dr Eagle Clarke has recorded that a small number had colonized some rough rocky slopes near Lochan Coire an Lochan, at an altitude of 2250 feet, where, however, they were exterminated by severe snowstorms in 1916.

Or take the case of our Peregrines and Buzzards: at the present day they mostly nest in cliffs, but in other countries they breed freely on trees, and probably once did so in Britain, for Sir H. Ellis mentions a nesting-place in Sussex woods belonging to Battle Abbey, founded by William the Conqueror,—"iii nidi acceptr' in silva." It is not likely that nests of the Sparrow Hawk which were common, would have been thought worthy of mention, so it is probable that the remark refers to eyries of the Peregrine or perhaps of the Goshawk.

The Influence of Houses

Other creatures have chosen a new type of domicile, which, one might have imagined, was fraught with danger to their welfare—the habitations of men. The Stork builds on the chimney tops of Europe, the shy African Thick-knee on the flat roofs of buildings in Cairo Zoological Gardens, but the most striking cases are those of the House Martin, the Swallow and the Swift. Swifts have been known to frequent their original type of nesting-site, building their loose beds of grass in the chinks of a cliff-face, but it is seldom indeed that they forsake their adopted place in the holes and crannies of human habitations. House Martins and Swallows so universally build in the shelter of houses, that it is rare to find any trace of their original habits. Yet occasionally a Swallow, building its nest in a tree or on a sea-cliff, or a

House Martin making its home in the darkness of a cavern, betrays the secret of its original haven, before man led its kind to alter their ways by supplying new building sites. That the preference for man's dwellings seems to be almost a primitive instinct of the Swallow's nature is shown by the unanimity with which different species of Swallows in all parts of the world—in Europe, China, India, South Africa, Australia and America—seek out the habitations of men for their nesting-places; and by specific cases, such as that mentioned by Dr Richardson as occurring in 1825, when Cliff Swallows (*Petrochelidon*) of North America, left their native cliffs to build under the eaves of a house at Fort Chapewyan —"the first instance of this species of Swallow placing itself under the protection of man within the widely extended lands north of the great lakes." Sometimes necessity compels a reversion from even so well established a habit: in the deserted towns of France, where during the War all the buildings have been demolished, Swallows have taken to building in dug-outs and in trees—at least a dozen nests having been seen in a single standing poplar tree.

The strength of the new instinct in the House Martin at the present day is indicated by the numbers of nests which are annually occupied on favoured houses. On a labourer's cottage of brick and tile in a village near Stratford-on-Avon, the Martins' nests were all destroyed in 1915, but in the summer of 1916 the birds returned in such numbers that the cottage sheltered 86 nests, so closely packed that the parent birds had in some cases to feed their young in turn from the door of their neighbour's nest. Mr O. H. Wild has drawn my attention to a similar congregation of Martins' nests on a small house on the southern slopes of the Pentland Hills at Baddingsgill in Peeblesshire. There during the past six years, nests have varied in number from 56 to about 70, in 1916 the occupant counted 74, and in the spring of 1917, before the year's migrants had arrived, I saw the old nests plastered thickly under the eaves, from five to seven being occasionally crowded together between the ends of a couple of rafters.

This curious change of habit in the Martin is interesting as an illustration of the subtle influence of civilization on the choice of the bird, but the examples just cited indicate also to how great an extent the building of man's tenements must have increased the opportunities for nesting and the numbers of the house-nesting birds. And this has led to another circumstance of some economic interest, for let us suppose that on an average each nest at Baddingsgill sheltered but four chicks and that the average number of insects caught daily to feed each chick was only 500 (a low estimate, for Mr O. H. Wild tells me that he has counted 400 to 500 insects in the gullet of a Swift—a single meal), then the daily catch for this colony of 280 youngsters (the 140 adults being omitted) during the nesting-season would average 140,000 insects,—Gnats, Midges, Weevils and Greenflies,—a work of no small service to man, the tiller of the soil.

House-sparrows, too, though a proportion of their number still retains the primitive tree-nesting habit, have to a very large extent developed an inordinate love for the chimneys and water-rhones of houses—an apparent affection for man's habitations commemorated in their Linnaean name of *Passer domesticus*.

Shelter and warmth and easy supply of food have all contributed to induce many creatures, which must originally have dwelt in the open, to accept the unwilling hospitality of man's abode. Houses have, indeed, a fauna of their own, the members of which have to some extent altered their habits in placing themselves under man's roof. The simplest case is that of the creatures which take temporary shelter in the warmth and security of houses in order to tide over periods of severe conditions out of doors. Take the swarms of Flies which appear year after year in some houses, covering the windows until, as Mr Hugh Scott described,

every pane of glass was densely covered with countless myriads of small flies; on the upper sides of the projecting cross-pieces of wood between the panes the flies rested in masses, literally crawling over each other, while all the part of the ceiling near the window was almost as thickly covered as the window itself.

Such swarms, which have recently been recorded from houses in Edinburgh and in Fifeshire, as well as from Cambridge, consist mainly of Two-winged Flies—*Limnophora septem-notata* being most common, although in one case *Chloropisca ornata* was abundant. Careful investigation by Professor J. H. Ashworth has shown that these thousands of flies, almost without exception females, which have sought shelter and been preserved through the inclemency of winter, are ready in the spring-time to lay fertile batches of eggs from which the hordes of a new generation would arise. Another source of attraction lies in our food supplies, an acquired taste for which induces flies, such as the Common House Fly (*Musca domestica*) the Blow-Flies or "Blue-bottles" (*Calliphora erythrocephala* and *C. vomitoria*), to spend the greater part of their existence indoors. The wide adoption of this new habit is well attested by the multitudes of fly-papers, fly-traps, and other death-dealing devices whereby man endeavours to protect himself from his detested messmates.

Flies have been followed into houses by Spiders, several of which have adopted the domestic habitat, spite of Titania's remonstrance

> Weaving spiders, come not near.
> Hence, ye long-legged spinners, hence.

The Common House Spider, *Tegenaria derhamii*, shows preference for dwelling-houses, while others, such as *Leptyphantes nebulosus*, are plentiful in some warehouses, and a few, as the introduced *Theridion tepidariorum*, frequent warm greenhouses in Scotland.

But the fauna of houses is almost without end—a wingless (Apterous) insect, the "Silver-fish" (*Lepisma*), hides in disused cupboards, chests of drawers, or amongst sugar, sometimes in the company of a Beetle *Niptus hololeucus*; other Beetles make their home in household properties—*Anobium paniceum* in bread and stored goods, and Furniture Beetles, *Xestobium tesselatum* in rafters and lintels, *Ptilinus pecticornis* and *Anobium domesticum* in furniture, where the latter, the "Death Watch," ticks its false portent of death in the still hours of the night, as Dean Swift humorously put it—

> Then woe be to those in the House who are sick,
> For sure as a Gun they will give up the Ghost,
> If the Maggot[!] cries click when it scratches the post.

There are besides, the Bugs of old houses (*Cimex lectularius*) and the Common Flea (*Pulex irritans*), bred in dust and dirt sometimes in such incredible numbers that the Rev. James Waterston has observed "a steady stream of larvae (chiefly) and adults falling from a cracked ceiling above which the main breeding-place lay." There are also the Crickets of the Hearth (*Gryllus domesticus*), and the Cockroach (*Blatta orientalis*), both of which shall be referred to in another connection; and last in our perfunctory list, but not least, the Common House Mouse (*Mus musculus*), which has forsaken the fields for the comforts of domesticity. The dwellings of man have induced in all of these a new habit of life.

The Influence of Towns

In a variety of ways towns seem to have told upon the habits of some creatures, and especially of birds. True, they have driven many a bird from its old haunts, for the migrants dare not penetrate the smoke cloud which veils our busy cities; but in other birds town life has bred a boldness of character which one is not accustomed to associate with their wild natures. Not so long ago the Rook was accounted one of the wariest of birds, so shy that a single Rook sitting upon a house-top was considered a sure portent of death amongst the inmates. Yet many Rooks have taken to city life. In Cheltenham a rookery lines one of the main streets. In the heart of Edinburgh there are rookeries, in one of which I counted over 100 birds in the autumn of 1917; and even busy London prides itself on their abundance within its boundaries. In some places the Raven is losing its dread of man, and is returning from the cliffs of the wilds to live nearer the centres of habitation, for, writes "Cheviot" in *The Field* of 20th May 1917,

there is a tribe of him that nests year by year nearer the towns and in the suburbs, where he has found he can live foully on live and dead out of reach of the game-keeping gun, and where his long-drawn and savage krar'! krar'! sounds strangely close to open windows and the quiet of garden trees.

So the Raven returns to its old trade of town scavenger—a trade long lost and entirely forgotten during ages of persecution.

The House Sparrow is a bold bird, but contact with the bustle of business removes its last traces of shyness, and with the size and activity of the city its familiarity and impudence increase. Other birds, once banished before the advance of civilization, are gaining a new confidence. Black-headed Gulls troop up the Thames to London each autumn in ever-increasing numbers, a habit said to date from the great frost of 1895. Of recent years, too, large numbers of Wood Pigeons have settled in London, nesting even in Trafalgar Square and feeding with the utmost composure amongst the human frequenters of the public parks, just as they pick grain from the very doors of warehouses in our busiest seaports. Nor is this familiarity confined to birds which may be regarded as common. So rare a bird as the Great Crested Grebe has nested for several years in the Penn Ponds of Richmond Park on the outskirts of London. In the streets of Edinburgh the hoot of the Brown Owl may be heard on many an evening when the din of the traffic and tramway cables has ceased; and flights of Pied Wagtails, containing scores or even hundreds of individuals, roost on spring and autumn evenings in the eaves of the British Linen Company's Bank in St Andrew's Square and on the ornamental stonework of the roof of the General Post Office, a coign of vantage which they have occupied for at least thirty-five years, and at which they arrive towards dusk with a rush of wings to be heard fifty yards away. Many of the birds of the town garden, the Blackbirds, and Thrushes, the Tits and Starlings, feed unconcernedly almost within arm's length of an observer, and Rooks have become well-nigh as impudent as Jackdaws.

One cannot but mention in this relation, though it has no special connection with the influence of towns, the confidence which the Robin develops in association with man. Most country bairns and many town children are familiar from experience with the Robin which sits upon the spade-handle, while the gardener (as is his custom) rests; and Robins have been known to accompany shooting parties in order to pick Fleas from the Rabbits killed, confidence in man overcoming, in this case, dread of the gun and its noise. This confidence is not an attribute of special individuals, for I know a cottage near West Linton, which, in the snowy spring of 1917, was invaded in less than a week by as many as seven different Redbreasts in search of food, all the birds being kept together in a cage till the storm was over.

TOWNS AND SONG

In another way, towns have changed the habits of their bird frequenters, for the spring song of town and city dwellers as a rule commences at an earlier date than that of their country cousins. The Thrush is a case in point, for although its song rarely commences before mid-February in the country, it is no uncommon thing to hear it in town before the close of January. The reason probably is that the town atmosphere in winter has a more equable and higher temperature owing to the abundance of fires, and the radiation and refraction of warmth from the walls of many houses. The ground, therefore, is less often frozen, the frosts and snows are less severe, and last for shorter periods than in the surrounding country. As a consequence, natural food, to which man adds his kitchen scraps, is more accessible in towns throughout the winter, and the signs of spring begin to show while country vegetation is still asleep. So food, warmth, and symbols of returning spring all tend to deceive the town bird as to the march of the seasons, and he gives way to his awakening energies in early bursts of song.

TOWNS AND NESTING

Now the spring-time song is part of the great mating enthusiasm of the opening year; and its early awakening has close connection with an interesting sequel—the early nesting of town birds. The House Sparrow, for example, is not an early nester in country districts where it has not become an overwhelming nuisance. Its activities in this respect belong to the second great wave of building in the latter half of April, some time after the first wave, which includes such birds as the Heron, Raven, Thrush, Blackbird and Robin. Yet in towns, Sparrows have forsaken the country habit and may be seen gathering the miscellaneous rubbish for their nests even from the beginning of the year. Starlings, too, which, like the House Sparrow, naturally take their place in the second wave of building have like it also been undergoing a process of speeding-up, and now instead of waiting for the ides of April, set to work in towns a good fortnight earlier. This curious change of habit is no doubt due mainly to the peculiar conditions of town life, but it may also have been induced in part by the increasing competition for nesting sites caused by fresh influxes of rivals or by the inordinate multiplication of the native birds.

CHANGES IN FOOD HABIT

In no way has the spread of cultivation and civilization so clearly influenced the habits of animals as in turning their tastes to new sorts of food. Two causes have fostered this tendency. In the first place the cultivation in large quantities of plants of native kinds and the introduction of new plants have placed before the vegetarian animals of the country a choice such as no native of uninhabited Scotland could have had. The result has been the attraction of many creatures to new foods. In the second place, the overwhelming increase of some species of animals under the conditions of civilization has created a fresh competition for food between races which depend on the same material, the result being that individuals or races are compelled to seek new sources of supply.

The changes of habit show several degrees of complexity. They may be simple alterations from one species of food plant or animal to another closely related species; or, more complex, from one kind of plant or animal to another entirely different kind; or, most fundamental change, from animal food to vegetarian food or *vice versa*. Examples of these different degrees of modification of habit will probably occur to most readers, so that only a few typical illustrations need be given..

Many animals have shared in the most natural change of all—the turning to a different variety of the same kind of food. The caterpillar of the Death's Head Hawk Moth feeds naturally upon the Deadly and Woody Nightshades—wild members of the Solanum family (though it has been known to occur on a few other plants). But with the introduction to Europe of the Potato—a species of *Solanum* native in America—the caterpillar, attracted by its abundance, has turned to this food plant. It is well to remember, however, that suitable food is but one item making for the success of an animal, and that although every year brings to Scotland fresh Death's Head immigrants from the Continent, so that they have been found in most counties, from the southern borders to the Orkney and the Shetland Islands, and although these immigrants lay eggs which hatch into caterpillars and reach the chrysalis stage, yet for lack of suitable climate the adult moth seldom emerges in Scotland under natural conditions and the mummy shrivels within the pupa case.

The adult or imago "Death's Head" shows a second interesting change in its habits of feeding. Naturally, it may be seen sipping honey while it hovers in front of some evening flower; but since man contrived that Bees should collect honey in hives, the "Death's Head" frequently chooses the easier way of creeping within the hives and stealing honey from the cells, a strange habit well known in Southern Europe, a case of which near Edinburgh has recently come to my notice.

The extraordinary spread of the Colorado Beetle in America was due to a similar simple change of food through cultivation. In its native haunts in Colorado it fed harmlessly on the Sand Bur, a wild species of the potato family, until the cultivated Potato beguiled it, and led it to devastate a continent.

The Colorado Beetle, however, widened the break with former habits by changing from one type of vegetarian food to new and different kinds—a change which can be traced to the competition engendered by the rapid multiplication of the beetles themselves. Before it had spread to the Atlantic coast, many individuals were found to forsake the potato family and to browse upon such plants as cabbages, Smartweed (*Polygonum*), Pigweed (*Amaranthus*), and Hedge Mustard (*Sisymbrium*).

Not only so, but its numbers induced other creatures to change their old established feeding habits, though not always without hesitation and much experiment. For several years Ducks were the only domesticated birds which turned to the Beetle for food, but after a few years Fowls learned to eat, first the eggs, then the larvae, and finally the adult beetles, so that as many as 30 and 40 of these were found in the crops of single chicks, and this not because other food was lacking, but from a newly acquired preference.

A similar development of habit occurs in our Starlings, which may frequently be seen perched on the backs of Sheep, having forsaken for the nonce the insects of the cultivated ground for the "ticks" parasitic on the animal. In this same category must be placed many introduced animals such as the Common Pheasant, which, first brought from the banks of the Phasis in Asia Minor, has had to satisfy itself with the strange products of our western lands. And what of the pests of our food and stored goods—the Meal Worm, hateful larva of a Beetle (*Tenebrio molitor*), the Flour Mite (*Aleurobius*), the Biscuit Beetle (*Anobium paniceum*) whose larvae, as sailors well know, riddle the biscuits carried on long voyages, the Cheese Mite (*Tyroglyphus*), the Clothes Moths (Tineidae) which destroy clothing, tapestry, furs and wool, the "Book Worm," a Scolytid Beetle, whose curious habit has earned it the learned name of *Hypothenemus eruditus*, the still more curious Tobacco Beetle (*Lasioderma testacea*), the grubs and adults of which devour cigarettes or any other form of prepared tobacco—and how many more? Surely the ways and inventions of man have modified the feeding habits of these to no small degree.

Yet still more marked is the break with custom noticeable in such creatures as forsake a vegetarian for a carnivorous, or a carnivorous for a vegetarian diet. Of these the most notorious is that dull olive-green parrot, with red underwing, streaked with blue and yellow as it flies—the Kea (*Nestor notabilis*), bird of the solitudes of the great snowy mountains of New Zealand. It is chiefly nocturnal in habit, and before man tempted it from the way of uprightness, it depended mainly on the berries of the trees and shrubs which grow in the sheltered gullies of the hills. Now it has acquired the despicable habit of settling on the back of a living Sheep and digging with its sharp beak through skin and flesh until it reaches and devours the succulent fat encasing the kidney. As many as 200 Sheep have been killed in a single night on a station at Lake Wanaka through these cruel attacks. It has been surmised that the Kea learned its gruesome habit through alighting, as does the Starling, on the backs of Sheep, in order to eat the parasites abounding there, and that by chance, perhaps in following the burrowing maggots of some fly, it happened upon the dainty concealed in the living flesh.

This direction of change, however, from vegetarian to

carnivorous diet does not seem to have been the usual one. It is true that the abundance of the Colorado Beetle induced that handsome American finch, the Rose-breasted Grosbeak (*Guiraca ludoviciana*) to give up its staple food of berries for easier gluttony upon the Beetle, with the result that superabundance of food led to the increase in numbers of this as of other birds in the Eastern States. But this is not the rule, at any rate in our own country. Here the tendency is distinctly that insectivorous birds should develop vegetarian leanings, a change of habit to be traced to the plenteous variety of the offerings of field and garden.

Fruit crops have had a predominating influence in this respect, for such insectivorous birds as Blackbirds, Thrushes and Starlings become more and more the pests of orchard and garden, finding in gooseberries, currants, strawberries and plums a diet which satisfies at one and the same time, their need for food and drink. Yet the change of habit has not developed to an equal extent in all these birds, for though Blackbirds and Thrushes are equally destructive in dry seasons, in wet seasons when moisture is plentiful, the Thrush is more content to follow its natural pursuit of worms and slugs while the call of ripe fruit seems always to appeal to its dusky relative. Even Wasps have learned to give up their hunt for Greenflies and other insects when plums are at the ripening. At Hilston in Cornwall the common Red Squirrel has recently adopted the fruit habit. Here the Squirrel used to be a rare creature, according to a correspondent of *The Field*, but in the few years before 1913 it increased rapidly in numbers till it could be counted by the score. In 1913 it began to enter gardens, tearing the nets to get at strawberries; from these it passed in succession to raspberries and ripe plums, till it finally attacked hard green peas, when the patience of the gardener giving out, a warfare of extermination was begun.

Field crops also have proved a temptation to many a bird. The attraction of potatoes for the Rook is an old story, but its raids on grain are by no means negligible, as the contents of its crop too often show. One would scarcely expect Wild Ducks to be induced to forsake the worms and insects of the marsh for the produce of the fields, yet in the Tweed valley on autumn evenings numbers of Mallard visit the stooks of oats in the harvest fields for their grain. Again, Mr O. H. Wild tells me that he has seen Mallard flying deliberately and regularly inland to potato fields where they consumed sufficient to damage the crop appreciably, and Rev. J. M. McWilliam relates a similar story of the Wigeon. Gulls, too, especially the Black-headed and the Common Gull, have left the seashore and the marsh to follow the plough, and may be seen on summer evenings chasing moths in pasture land. It is generally supposed that in the fields they confine their attention to grubs and worms, but while these no doubt form the larger share of their new diet, examination of their gizzards convicts them of an unsuspected taste for grain and even turnips. In North Ronaldshay, in the Orkney Islands, turnips have formed a staple part of the autumn food of Gulls for close on forty years, and during the last dozen years, the turnip-eating habit has become common in parts of Aberdeenshire.

A notable change in the feeding habits of the Grey Lag Goose is clearly traceable to the influence of cultivation. In North Uist, Mr F. S. Beveridge has recorded, towards the end of August when oats have begun to ripen, the Geese appear in considerable flocks, select the most convenient field and proceed utterly to demolish it. If the grain be too high to be reached comfortably, the Geese beat the crop flat by settling on it in closely packed flocks, or by beating it with their wings. After the grain has been harvested, they turn their attention to the potato crop, if the weather be severe and more natural food unobtainable.

One little change of habit, a local migration, may be noted in closing this account of the influence of civilization and cultivation on habits of feeding. Every spring there are reared in the safety of the large towns millions of Sparrows, which, so soon as the grain harvest in the neighbouring country approaches, forsake the habitations of men and seek the fields in countless hordes. On dry summer days, clouds of town Sparrows, recognizable by their dusky plumage, may be raised from any cornfield within miles of a great city, and there they spend the summer and autumn taking heavy toll of the ripening ears, and returning to the warmth and abundant food of the city in winter. These migrations between city and surrounding country are now carried out regularly and on a much greater scale than the somewhat similar movements of Finches and Buntings, which in winter leave the barren hedgerows for the bounteous farmyard.

FAUNAS OF CIVILIZATION

Changes of habit have been induced in individual animals and species by the seductions of man's cultivation and civilized ways, but new and peculiar animal associations, or faunas on a small scale, have actually been brought into existence through his intervention. A couple of instances will illustrate this curious creation of new worlds for some lesser organisms.

The Fauna of Waterworks

It is notorious that the sewers of great cities harbour armies of Rats, which feed upon the garbage underground; but many water supplies, also, have their own peculiar tenants. Natives of the streams, or immigrants to the reservoirs from which the water supply is drawn, find their way, sometimes without disguise, sometimes in the form of minute spores or eggs, into the pipes of the water system, especially where filtration is deficient. In the pipes regular colonies may thus be formed which live there year in, year out, unknown to man, until their gradual accumulation forces itself upon his notice by interfering with the water supply.

It is scarcely to be expected that creatures so large as vertebrates should form part of the waterwork's fauna. Yet more than one household has been agitated at finding a living Eel issue from its water-tap. Not long since, in fact, the water-tap Eel was a topic of too frequent occurrence in the London newspapers, and the persistence with which it choked the pipes of the Hamburg supply led in 1886 to a careful investigation of the animal content of that system. A sieve-like apparatus was attached to the mains at various points throughout the city, and the water was allowed to pass with moderate force through this collecting apparatus, with the result that the animals contained were filtered out. The results were surprising. Worse than that, they were remarkably uniform at each of the points examined—an indication that the haul was no chance capture but represented a definite association of animals more or less constant throughout the entire water system. It is impossible to detail the species of the fifty distinct genera of creatures thus captured. Eels were common, sometimes as many as six in a single sample, so that the investigator estimated that thousands must have been present in the system. Sticklebacks, a Flounder, and a Burbot were entrapped. Living shell-fish were frequent; not only the freshwater Pond Snails, but even species of much larger Freshwater Mussels. Worms were common, and some of the smaller Leeches were present "in almost incredibly large numbers." The water was peopled by swimming crustacea—Water-fleas, the Freshwater Shrimp, and such like—and in almost every sample, hundreds, even thousands, of the Freshwater Slater (*Asellus aquaticus*) were found "in ugly crowds." The attached, inactive inhabitants of the water-pipes were even more numerous and of more economic importance than this enormous army of active swimming organisms. A Freshwater Sponge and a Hydroid Zoophyte were common; but more serious were the growths of "Leitungsmoos"—plant-like colonies of animal moss or Polyzoa, which in Antwerp were found to form all round the inside of a 60 cm. (about 24 inch) pipe a coat nearly 10 cm. (over 4 inches) thick. From about 220 yards of this pipe, two cart loads of the animal obstruction were removed.

Other cities, even if they yield first place to Hamburg, have still sufficiently remarkable water-pipe faunas. From the water-conduits of Paris, there were described in 1913 no less than forty-four species of molluscan shell-fish alone; in Ypres (before the War), Dr A. Kemna found that the opening of a hydrant was a signal for the appearance of masses of shells, Polyzoa, and Worms in the jet of water.

Britain also has its waterworks' faunas. In 1910 the irregularity of the Torquay water supply and the choking of taps and meter strainers, led to the discovery that the mains in every part of the district of supply were coated and choked with living growths of the Polyzoon, *Plumatella emarginata*. Freshwater Sponges (*Spongilla lacustris*) have been found growing in luxuriance in the pipes of Cardiff waterworks. Eleven species of Mollusca have been recorded from a water-main at Poplar, and 90 tons of the introduced Zebra Mussel (*Dreissensia polymorpha*) (Fig. 68) are said to have been removed in 1912 from a quarter of a mile of water main, conducting unfiltered water, at Hampton-on-Thames. There the solid growths of the shell-fish on the inner surface of the pipe had reduced the diameter from thirty-six inches to nine inches. From various supplies in Scotland, I have seen odd samples of Hair Worms (*Gordiidae*), Freshwater Shrimps, and Water-fleas, while tangled masses of *Paludicella articulata*, some pieces of Freshwater Sponge (*Spongilla*

Fig. 68. Zebra Mussels—from a mass of some 90 tons removed from a water-main at Hampton-on-Thames. Almost natural size.

lacustris), a few young specimens of a Freshwater Snail (*Limnæa peregra*), and some larvae of Gnats (*Chironomus*) were received by Dr S. F. Harmer from Aberdeen waterworks in March 1913.

THE FAUNA OF A COAL PIT

Coal pits are as artificial in their origin as are waterworks, yet they also have become tenanted by a fauna of their own, whose members live and multiply and die at great depths, far from the light of day. In some respects a pit fauna stands on a par with a cave fauna, for many conditions of moisture, light, food and shelter are somewhat similar in both, but while the cave possesses an assemblage of natural immigrants, the coal pit fauna has been introduced by man along with pit-props, with hay and straw for the pit horses, and with other stores. Probably many of the creatures thus forcibly carried underground soon disappear, but several become established and carry on the tradition of their race under new conditions in a new world. Of thirteen species of animals which I recently recorded as inhabitants, at a depth of 750 feet, of a coal pit at Niddrie in the Midlothian Coal Field, there is little likelihood that such visitors as a Sparrow, the Beetle *Thanasimus formicarius*, and perhaps the Two-winged Fly *Phora rufipes*, were permanent tenants; on the other hand there is clear evidence that Rats and Mice, Slugs (*Limax maximus*), Cave Spiders (*Lessertia dentichelis*), the "Clocker" Beetle (*Quedius mesomelinus*), Spring-tails (*Tomocerus minor*), two species of Earthworms, a Mycetozoon and perhaps the Moth Fly (*Psychoda humeralis*) had made themselves permanently at home, for the many individuals in the pit included young at various stages of growth or development.

In this underground world habits were modified less than might have been expected: the Spiders fared sumptuously upon the Insects that frequented the workings, for their webs contained many wings and remains of bodies, the Earthworms were forced to swallow coal dust in place of earth for the organic matter it contained, and the Slugs were driven to a diet of fungus. Solely to the workings of man this little association of animals owed its strange existence in the moistness and perpetual night of a coal pit.

CHAPTER VIII

CAMP-FOLLOWERS OF COMMERCE, OR ANIMALS INTRODUCED UNAWARES

> Brought from under every star
> Blown from over every main.
>
> We sail'd wherever ship could sail,
> We founded many a mighty state.
> TENNYSON.

SINCE nomadic man first began to move his tents from one fertile valley to the next, he has taken his place with the beasts of the field and the fowls of the air as an unconscious factor in the dispersal of living things. The effects of this unintentional transference of plants and animals, limited by the extent of human migrations, may in the earliest days of our race have been of small moment; yet even in the Old Stone Age they must already have had considerable significance, for no fact is clearer than that succeeding periods of climatic change in Europe brought with them from lands unknown, their own characteristic peoples.

We know that the people of the New Stone Age did actually convey with them from western Asia, domestic animals strange to the fauna of Europe, but what strange bed-fellows and camp-followers accompanied them willy-nilly, we can only guess. With modern extensions of travel and commerce, and perfected means of intercommunication across land and sea, the unintentional transference of plants and animals has entered a new phase, and has already become a matter of grave economic importance, which has demanded and has received, in countries whose scientific eye is undimmed, the attention of governments and of law-makers.

Plants are less mobile than animals and therefore afford a more striking illustration of the efficiency of man as an agent of dispersal. When the little lonely island of St Helena, lying in the middle of the Atlantic Ocean, 1000 miles from nearest land, was discovered in 1501, it was described as possessing about 60 species of plants, all of which, with very few exceptions, were unique. But before the days of the Suez Canal, St Helena lay on the great ocean highway from Europe to the East. The result of constant traffic is clearly perceptible in the change which took place in its flora; for in less than two and a half centuries, a little design and much accident had brought its plant species up to the number of seven hundred and fifty. So also the settlers who formed the first British colony in New England were astonished to find weed after weed, known to them in the old country, spring up in the neighbourhood of their new cultivation, until they had more than thirty familiar weeds they had thought (and had hoped) to have looked on for the last time on the other side of the Atlantic.

How do these unsought aliens find their way across land and sea? They seem to utilize any means of transport that man has devised for his own furtherance, concealed often in curious corners and crannies until they are dumped in the fertile soil of a new country. Napoleon's troops on their return to France from the Russian campaign of 1814 brought, in the stuffing of their saddles, in fodder and in other unsuspected ways, seeds of plants which had bloomed on the banks of the Dnieper and deposited them in the valley of the Rhine; they even introduced the plants of the steppes to the heart of France. In similar ways, many plants from Algeria and other countries entered France with the fodder imported during the Franco-German war of 1870, and successfully established themselves.

Animals are more alive than plants, and while their passive stages—resting spores, full-fed larvae, pupae or hibernating individuals—may be transferred from one country to another almost with the ease of insensate seeds, they themselves are also liable to be carried unintentionally, since many find concealment amongst stores and material for transportation.

Regarding the general effects in a native fauna of these invading animal stowaways, there is little to be added to the discussion of the effects due to deliberate introductions by man, outlined in another place (p. 243): for both deliberate and accidental introductions give rise to the same new competition for food and room, which leads to the weakening of one or other of the competitors. But this may be said, that since the accidental introductions owe their presence to concealment, they are as a rule smaller in size and less conspicuous than man's imported favourites; and also, that the former are on the whole more troublesome and pestilential, for while the deliberate introductions do damage mainly because they have betrayed the character man had attributed to them, there can be no question about the character of the creatures which have escaped man's vigilance—their chief end seems to be to work mischief. In proportion to their numbers the alien stowaways which become established in a country include more economic pests than the native fauna they invade. But after all this is exactly what is to be expected: for a native association has been brought by long trial and struggle to the dead-level of old age—Nature's balance of power; while the newcomers, freed from the enemies which had developed along with them in their former home and had kept them in check, and finding in the country of their adoption no similar restraint, increase and do damage to their heart's content. So the evil continues until the very numbers of the pest induce some member of the invaded fauna to turn upon them as a convenient food supply, or until man devises some cunning means for their destruction, or counters the foreign interloper by deliberately introducing its foreign enemy (see Counterpests, p. 259).

In the following pages are indicated the modes and results of the unforeseen introduction of animals to Scotland.

VIII. 1

HANGERS-ON OF MAN AND HIS DOMESTIC ANIMALS

THE accidental importation of animals is of more fundamental importance in the make-up of our fauna than would at first seem possible or probable. It is difficult to realize, for example, that the parasites of man and of his domestic animals have been distributed to the ends of the earth through the migrations of the human races, and that almost all of these were absent from our fauna before the advent of man. It is true, of course, that some of our modern tormentors may have occurred naturally on one or other member of the native fauna, and may have turned their attention to man only after his appearance in this country. The Human Flea (*Pulex irritans*) is a true parasite of the Badger in Britain and man may have derived it from this source. But there can be no doubt that the virulent Plague Flea (*Xenopsylla cheopsis*), responsible for the transmission of the deadly bubonic plague, which in the fifteen years between 1896 and 1911 caused over 7,000,000 deaths in India alone, is an undesirable alien occasionally introduced into our large seaports from the East. It is almost certain too that some, at any rate, of the Fleas of the Brown Rat, of domestic dogs and cats, and of poultry,—animals which owe their presence to the influence of man,—have been brought with their hosts from foreign lands. Just so the deliberate introduction of the European Starling to Australia involved the importation of a Fowl-Mite (*Liponyssus bursa*), which proved troublesome to the human inmates of a house in Sydney in which Starlings were nesting.

It is unnecessary, as it is impossible, to trace the history of the introduction of human parasites to Britain and Scotland. Most have been so long established that their migrations, like those of man himself, are lost in the mist of ages. But a few cases may be mentioned to illustrate the general theme.

The original home of the Bed-bug (*Cimex lectularius*), a creature known to Americans by the elegant and apposite name of "Mahogany Flat," has long been a subject of bitter recriminations. Englishmen were too ready to attribute it to America; Americans forcibly stated that it came to the New World from England with the early colonists. Now, in an *entente cordiale*, for which there are admirable scientific grounds, both nations have agreed to regard it as an alien, native of the distant eastern portion of the Mediterranean basin. Thence with man it has become cosmopolitan, a gift of civilization to new worlds. Unknown in Britain before the sixteenth century, so far as records reveal, it had certainly become sufficiently common in 1583 to cause alarm amongst ladies of high degree, whose habitations at Mortlake, so Moufet informs us, it had unceremoniously invaded. In the early days of its appearance, influenced by the movements of commerce, it was most common in seaports, as was to be expected, and even as late as the first quarter of the eighteenth century it was said to be a rare object in places inland.

A near relative of the Common Bed-bug,—the Tropica Bed-bug (*Cimex hemiptera*), a native of the warmer regions of Asia and Africa, has taken part, like the other, in the wanderings of man, for it has become established in Australia and North America, in the Antilles and Brazil, though its predilection for a high temperature tends to restrict its range.

Fleas, especially *Pulex irritans*, and various forms of Lice—the Head-Louse (*Pediculus capitis*), the Clothes-Louse (*P. humanus*), and the Crab-Louse (*Phthirus pubis*)—have been carried everywhere with man; the Itch-Mite (*Sarcoptes scabiei*) is also universal, and even the Spiny Ear-Tick (*Ornithodoros megnini*), which sometimes forsakes its usual habitat in the ears of American domestic animals to take up its abode in those of man, has been transported to England in the ear of an American visitor.

Internal parasites—"Flukes," Round-Worms, and Tape-Worms are even more deeply committed to human migrations than their external confederates. The Beef Tape-Worm (*Taenia saginata*), the Pork Tape-Worm (*Taenia solium*), the Dwarf Tape-Worm (*T. nana*), and the Broad Tape-Worm

(*Bothriocephalus latus*), parasites of the food-canal of man, have been carried far and wide. Round-Worms, including the Maw Worm (*Ascaris lumbricoides*), the common *Oxyuris vermicularis*, the Miner's Worm (*Dochmius (Ancylostomum) duodenale*), the Whip Worm (*Trichocephalus dispar*), and the Pork Thread Worm (*Trichinella spiralis*), have become cosmopolitan.

The kind of way in which many of these noxious infestations have spread in times past is still well illustrated in the case of a dangerous parasite of the blood-vessels of man—*Bilharzia haematobia*. This species, which is swallowed by man in impure drinking water, occurs over Africa from Egypt to the Cape of Good Hope, and it has been found that Mecca is a permanent centre of dispersal, whence Mohammedans, returning from their pilgrimages, carry the parasite to widely distant parts of the Old World.

PARASITES OF LIVESTOCK

The introduction of domestic and other animals from foreign lands has also been the means of bringing a varied and undesirable assemblage of aliens into our country. Like man, these too have their external and internal parasites, their temporary and permanent hangers-on. Without the presence of the Pig, man would escape the Pork Tape-Worm (*Taenia solium*), and the Pork Thread-Worm (*Trichinella spiralis*), for these find a temporary host in the domestic animal, (as many as 12,000 to 15,000 of the latter having been counted in a gramme of muscle). But wherever the Pig has gone they have accompanied it.

The Round-Worm of the Horse (*Ascaris megalocephala*), a creature sometimes 17 inches long, often found in great numbers; the Tape-Worm fauna of the Dog's interior, which includes at least a dozen distinct species; the Liver-Fluke of the Sheep (*Fasciola hepatica*); the Beef Tape-Worm (*Taenia saginata*), whose intermediate host is the Ox; and many other internal parasites of domestic animals,—Tape-Worms, Round-Worms and "Flukes"—have been carried wherever civilization has penetrated with its domestic stock.

The domesticated and other introduced animals are, besides, responsible for the transportation throughout the world of many external parasites, the majority of which belong to the insect tribes or to the arachnid Mites and Ticks. Amongst the former are such pests as the Ox Warble (*Hypoderma*), which ruins over 32 per cent. of the hides of cattle in some parts of America; the Bot-Fly (*Gastrophilus equi*) of the Horse, the "Spider-Fly" or "Tick-Fly" of the Horse (*Hippobosca equina*), the so-called Sheep-Tick (*Melophagus ovinus*),—a two-winged fly adapted to creep on the skin beneath the wool of the fleece,—and the Green-bottle Fly (*Lucilia sericata*), an insect which has become notorious as a destroyer of Sheep, on which the eggs are laid and the larvae feed. There can be little doubt that this species has been carried far and wide with live Sheep or with fleeces; of recent years it has done much damage in the Netherlands, having been introduced from southern Europe or Asia Minor, though some say from Britain. It has also appeared in Australia, where it "blows" Sheep just as in this country, and in the United States of America, where it has become one of the nuisances of slaughter-houses.

Amongst Arachnids the introduced pests include such forms as the red Chicken-mite (*Dermanyssus gallinae*) of domestic poultry, which is common in both Old and New Worlds, the true Sheep-Tick (*Ixodes ricinus*), now scattered from China to the British Isles, and throughout North America, and the Asiatic-Tick (*Dermacentor reticulatus*), a transmitter of biliary fever in Horses, which occurs in Europe and northern Asia on most kinds of domesticated animals, and has also been found in England and in Wales on Sheep.

Other introduced animals have brought their parasites with them, but the significance of such miscellaneous importations is of much less account than the deliberate, sustained and universal dispersal associated with domestic animals. One of the most important of the odd carriers is the Brown Rat (*Mus decumanus*), which has borne with it to the uttermost bounds of the earth one or more of the twenty-six species of Fleas which find it a happy hunting-ground, and which include the transmitter of the deadly bubonic plague.

Even our seas have been enriched by strange aliens

which have clung through thick and thin to their hosts during the vissicitudes of transportation. An interesting recent example is furnished by the appearance of the American Slipper Limpet (*Crepidula fornicata*) in the Thames estuary. The first sign of its presence there was a dead shell found on the shore at St Osyth in 1891, although a fisherman had recollections of the "Crow Oyster" extending back some fifteen or twenty years. In 1893 a living example was found amongst Oysters from the River Crouch, and thereafter records came with ever-increasing frequency, until it was discovered that the Slipper Limpet, from being a rarity, had become a pest. Its numbers on the oyster beds became so troublesome that endeavours were made to eradicate it, a special crushing apparatus being arranged for converting into manure the "Limpets" dredged from the bottom. About 1911, the Blackwater Fisheries alone yielded 35 tons of Slipper Limpets in four weeks; and since then the multiplication of the alien has been even more rapid, for in twelve months in 1914–15 upwards of 1000 tons, dredged chiefly from the estuaries of the Blackwater and the Coln, were crushed and used for manure by the farmers of the district. The precise relationship between the Slipper Limpet and the Oyster is unknown, but whether the former be a semi-parasite or only a constant messmate, there seems to be little room for doubt that it was introduced with foreign and probably American Oysters, brought for relaying in the oyster beds of the Thames estuary.

VIII. 2

STOWAWAYS ON SHIPS

That we have set our habitations in an island implies that our introduced animals have all been sea-borne, except such as entered the country with Neolithic man, while yet the English Channel was a dry valley, and the North Sea a marshy plain; and since the unforeseen importations have all made the voyage in secret and in hiding, it follows that all might be regarded as stowaways. I reserve this section, however, for a few free-lances, which have sought shelter on ships as it were on their own account, having no close or essential association with any particular cargo. The more restricted travellers I shall consider later, in relation to the materials in the company of which they reached us in concealment.

THE BLACK OR SHIP RAT

Rare though this interesting rodent is nowadays in Britain, it was no unsuccessful stowaway in the days of its prosperity, for its lightness and skill in climbing make it outstandingly the "Ship Rat." The place of origin of the Black Rat (*Mus (Epimys) rattus*) was probably in southern Asia whence the rodents spread through Arabia, northern Africa and southern Europe. At what period it reached Britain is uncertain, but Messrs Barrett-Hamilton and Hinton consider that its apparent absence in Europe in the eleventh and twelfth centuries, and its firm establishment in western Europe shortly afterwards, indicate that it must have been brought to our shores with the navies of the Crusaders on their return from the Holy Land—a direct, if undesirable tribute to our predominance on the seas, for it did not appear in Germany till a much later date.

In any case it was sufficiently common in England in the fourteenth century to be a pest demanding stringent measures, witness Chaucer's comment:

> And forth he goth, no lenger wold he tary,
> Into the toun unto a Pothecary,
> And praied him that he him wolde sell
> Som poison that he might his ratouns quell.

For the earliest reference I know to the presence of the Black Rat in Scotland I am indebted to Mr A. O. Curle, and quote it from the sixteenth century manuscript of the Moray Papers by permission of the Right Hon. the Earl of Moray. The writer quaintly described the death and burial of an old horse, which "made his testament on ye castell hill and put debait amangs ye doggis and retrings[1]."

Yet early Scottish references are few and are mostly confined, not to recording the presence of Rats, but to marvelling at their absence from certain areas—a clear indication of their general distribution and abundance. Thus Bishop Leslie in 1578 recorded "a wondir, the rattoun lyues not in Buquhane," or Buchan, a district in eastern Aberdeenshire: "In this cuntrey na Rattoune is bred, or, brocht in frome ony vthir place, thair may lyue [live]." In the seventeenth century, 1630 to be exact, a similar reference to Sutherland, contains a definite statement of the means by which the Rats were introduced:

There is not a ratt in Sutherland, and if they doe come thither in shipps from other pairts (which often happeneth) they die presentlie, how soone they doe smell of the aire of that cuntrey. But they are in Catteynes [Caithness], the next adjacent province, divyded onlie by a little strype or brook from Sutherland.

At the same period, a like tradition held regarding Ross-shire, much to the mystification of Richard Franck, an ex-Cromwellian trooper, when in 1656, his journeyings in Scotland brought him to the northern counties. For, said he,

The inhabitants will flatter you with an absurd opinion that the earth in Ross hath an antipathy against rats as the Irish oak has against the spider. And this curiosity if you please to examine you may, for the natives do: but had they asserted there were no mice in Ross, every tongue had contradicted them. Now mice and rats are cousin-german, everybody knows that knows anything, and for the most part keep house together. But what difference has hapned amongst them here as to make such a find in this country of Ross that the rats of Ross should relinquish their countrey and give possession wholly to the mice; This is a mystery that I understand not.

He added that

to the best of my observation, I never saw a rat; nor do I remember of any one that was with me ever did; but for mice, I declare, so great is their plenty that were they a commodity, Scotland might boast on't.

[1] *retrings* I take to be "rottens," rats.

The parish of Dunnet in Caithness, and also Liddesdale in the Lowlands boasted a similar immunity, and earth from these districts was in great demand for flooring barns, which thereby became, it was believed, ratproof.

Unfortunately all the stories do not testify to the absence of Rats, and Martin in 1703 gives a curious account of their abundance in the Outer Hebrides, which will serve at the same time to indicate how great a plague the troublesome immigrants from overseas had become.

> I have seen a great many Rats in the Village Rowdil [in Harris], which became very troublesome to the Natives, and destroyed all their Corn, Milk, Butter, Cheese, etc. They could not extirpate these Vermin for some time by all their endeavours. A considerable number of *Cats*, was employed for this end, but were still worsted, and became perfectly faint, because overpower'd by the *Rats*, who were twenty to one; at length one of the Natives of more sagacity than his Neighbours, found an expedient to renew his *Cats* Strength and Courage, which was by giving it warm Milk after every Encounter with the *Rats*, and the like being given to all the other *Cats* after every *Battle* succeeded so well that they left not one *Rat* alive, notwithstanding the great number of them in the place.

Since these days the Black Rat has disappeared from many places other than Harris. With the appearance of the Common or Brown Rat in the middle of the eighteenth century, its villany was out-villained, and it gradually sank in spirit and in number as it was ousted by its pushing rival first from the seaport towns and finally from places inland. It was common in rural Aberdeenshire till 1830, was "not very uncommon" in Keith in 1838, and had a colony at Cairnton of Kemnay in 1855. In 1813 it was the only species known in Forfar. In Moray it occurred in 1844 but had disappeared shortly afterwards. In Edinburgh it still lingered in 1834, driven like the human outcasts of fortune to "the garrets of the high houses in the old city," while its successful relatives battened in the cellars and the sewers.

Now the Black Rat has almost gone[1]. Some still exist in South Ronaldshay in the Orkneys, and occasional fresh stowaways land and attempt to gain a footing in the great seaports where they once swarmed (there are recent Scottish records from Leith, Greenock, Glasgow, Paisley, and Torry,

---

[1] Although recent reports (Nov. 1919) indicate that it has again got a foothold and is increasing in some seaport towns of England.

Kincardine), but in this country the Black Rat has never been able to face the competition of the Brown, and disappears as certainly and almost as rapidly as it comes. Yet in the days of its abundance the influence of the Black Rat was more important than mere numbers would suggest, for, as I shall discuss in another place, its presence was a predominant factor in the occurrence and distribution of the plague which ever and again ravaged the great cities.

THE ALEXANDRINE RAT

Closely related to the Black Rat, indeed usually regarded as a brown-coloured tropical race of the dark species, the Alexandrine Rat (*Mus (Epimys) rattus alexandrinus*) ranges from south-eastern Europe and northern Africa to India. It has long been known to pay passing calls at our ports, on board ships from the East, and Dr Eagle Clarke and Mr William Evans have recorded many specimens from steamers in Leith. Yet it seems never to succeed in establishing colonies, though recently specimens have been found in company with the Common Rat in the stores of the Zoological Park at Corstorphine near Edinburgh. Whether the two species will succeed in living there in harmony, in view of the abundance of the food supply, is a matter which only years, or the ratcatcher, will settle.

THE COMMON OR BROWN RAT

Most successful of the stowaways has been the Common or Brown Rat (*Mus (Epimys) norvegicus*), which from its native home on the steppes of Asia, in the regions which lie between Lake Baikal and the Caspian Sea, has spread from country to country till the whole of the civilized world has become its playground. Romance of a sort attaches to the story of the colonization of Europe by the Brown Rat, partly because of its extraordinary success. We cannot but wonder at the first hordes of the invading army, which in 1727, Pallas tells us, compelled by the dearth caused by an abnormal year of multiplication in their native territory, burst their bounds and like a living river, flowed westwards from the Caspian region across the Volga, whose bed was choked with their dead bodies, and away to the west and the north, over the vast plains of Russia. This great migration speedily made Russia, especially the region of the Baltic ports, a centre of dispersal whence shipping carried the newcomers to all parts of Europe. While this migration was populating central Europe, and even before it had begun, the Brown Rat, occasionally carried to seaports with cargoes from southern Russia, was making here and there experimental colonies on new coasts.

To this latter mode of dispersal it is supposed that Britain owed its first Brown Rats, the date of their arrival being generally set down as 1728 or 1729. When it first appeared in Scotland is uncertain, but the importance of the trade carried on in the eighteenth century between our great seaports and the countries of the Continent, suggests that the Brown Rat as an immigrant cannot have lagged much behind its English brethren. Probably it was well established before the middle of the eighteenth century. At any rate, Professor John Walker, whose essay on *Mammalia Scotica*, published in 1808, was probably written between 1764 and 1774[1], speaks of it in a way which would suggest that it was common and well established when he wrote.

Further, Walker actually describes the arrival and subsequent establishment of the Brown Rat in the district of the Solway, some twenty years before he wrote, that is to say between 1744 and 1754. His remarks are well worth translation and reproduction. Says he, in effect, of his "*Mus fossor*...The Norway Rat: ...First brought, as they say, to Scotland in ships from Norway."

Wheresoever it pitches its abode, it pitches out the Black Ratten utterly.
The Black Ratten, the Water Ratten and the Norway Rat were previously entirely unknown in Annan [a district on the Solway in Dumfriesshire]. Because, according to tradition, these animals were unable to live in that district. For the which reason, the soil of Annan, was carried to districts afar off, with great care, and with none the less folly, for the ruination of Rats. However about twenty years agone, the Norway Rat was cast on

---

[1] The late Mr Barrett-Hamilton and Mr Hinton, in their fine *History of British Mammals* (1916, p. 609), err in saying that Walker attributed the *arrival* of the Brown Rat to the period between 1764 and 1774—this is apparently a misinterpretation of the correct statement made by Mr Evans in his *Mammalia of the Forth*, 1892, p. 73. None of these authors, however, has referred to the actual instance of the establishment of the Brown Rat mentioned by Walker, and translated above.

shore from ships driven to the mouth of the river Annan, and now is scattered through almost the whole region of Annan. But the Black and the Water Rattens have not yet put in an appearance in Annan, a district encompassed as with ramparts, by the ocean, by alpine regions and by deep rivers.

From Walker's actual statement we learn that the Brown Rat appeared in Annan about 1750. This case he mentioned for a special purpose—to contrast it with the traditional absence of Rats from the district concerned; but that this was not the earliest establishment known to him is clear from the remark in which he assigns the first coming of the Rat from Norway to an earlier period, to which tradition alone bore witness. On these grounds it is legitimate to attribute the first introduction of the Brown Rat to Scotland to a period previous to the middle of the eighteenth century, perhaps in the region of the thirties.

At first the Brown Rat settled mainly in the seaports and seaboard towns, where it found garbage in plenty for its sustenance. Soon its colonies, outpacing the food supply, overflowed into the country, so that by the beginning of the nineteenth century, it had obtained a hold in most districts. The New Statistical Account of the Peeblesshire parish of Newlands gives an interesting description of the colonization of that part of Tweedside. This I quote to illustrate the influence of peopled valleys in determining the direction of the Rat's dispersal:

> The brown or Russian or Norwegian rat...a good many years ago invaded Tweeddale, to the total extermination of the former black rat inhabitants. Their first appearance was in the minister's glebe at Selkirk, about the year 1776 or 1777, where they were found burrowing in the earth, a propensity which occasioned considerable alarm, lest they should undermine houses. They seemed to follow the courses of waters and rivulets, and, passing from Selkirk, they were next heard of in the mill of Traquair; from thence, following up the Tweed, they appeared in the mills of Peebles; then entering by Lyne Water, they arrived at Flemington Mill, in this parish; and coming up the Lyne, they reached this neighbourhood about the year 1791 or 1792.

Nowadays there are few places which have escaped its detested presence, and the Brown Rat has become one of the most abundant and most destructive members of our fauna. In many areas the plague of Rats has become a menace to agriculture, so that we find the farmers in East Lothian clubbing together in 1909, and engaging four ratcatchers to "work" the county. From that time till 1914 ratcatchers varying in number from two to seven were constantly employed, with the result that in the six years 116,857 Rats were killed. This is nothing, however, to the massacres which have taken place elsewhere, as in 1901 on a 2000 acre farm near Chichester, where, in a season, systematic destruction by poisons, traps, and ferrets accounted for 31,981; while in addition more than 5000 were killed at the thrashing—an astounding total of over 37,000 for a year[1].

Suppose we allow a single Rat on an average to each acre of cultivated land in Scotland, our stock would be approximately 5,000,000 in number, and suppose we allow that each Rat on an average contents himself with food to the value of one farthing per day (a low estimate, for actual trial has shown that the cost is more than one halfpenny, but some Rats feed on garbage and refuse), then the bill for food alone would amount to £1,875,000 a year. Add the waste and damage done to houses and furniture, game and poultry, flowers and bulbs, and we find that the harbouring of this unsought alien must cost Scotland roughly £2,000,000 a year. On a similar basis, Boelter calculated that the food of the 40,000,000 rats of the United Kingdom must cost over £15,000,000 a year. In the United States in the great centres of population of over 100,000 inhabitants, the annual loss has been reckoned by Prof. David E. Lantz at £4,000,000. In 1904 Rats cost France an estimated amount of close on £8,000,000; Denmark it has been stated, loses £600,000 a year, and Germany at least £10,000,000. To such financial losses must be added the even more important influence of the Rat in disseminating disease among men, a subject to be referred to again (p. 507).

Yet in all these countries which suffer so greatly through its ravages, the Rat is no more than a disreputable alien, a rascally stowaway from overseas.

But the ravages of Rats are not confined to men's goods,

---

[1] So serious has the Rat plague become in Britain that the Government has been compelled to take action; and while the Board of Agriculture and Fisheries successfully carried out in October 1919 one national rat-week for England and Wales and another in December, Parliament has endeavoured to enforce destruction by means of the Rats and Mice (Destruction) Bill, which came into force on 1st January, 1920.

although the presence of man, with his stores of food and litter of garbage, is the great determining factor in the distribution of the common species. Wherever the carcase is, there will the Rats be gathered together, and so they occur in great hordes far from man's habitations, infesting the cliffs of Caithness and other parts, and, as Mr O. H. Wild tells me, many isolated and uninhabited islands off our coasts, as Craigleith off North Berwick, the Gaelic-named "Home of the Bird" off Isle of Muck, Puffin Island near Anglesey, and others; for Rats are excellent swimmers and have made their way to many a solitary isle. In these cases the Rats depend on the eggs, young and even adults of the birds which frequent the coast or islands, for they live in the burrows of the Puffins, enjoying rich fare during the breeding season and managing to scrape along during the winter on the molluscan and crustacean shell-fish and the refuse of the seashore, until spring brings again birds and plenty.

Even during the temporary breeding season, however, the Rats must have an appreciable influence in restricting the bird-colonies, and it is possible that their introduction may have had more to do with banishing and exterminating ground-nesting birds from closed areas like islands than one generally attributes to it. Even the smaller ground mammals may suffer from the Rat's voracity, for it is one of the worst foes of the Rabbit, hunting down the young in their burrows with the savage pertinacity of a ferret.

There is a type of ship-voyager, very different from the Rats, and perhaps of not much account in the make-up of a fauna, which clings to the exterior of a vessel and with it is taken from sea to sea. I think of those borers in the timbers of wooden ships—the Teredo or so-called Ship-Worm—a bivalve mollusc, and of the creatures which form a crust or a coat upon the external surface of hulls whether they be of wood, copper or iron—stony Barnacles and stinging Hydrozoa. It is almost impossible now to disentangle from the complicated net of Nature's facts the threads of influence upon the spread of these marine creatures created by the trafficking of ships; but many of them have a suspiciously wide distribution—the Common Acorn Barnacle of ships (*Balanus tintinnabulum*), for example, has been found all the world over. Occasionally, however, a stray piece of evidence comes forward to show that ships do actually contribute to our fauna by such transportations. There is an American Oat-pipe Coralline, *Tubularia crocea*, which, widely distributed on both the Atlantic and Pacific coasts of the New World, was unknown in British waters until, in 1897, it was discovered in Plymouth Sound attached to the stern of a large three-masted sailing-ship, the *Ballachulish* of Ardrossan. There could be no doubt as to the original provenance of these fine colonies of Hydroid Zoophytes, for the *Ballachulish* had made the voyage direct from Iquique, Peru. The significance of their presence lies in the fact that when they were examined in Plymouth by Mr E. T. Browne, the colonies were fully developed and ripe, and were setting free in the Sound great quantities of their tiny, jellyfish-like young, which, if conditions favoured, would give origin to an alien stock in the marine fauna of the English Channel. How many myriads of similar cases of involuntary transportation have gone unrecorded since man first went down to the sea in ships?

VIII. 3

SKULKERS IN DRY FOOD MATERIALS

OF all the means which the lesser creatures have unconsciously employed in insinuating themselves into new countries none has found greater favour than concealment in food materials, and none has added greater numbers and variety of animal life to our fauna Most of the creatures thus imported are small in size, and although practically all of them belong to the groups of Insects or of Mites, it is surprising under how many different guises they come. Many lurk amongst grain, a few in flour, some hide in tunnels of their own making in biscuits, others in holes drilled in peas, some are concealed amongst sugar, others amongst copra or the seeds of cotton. The secret of these preferences lies in the simple fact that some of our food materials happen to be theirs also, so that these lesser things, each engrossed in feeding on its own particular food, are carried wheresoever the food material goes. But the very fact that they come in secret and hiding, makes it almost impossible to trace their dispersal in countries where international commerce has long held sway.

Foremost amongst these passive immigrants, by reason of their ancient naturalization, are the Common Cockroach and the House Cricket, two Orthopterous insects.

THE CRICKET ON THE HEARTH

The latter, *Gryllus* (*Acheta*) *domesticus*, is by repute an introduction to Britain, but so long is it since commerce carried it over the world that even its original home cannot now be traced. Two deductions may, however, be made from its habits; first, that it is no native of Britain, else why should it exist only in the warmest corners of houses? In Scotland at the present day, it seldom occurs even in kitchens, but is often found about bake-houses in Edinburgh, Glasgow, Paisley, Dumfries and other towns. Occasionally it obtains a footing out of doors, as in a quarry at Slateford, where Mr W. Evans discovered a colony in 1907, but such colonies are always associated with material cast out of houses, and seldom survive for long. Secondly, the distinct preference shown by the House Cricket in Britain for artificial warmth hints that originally it belonged to a tropical climate.

An early Scottish reference to this insect is Sibbald's record (1684) of *Grillus Focarius*—the Cricket on the Hearth. Of recent years it has been carried to America in shipping and has become firmly established, being especially common in Canada.

COCKROACHES

The belief is almost universally held that the Common Cockroach (*Blatta orientalis*) (Fig. 69, 4, p. 437), came to Europe and to Britain in comparatively recent times from tropical Asia. At the present day in Scotland it is common where warmth and abundance of food offer it congenial conditions, in kitchens, bake-houses, and, as Mr William Evans has recorded, even in the warm underground workings of coal-pits, as at Bo'ness and Dalkeith. But this general distribution is a recent development, for the Cockroach when first referred to as a British resident, in 1634 by Moufet, lived in wine-cellars and flour mills and was probably confined to London and perhaps a few other great ports. Fifty years later Sibbald recorded it, as "Blatta, the Moth Fly," from Scotland. Its mode of arrival is tolerably certain, for Moufet tells how, even in the sixteenth century, Sir Francis Drake, on capturing a ship, *Philip*, laden with spices, found a vast multitude of winged Cockroaches on board; but whether this was the Common or American species cannot easily be decided. As befits an ocean voyager the Cockroach first populated the seaports. Thence its spread countrywards seems to have been a slow one, for it was not till about 1790 that Gilbert White found it in his own house at Selborne, 50 miles from London. In America the distribution of the "Black Beetle" is also clearly due to commerce, for it is abundant in all the eastern and Mississippi valley States and thins out westwards as it reaches the great plains.

In more recent years other Cockroaches have made

landings in Britain and have attempted to gain a foothold. The most common of these is the American Cockroach (*Periplaneta americana*) (Fig. 69, 2) about 1½ inches long, a much larger species than the common "Black Beetle," and light brown in colour. It is a native of tropical America, and according to Professor V. L. Kellogg occurs in such numbers on sailing ships entering San Francisco after their long half-year voyages round the Horn, that the sailors while sleeping wear gloves to prevent their finger-nails from being gnawed by the voracious hordes which tenant the whole ship. In some cases the opening of boxes and barrels of biscuits after a voyage has led only to the horrible discovery that the biscuits have been entirely replaced by crawling masses of Cockroaches. It is not surprising to find that this pest should have travelled with traffic from its native home in Central America and Mexico, far northwards in the United States of America, and it is little wonder that the Ship Cockroach should appear frequently in our seaports, though it seldom succeeds here in establishing permanent colonies. It is common in the London Zoological Garden in Regent's Park, but in Scotland thrives only in greenhouses, as at the Royal Botanic Garden, Edinburgh, the Glasgow Botanic Garden, Woodside Conservatory, Paisley, and elsewhere, although spasmodic occurrences have been reported from various parts of Edinburgh and Leith.

The Australian Cockroach (*Periplaneta australasiae*) (Fig. 69, 5), a native of the Australian Pacific region, has also followed the trail of commerce throughout the world, though its tropical preferences have limited its power of colonizing temperate lands. It differs from its American relative in being deeper in colour, and in having a yellow stripe outlining the shield-shaped portions of the back, behind the head (pronotum). Like the trans-Atlantic species it is found in Britain only in warm glass-houses, some of its more definite stations being in the Botanic Gardens of Kew and Cambridge, in Belfast, in the Royal Botanic Garden, Edinburgh, and in Paisley; but in the United States of America, it has taken a firm hold and is common in Florida and the other warm southern States.

Much less in size, only half an inch long, but as a colonist in Britain ranking next to the Common Cockroach, is the

SKULKERS IN DRY FOOD MATERIALS 437

Fig. 69. Some alien Cockroaches introduced into Scotland. Natural size. 1. German Cockroach. 2. American Cockroach, male with extended wings. 3. Wingless female Cockroach, brought to Aberdeen with bananas from Jamaica. 4. Common Cockroach. 5. Australian Cockroach.

pale yellowish brown German Cockroach (*Phyllodromia germanica*) (Fig. 69, 1), easily recognized by a couple of dark bands along the shield-like plate which lies in front of the wings (the pronotum). Whence this species originally came is not known, but as its name implies, it has long been especially abundant in Germany, from which centre of dispersal it has endeavoured to dominate the world. Its secret diplomacy has met with no little success, for it has insinuated itself into almost every country in Europe, and in some localities, as in Vienna, it seems actually to be displacing the formidable "Black Beetle." In Britain it has formed well established colonies in many towns, in London hotels, in a baker's shop in Leeds, to which town it is said to have been brought by soldiers returning after the Crimean War. In Scotland it has been found in abundance in an hotel and a restaurant in Edinburgh, is said to occur in all quarters of Glasgow and has been recorded from Paisley. In many parts of the world the German Cockroach has obtained foothold, and its increase in various parts of Europe and America suggests that in Britain also it may yet prove more of a pest than it is to-day.

IMPORTS WITH WHEAT

Since secrecy is of the essence of pest distribution, many of the imported skulkers had become widely spread before their identity or even their existence was recognized, so that definite trace of their original place of habitation has been lost, and we can only urge their known habits and their wide distribution in plea of the thesis that they have been introduced and dispersed by commerce. The number of these introductions is so large and their own size and interest is as a rule so small, that it is impossible here to do more than indicate a few of special interest, either because of the importance of their human relationships, or because of the probability with which their history can be traced.

Animals which feed upon wheat may be introduced in several ways and at several particular stages of their existence. Some insects slip into new countries as full-blown adults, some as active larvae, others as resting pupae; some come, making no pretence of hiding, lying or wandering openly amongst the grains they feed upon, while others bore into and are concealed within individual grains of wheat.

The Hessian Fly (*Mayetiola* (*Cecidomyia*) *destructor*), a notorious destroyer of growing wheat, is an excellent example of introduction through insignificant passivity. The damage is caused by the Fly in its grub stage, the maggots attacking the stems of growing corn at one of the lower joints or knots. Now this being so, it seemed likely that the insect had been introduced in the maggot stage with straw, but careful and prolonged examination of imported straw revealed no trace of the foreign pest. The resting period of the fly, which succeeds the maggot stage, is passed in a dark brown oval pupa case, resembling, in both shape

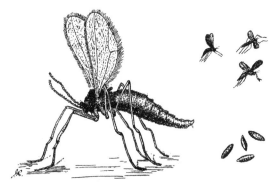

Fig. 70. Hessian Fly (*Cecidomyia destructor*) natural size and enlarged. The seed-like puparia, in which form the Fly was probably introduced to Britain, are shown as of natural size.

and colour, a rather small flax-seed. There can be little doubt that to this resemblance we owe the Hessian Fly—that it was first brought to Britain in cargoes of foul grain, where the pupa cases of the insect passed unnoticed amongst the real seeds of flax and of a dozen other weeds.

Whence it reached Britain has never been satisfactorily determined, but the fact that the insect parasites bred from the imported pupae were identical with those which had been found in native Russian pupae, suggested very strongly that the Hessian Fly came to Britain from Eastern Europe. But wherever its original home may have been it was not long in advertising its settlement in our Islands. First discovered in 1886, it spread in the following year over the whole country, from Kent to Cromarty, causing incalculable loss of wheat and barley by its devastations. Since then there has been no serious or general attack in this country, but the Hessian Fly is still with us, and may be "biding its time." Commerce has carried the Hessian Fly to many lands. In the United States of America, where its name originated on the supposition, probably erroneous, that it was brought in straw from Europe by the Hessian troopers of the Revolutionary War, it has been the cause of greater destruction than even in Europe. Millions of bushels of the wheat crop of 1915 were ruined by it, and the annual damage to crops in the United States caused by this tiny insignificant insect has been estimated at about £2,000,000.

Probably much the same mode of transportation as gave the Hessian Fly to the world accounts for the wide distribution of its near relative, the Wheat Midge (*Contarinia (Cecidomyia) tritici*), the "Red Maggot" of which does a considerable amount of harm in this country as well as in Europe generally and in North America. With it, as a creature of uncertain provenance, may be placed the Corn Aphis (*Macrosiphium granarium*) which has caused an appreciable amount of damage in southern Scotland, is common throughout Europe, was unrecorded from only ten of the United States of America in 1916, and has been found in East Africa—clearly a camp-follower of commerce.

A minute yellow ant (*Solenopsis molesta*), and the so-called White Ant of the United States, a Termite (*Leucotermes flavipes*), have been transported with food materials from America to Europe, where both have done much damage. The former, occurring amongst stores, is said to have proved very troublesome to English housekeepers. In Scotland, another American species, the Small Red House Ant (*Monomorium pharaonis*), a common inhabitant of London eating-houses, has been found in great numbers in Edinburgh, as well as in Roxburghshire and Aberdeenshire. In one place in Edinburgh where fruits, cake and confectionery were stored, Dr R. S. MacDougall has recorded that it was found necessary to employ a man whose chief work was the destruction of these ants.

Many beetles find their way to Scotland in cargoes of

wheat, either in the adult or in larval stages. African Grain Beetles, *Tribolium castaneum* or *ferrugineum* and *Tenebrioides mauritanicus*, have been found in wheat imported from Egypt to Glasgow. Both species, as well as *Tribolium confusum*, are thoroughly naturalized under cover in Britain, and the first has already been carried to India and the United States as well as to Europe. The fact that all three have been found damaging army biscuits suggests

Fig. 71. Granary Weevil (six times natural size) and destroyed wheat (nat. size).

that the Great War may be an important factor in distributing them throughout the world. The most destructive and widely spread of Grain Beetles, however, are probably the tiny brownish-black, long-snouted Corn or Granary Weevils, *Calandra granaria*, the adults of which gnaw the grains from the outside, while the grub spends its whole life within a grain, devouring the inside until only an empty shell remains. Whole cargoes of wheat and barley, worth many thousands of pounds, have been completely destroyed during a voyage by the ravages of Corn Weevils. *Calandra* probably belonged originally to a warmer country than ours, but now commerce has added it to the faunas of almost all parts of the world. In the British Isles it is frequently found amongst wheat or barley in mills, granaries or breweries, and its presence has been recorded and its ravages deplored in Canada and the United States, in Europe, Asia, Africa, and even in far Australia.

So often it happens that the dispersal of a species has been completed by man before the stages of its progress could be observed, that particular interest attaches to a form which is just beginning to find a place in a new country, and which still is treading the progressive steps which lead to universal distribution. Of Beetle species which by long standing naturalization have become part and parcel of our fauna are such household tenants as the Store Beetles, *Niptus hololeucus* and *N. crenatus*, but their congener *Ptinus tectus*, regarded as hailing originally from Tasmania, is a recent arrival in this country. The earliest specimens were found in 1901, and since that time it has been discovered in bakers' shops, granaries, and store-rooms in several places, including, in Scotland, a meal-mill at Dunfermline (1905) and a bake-house at Stromness in Orkney (1905). So far this invader has occurred predominantly at seaports or in the immediate neighbourhood of ports.

Because of its recent discovery in Scotland the occurrence of a small pale greyish-brown moth much resembling a Clothes' Moth in appearance, may be noted. The Angoumois Grain Moth (*Sitotroga cerealella*) infests Indian corn, wheat and barley, and in such cargoes has been carried over the whole world; for it is regarded as a pest in Africa, from Algeria to Nyassaland, as much as in Europe, and in Australia as much as in Canada and the United States. Dr R. S. MacDougall states that it has been taken in Scotland as well as in England, and some idea of the abundance of such pests and of the destruction wrought by them may be gained from a description quoted by him of a whole cargo of Indian corn, recently condemned in an English port

on account of its being infested with the parasites. The warehouse was a crawling mass of them. There were millions of small grey-white moths, and nearly every grain contained a weevil.

IMPORTS WITH FLOUR

Grain escapes the attacks of some of the pests only to be devoured in its more specialized forms as flour or biscuits. The American Meal Worm, the hard-skinned waxy-looking grub of a brownish, half-inch long Beetle, *Tenebrio obscurus*, is now almost as common in our flour and meal, in stables, stores and pantries, as its relative, the original European Meal Worm (*Tenebrio molitor*); and both species have been dispersed throughout the world by the traffic in food stuffs. A similar fate has befallen the Meal Snout Moth (*Pyralis farinalis*), owing to the success with which its caterpillars conceal themselves in stored cereals, flour or biscuits. But the worst of the introduced flour pests is another Pyralid, the Mediterranean Flour Moth (*Ephestia kühniella*), of pale leaden-grey colour with wing expanse less than an inch across, for this in the course of some fifty years has established posts throughout the world. First recorded in a flour-mill in Central Europe in 1877, it appeared in England ten years later, and since 1888 has sporadically caused damage there and in Scotland as far north as Aberdeen. Now, it is well known throughout Europe, and in the New World, where it appeared in 1889, it has spread from Canada to Chili. The damage caused by the lumping of flour on account of the silken galleries woven in it by the caterpillars is reckoned at many thousands of pounds a year. Destruction of a similar kind is caused by the European Grain or Wolf Moth (*Tinea granella*), which has become established in the New World.

The presence of so many well-fed inhabitants of grain and flour has led to the introduction of animal types different from themselves, which find in them an easy and abundant food supply. Thus the carnivorous Grain Mite, *Pediculoides ventricosus*, has been carried with the soft larvae it devours to many a new country, to the annoyance of mankind, for the mite readily transfers its attention to the human body and gives rise to an irritable skin eruption known as grain-itch. So troublesome is the red rash due to the mite attacks, that, when in 1913, several cargoes of grub-infested cotton-seed arrived in London from Alexandria in Egypt, it was found necessary to raise by 50 per cent. the wages of the labourers discharging the cargoes in order to keep them at work. In several widely scattered places in Europe, Africa and America the presence of the Grain Mite has been betrayed by its evil deeds.

ALIENS IN BISCUITS

How many voyagers on ships, rejecting "weevily" biscuits put before them, have realized that they were helping

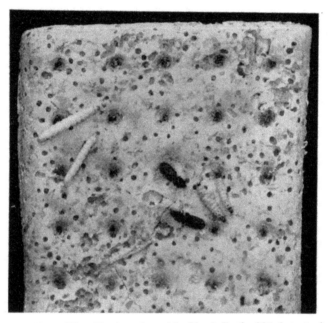

Fig. 72. Piece of Dog Biscuit perforated by Biscuit Beetles (*Sitodrepa panicea*). Lying on the surface are other insects found amongst biscuits. Natural size.

to add new elements to old faunas? It is matter of common knowledge that biscuits carried on board ship during a long voyage become riddled by the grubs and adults of Biscuit Beetles, especially *Sitodrepa (Anobium) panicea*, and this even although the greatest precautions are taken in packing and preserving the food. Many species of Beetles, and of Moths as well, have been distributed over the world through the unconscious wanderings of biscuits. Even before the Great War, the ration biscuits supplied to the widely scattered outposts of our Empire were frequently found unfit for food, notwithstanding that they were securely packed in tin boxes in the factories. A chance lot of unopened biscuit tins, drawn at random from stores at various foreign stations, revealed the fact that the biscuits harboured four different species of Moths[1] and eleven distinct species of Beetles[2]. The authors of this investigation, Messrs Durrant and Beveridge (1913), came to the conclusion that these pests had gained access to the biscuit box before it was sealed up in the factory. But the pests were not necessarily natives of the country where the biscuits were manufactured, for they may have come thither in flour from the ends of the earth. It is no simple chain of events which has distributed Biscuit Beetles and Moths throughout the civilized world.

SUGAR AND TOBACCO

Other materials have brought their own pests. Raw sugar imported to this country swarms with minute mites (*Glycyphagus*), which set up a slight irritation, known as "Grocers' Itch," on the skin of persons handling the sugar in quantity. Wherever tobacco or cigarettes may go, only too often the Cigarette Beetle (*Lasioderma serricorne*) accompanies them.

PEAS AND BEANS

Peas, Beans, and possibly Lentils also have been the means of contributing to the fauna of Scotland, for each of these harbours its own peculiar pest. While pea-pods are green in the fields, the Pea Beetle (*Bruchus pisorum*), a small beetle one-fifth of an inch long, of rusty black colour broken by a white spot on the thorax, lays its eggs upon the pods. The grubs, so soon as they hatch, bore their way through the pod and into the peas, and there they feed, hidden from the eyes of the world. As the pea ripens the larva still continues to feed within, until only a shell, concealing a mature larva, remains. In this safe shelter, the Pea Beetle, a native of North America, has been transported to most civilized countries. It has been reported from several

---

[1] *Ephestia kühniella, E. cautella, E. elutella*, and *Corcyra cephalonica*.
[2] *Silvanus surinamensis, Trogoderma* sp., *Sitodrepa panicea, Lasioderma serricorne, Rhizopertha dominica, Ptinus tectus, Tribolium castaneum, T. confusum, Tenebrioides mauritanicus, Calandra oryzae*, and *C. granaria*.

Scottish localities, and although its ravages are not so noticeable here as are those of its near relatives, the Bean

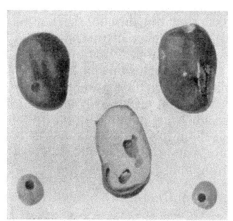

Fig. 73. Beans and Peas damaged by larvae of Bean and Pea Beetles. Natural size.

Beetles, infested peas may sometimes be seen in shops. The habits of the Bean Beetles and of the Lentil Beetle are essentially the same as those of the Pea Beetle, except that the adults select, and the grubs feed upon different legumes. Bean Beetles (*Bruchus rufimanus*, Fig. 74, and *B. obtectus*), originally members of the North and Central American faunas, are brought to Scotland every year with imported beans, and "weeviled" beans exposed for sale may be found without difficulty, those of the Seville and Aquadulce varieties, Dr R. Stewart MacDougall states, being specially liable to infestation. Of recent years, Lentil Beetles have also been found in the seaport towns of Britain—*Bruchus lentis*, having been identified by Messrs J. Edwards and E. C. Champion from Egyptian seeds at Gravesend and Birchwood, and the Chinese Lentil Beetle (*Bruchus chinensis*) in imported lentils at Dartford and Darenth Wood in Kent, Putney in Surrey, and New Forest in Hampshire.

Fig. 74. Bean Beetle. Seven times natural size.

VIII. 4

FOUNDLINGS AMONGST FRUIT

MANY and confused are the diverse types of creatures that have been imported unknowingly into Britain, yet the diversity ranges itself round definite main methods of dispersal. Each carrying agent, it will be seen, bears its own characteristic types of animals, so that from the nature of an imported animal it is possible to say, with close approximation to accuracy, with such and such a kind of cargo this traversed the oceans to our shores.

So we leave the Beetles and Micro-Lepidoptera of dry goods, for the varied assortment of aliens which have taken advantage of the transport of fresh fruit. The development of this trade is largely of recent date, and well illustrates the influence upon our animal life of commerce, which brings an influx of new animals with its every extension.

THE LIVING FREIGHT OF BANANAS

Bananas in their great clusters have been fruitful carriers of strange animals. From widely scattered towns in Scotland, from Aberdeen in the north, from Cupar, Edinburgh, Leith, Musselburgh, and from Dumfries in the south, I have seen many unexpected arrivals, exotic Jacks-in-the-box that appeared when cases of foreign fruit were opened. Most interesting of these was a Green Tree-Frog, a species of Hyla, which from the tropics, as it were with one bound, landed full of activity in an Edinburgh fruit shop. More terrifying are the appearances of that heavy and fierce-looking monster, a Bird-Eating Spider (Mygalid), which has appeared in Edinburgh, though it is fortunately a less frequent visitor than the delicate Snowy Tree Cockroach (*Oecanthus (Panchlora) niveus*) (Fig. 75, p. 448) of the East and middle West of America. The slender, half-inch long, ivory-white body, tinged with the palest green, and armed with long antennae and long legs, gives a tropical beauty to these Cockroaches, which are the most common of the banana's living freight. In addition to the above I have occasionally seen, forwarded to the Royal Scottish Museum, different species of Cockroach, a Locust, and on one occasion a Burnet Moth (Syntomidae or Zygænidae)—all of which were imported amongst bananas from Jamaica. Other records include such curious finds as Snakes, Lizards, and even a bird's nest with eggs intact, that has been said to have arrived safely in London hidden in a cranny of a banana cluster. In

Fig. 75. Snowy Tree Cockroach—frequently imported with bananas. Natural size.

addition, the packings of banana consignments carry a lesser but more dangerous cargo: old banana leaves are frequently infested with injurious Scale-Insects (such as *Hemichionaspis minor* and *Chrysomphalus aonidum*), and the Hilo Grass packing often brings with it pupae of the pestilent Mediterranean Fruit Fly (*Ceratitis capitata*).

The efficiency of the banana as an animal smuggler depends upon the generous concealment afforded by its rugged bunches of fruit. There is little room for concealment, one would imagine, on fruits which are picked singly and packed with moderate care, yet insects find a way.

FOUNDLINGS AMONGST FRUIT

THE APPLE AS A SMUGGLER

Take for example the apple as illustrating the different modes in which fruit may harbour and distribute pests.

Fig. 76. Branch of Apple-tree covered with Apple Mussel Scale.
Four times natural size.

Minute creatures may attach their almost invisible selves to the exterior of the apple, and so escaping careless examination may be carried to the ends of the earth.

The Apple Scale

The Mussel or Apple Scale (*Lepidosaphes ulmi* or *Mytilaspis pomorum*) (Fig. 76, p. 449), a tiny Hemipterous Insect or Plant-bug, is supposed to have belonged originally to the Old World, but wherever the apple has gone the Mussel Scale has kept it close company, travelling most frequently perhaps on the stems of transported apple-trees, but sometimes on the apples themselves. It is common in Britain, common indeed throughout Europe, northern and southern Africa have it, and it has reached far Japan; in the New World, Canada and the United States have publicly declared it a pest, and strive to intercept new arrivals, and Chili is concerned about its ravages; across the seas, Australia and New Zealand have been forced to make efforts to control it within their boundaries.

Mediterranean Fruit Fly

It is the habit of another series of animals to conceal themselves, still more effectively, *within* the apples which are to scatter them abroad. Of such none offers a better illustration than the Mediterranean Fruit Fly (*Ceratitis capitata*), a notorious pest of many fruits. The female of the Fruit Fly, a small two-winged fly with spotted yellowish-brown body and yellow wings, pierces the apple skin and lays a few eggs in each puncture. In from two days to over a week, according to the warmth or coolness of the weather, the larvae hatch from the eggs and feed upon the pulp of the apple for from one to three weeks, when they are full-fed and emerge from the fruit, ready for pupation. In a like period, the adult fly emerges from the pupa case and in from five to ten days thereafter is ready to commence the egg-laying which is to found a new generation. Now it is evident that this life-history and the impartiality with which the Fly attacks many kinds of fruit, afford much opportunity for the unnoticed dispersal of the Fruit Fly; for fruit in which the eggs have been placed in one continent may set free its cargoes of full-fed larvae in another. It is not surprising to find, therefore, that while the original home of the so-called Mediterranean Fruit Fly appears to be in the West Indies, the pest has established itself, thanks to the ramifications of commerce, in almost all tropical and temperate lands. Every year its larvae are imported into Britain in hundreds within Spanish oranges, to perish miserably in the marmalade pot—uncomfortable thought! Yet even when the Fly emerges in Britain, it does not seem to establish itself, owing probably to our temperate clime or to our weather. But it is common in southern Europe, and in Africa, from Tunis to the Cape; it has obtained a foothold in Australia and New Zealand, has been intercepted in California, and has been declared a pest in British Columbia. There are indeed few worlds left for the Mediterranean Fruit Fly to conquer.

Every fruit has its pest, and if the apple has carried the Mediterranean Fruit Fly whither it knew not, the Pear has performed the same service for another two-winged (Dipterous) Fly, the Pear Midge (*Contarinia* (*Diplopsis*) *pyrivora*). Since it was introduced to Britain, probably from Europe, many years ago, this insect has done much and ever increasing injury to the pear crop. It is distributed throughout Europe and has become a nuisance in the north-eastern parts of the United States.

Codlin Moth

Other pests utilize the apple much in the same way as the Fruit Fly; take for example that notorious destroyer of orchards, the Codlin Moth (*Cydia* (*Carpocarpsa*) *pomonella*) (Figs. 77 and 78, p. 452). See with what fine adjustment the life-cycle of the Codlin Moth has been arranged for its own success and man's distraction! The adult moth, a tiny creature less than three-quarters of an inch across the expanded wings, has deep grey forewings, with heavy brown lines and golden tips, and darker shimmering golden hindwings. It lays its eggs in the "eye" of a young apple and the caterpillar on hatching bores into the pulp, and spends the next three or four weeks of its life feeding comfortably and eating its way towards the centre of the apple, where it destroys seeds and core. It is now that the danger of transportation is to be feared, for the duration of the caterpillar stage within the apple is sufficiently long to afford plenty of opportunities for travel. In any case, if the apples be stored, the moth larvae are safely tucked away within them in sheltered cellars, and, unless they come to an untimely end in the meantime, eat their way out of the apple and form chrysalids in the spring. The Moths begin to emerge from the chrysalids about the end of May—in time to attack the new apple crop. By right of birth the Codlin Moth probably belongs to Europe, but Europe has dispersed the pest with

Fig. 77. Section of damaged apple containing larva of Codlin Moth. Natural size.

lavish carelessness, and from countries to which she sent it generations ago, it returns in fresh imports, to heap coals of fire upon her distracted head. Wherever it goes its presence is marked by widespread destruction. In the United States

Fig. 78. Codlin Moth. Twice natural size.

of America, this frail night-haunting species is estimated to do damage to the extent of £2,000,000 a year; in the New World it ranges from Canada to the southern States of South America; in the Old World, from Norway to South Africa, and from Great Britain to the far east; while in the midst of great seas, Australia, Tasmania and New Zealand denounce its presence.

Apple Seed Chalcid

And last of the types of smuggled goods transported by the apple, is one that, concealed in the secret of secrets, is almost certain to elude the watchfulness of man. For the larva of the Apple Seed Chalcid Fly (*Syntomaspis druparum*) is hidden within an apple seed. This little four-winged Chalcid, somewhat wasp-like in shape, bright green in colour with coppery or bronzy metallic reflections and brownish yellow legs, is only about three-tenths of an inch long, yet it has contrived to give its offspring an excellent start on a traveller's career. In July, where they are plentiful, the adult Apple Seed Chalcids may be seen flitting with a rapid zigzag course in the orchards. Here and there a female stops upon an apple, at this time grown to half an inch or an inch in diameter, pierces the skin with a slender ovipositor as long as its own body, pushes this delicate weapon home until its tip has penetrated a seed, and finally deposits one or more eggs in the seed's soft interior. The eggs hatch and the larvae begin their existence with a battle for life or death, for the six or seven which sometimes hatch in one seed are reduced to one by the old expedient of cannibalism. The grub having consumed all the store of food contained in the seed becomes full fed and moribund, and in this condition spends one, two or three winters in the apple seed, till the warmth of May restarts life activity, pupation takes place and the adult emerges in late June or early July.

The opportunities for unwitting transport furnished by this habit of life are obvious, and these have been taken full advantage of. The Apple Seed Chalcid is undoubtedly a European species. It was observed in Switzerland so early as 1803, in France about 1865, and in 1885 or 1886 the failure to germinate of forty pounds of apple seed which had been planted in Hungary, led to the discovery that the kernels had been devoured through its ravages. Its name occurs also on the list of British Chalcids.

The Apple Seed Chalcid was not noticed in the United States till 1906, but there can be little doubt that it had long existed unobserved in that country, for after attention had been drawn to its presence, it was found to be widely spread throughout the eastern and northern states. At what period it was carried to the United States we can only guess, but the early colonists, who are known to have planted apple-seeds from the old country, may have carried it thither. It may have passed from the settlers, with apples, into the hands of trading Indians, and been scattered by them, as they threw away the seeds, along the western trails. The westward spread of the apple, due chiefly to the Indians, may have meant also the spread of the lurker in the apple seed.

It is easy to realize how large and soft fruit may harbour pests, but less easy to see how evil can lurk in the heart of a tiny seed. Yet the seedsman's traffic, so it has been discovered, has been responsible for the transportation to this country of several injurious insects.

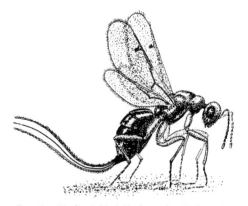

Fig. 79. Douglas Fir Seed Chalcid (female). Enlarged eight times.

DOUGLAS FIR SEED CHALCID

Take the interesting case of the Douglas Fir. The Columbia Red Wood or Douglas Fir is a native of western North America. For ninety years it has been grown in this country, first from seed brought across the Atlantic, and thereafter from seed ripened in Scottish woods. On a single estate, that of Durris in Kincardineshire, 300 bushels of good seed used to be gathered in a season, but in 1905 the forester, Mr Crozier, had to report that the seed was "not worth the trouble of gathering." And why? Because, as Dr R. Stewart MacDougall discovered, the kernels of the seeds harboured and had been devoured by myriads of grubs of a Chalcid Fly

(*Megastigmus spermotrophus*) (Fig. 79). In the spring time, from May onwards in Scotland, the tiny female insects, three-twentieths of an inch long, wasp-like in shape, brownish-yellow in colour, with clear wings, hover about the tops of the Douglas Firs, and have actually been seen inserting a long ovipositor between the scales of the fir cone and depositing an egg within the seed concealed there. There the grubs hatch and feed, invisible and safe, for no trace of a hole betrays their presence. Resting in their seed over winter, carried whither the winds or man determine, they

Fig. 80. Douglas Fir Seeds from Peeblesshire—showing, in the proportion in which they occur, seeds with escape holes of the Douglas Fir Seed Chalcid, an individual of which is issuing from a central seed. Enlarged slightly more than twice natural size.

finally emerge as adults, thousands of miles perhaps from the place where they fell asleep. This parasite has been found in Douglas Fir seeds imported from Colorado, and the obvious supposition is that it was originally brought in such seed to this country. Here it has taken increasing hold, for Mr Crozier records that though it was present on his transference to Durris in 1896, the damage it then caused was comparatively trifling. Nine years, however, sufficed to give its ravages "a serious aspect," a very large proportion of the seeds being destroyed. Since then it has spread widely in

Scotland, from Kincardineshire to Peeblesshire, for in a sample of Douglas Fir Seed of the 1915 crop which I have examined from the latter county, 51 per cent. of the seeds contained a tiny circular opening through which the adult Chalcids had escaped (Fig. 80, p. 455).

Several related Chalcid Flies have been brought to Scotland, concealed in the heart of foreign seeds. I have seen specimens bred from seeds of the American Spruce (*Abies grandis*) grown in Dumfriesshire, and from seeds of the Japanese Larch (*Larix Leptolepis*) imported from Japan.

Even the fragrant rose has a pestilence gnawing at its heart, and within its seed the Rose Seed Chalcid (*Megastigmus aculeatus*), whose grubs are found in both wild and cultivated varieties, has been scattered throughout Europe and North America.

VIII. 5

CREATURES CONVEYED BY PLANTS AND VEGETABLES

In recent times a great increase in the introduction of the pests of green plants has taken place. Even in the old days of sailing ships fresh or living vegetation could be transported from the Continent, and to this traffic we may possibly owe those pests of cultivation, the Common and Small White Butterflies, for who knows what past centuries may not have brought? But distance and time placed an embargo on the carriage of fresh plants from much further afield. The development of rapid transport has, however, broken the spell which protected us from many an undesirable alien, and America and Britain now exchange their pests of vegetation with disconcerting readiness.

Vegetables, stocks for orchards, and flowering plants for garden or greenhouse, have all encouraged this undesirable testimony to the extent of our commerce.

TYPES TRANSPORTED BY VEGETABLES

The transport of vegetables from Europe to Britain, especially for planting, may have in past times added new forms to our fauna just as it certainly now adds new numbers to old established species. It is no uncommon thing to find live Locusts in the sale-rooms of Covent Garden, escapes from consignments of cauliflowers brought from Naples—a hint of the carrying powers of green food.

The general resemblance which the fauna of Britain bears to that of Europe makes it difficult definitely to say that such and such an obscure European creature was unknown here until commerce introduced it. But no such disability applies to comparison of the inhabitants of the old and the new countries. So seldom are the native species of the New World, or of our colonies elsewhere, identical with those of Britain, that the naturalization credentials of a species common to both must be examined with suspicion. Speaking generally, the further distant two countries are from one another the easier is it to trace their commerce in living things.

Consider the influence of the introduction of cabbages or similar cruciferous plants upon the distribution of very different kinds of insects. A common British Butterfly—only too familiar—the Small White (*Pieris rapae*), a species widely distributed in Europe from the Mediterranean Sea to Scandinavia, was carried to Quebec by accident about 1860. Since then it has spread throughout the whole of Canada and the United States by an invasion, the progress of which, according to Professor Riley, can be traced step by step westwards, from the landing-place on the Atlantic coast. By a further lift it has also become established in the distant isles of Hawaii. It has long been the most serious of Butterfly pests in North America, and there its chance introduction has led to another addition to the fauna —the deliberate introduction of the tiny Hymenopterous Braconid (*Apanteles glomeratus*), whose grubs feed within and upon the caterpillars of the Cabbage Whites, and without whose active assistance cabbage beds in Britain would be wholly given over to ravenous caterpillars.

A very different insect, the Cabbage Fly (*Chortophila* (*Phorbia*) *brassicae*) is an equally notorious pest. The maggots of this small light-grey two-winged (Dipterous) fly, related to and somewhat resembling a House-fly, but on a much smaller scale, bore into the roots of cabbages, cauliflowers and other crucifers, destroying the plants, even to the extent of tens of thousands of acres in a year. This pest of European origin occurs in northern Africa and has been carried with cruciferous plants to North America. Here it has gradually subdued Canada and the United States, having reached the limit of the continent in 1914, when it did serious damage amongst cabbages, cauliflowers, turnips and radishes in Alaska.

Another example will suffice, for our purpose, to complete the tale of the cabbage and its introductions. In summer and autumn, the outer leaves of cabbages may often be seen wrinkled and covered with myriads of Greenflies —the Cabbage Aphis (*Aphis brassicae*), whose attacks finally stunt and deform the whole plant. The order of its going I know not, but this Aphis has travelled to the ends of the earth, probably from Europe where it is widespread and abundant. Southwards it has made a home in Africa, furthest east it is known in Japan, furthest west in Canada and the United States, and it has reached the isles of Hawaii in the midst of the Pacific Ocean.

Butterflies, Two-winged Flies, and Plant-bugs—the cabbage and its kind disperse them with equal impartiality, but occasionally an unlooked-for alien is smuggled in under the mantle of the cabbage leaves. In this way a couple of young Land Snails (*Achatina fulica*) were brought to Ceylon, and were thrown unwittingly into a garden. These refugees grew in size and multiplied, their descendants spread from their garden into other gardens and over a district, peopling in 1910 an area of three or four square miles with incredible numbers. In spite of the fact that a fully developed *Achatina fulica* has a shell $4\frac{1}{2}$ inches long, shaped like a "Buckie" or Dog-Whelk, 375 of these pests were counted in a garden in a space of four yards square, 227 were found in a single cluster on a coconut tree, and millions in all could be seen crawling upon the ground, or climbing walls, poles or the trunks of trees. They did little serious damage, but serve to prove the influence of the cabbage as a smuggler.

Many an old-established vegetable is accompanied in its travels by its own peculiar camp-followers. Asparagus has carried the European Asparagus Beetle (*Crioceris asparagi*), —a red, yellow and black pest which devours the crowns of young plants,—to Canada and the United States, as well as to the Argentine, in all of which countries it has become an established nuisance. And so with many other vegetables and their pests; but enough has been said to show the effects of a mode of transference, the operations of which in Scotland have been obscured by the passing of time.

TYPES TRANSFERRED BY NURSERY STOCKS AND LIVING TREES

Stocks and trees, like vegetables, have their own peculiar population: it is remarkable how large a proportion of the creatures introduced to new countries through the agency of living trees, belongs to groups of Plant-bugs—the

Greenflies and the Scale Insects. I have already mentioned the Mussel Scale as having benefited by the traffic in apples, but in all likelihood the trees are more to blame for its spread than the fruit. An equally clear case is that of the Mulberry Scale, called because of its shape, *Aulacaspis pentagona*, a pest upon great variety of trees and shrubs. The female Scales are unable to fly, and the larvae can move only a short distance, so that what progress the species has made in its distribution is due to the interference of man. Supposed originally to have belonged to Japan, it is now found in all tropical or warm temperate lands: throughout southern Europe, in Africa, in Canada and the United States, in Brazil and the Argentine, in Australia and in the lesser islands—West Indies, Seychelles, Zanzibar. There is no doubt as to the means of its dispersal: it has actually been intercepted over and over again in the act of being smuggled into the United States and the isles of Hawaii upon the stems of trees and shrubs. In 1898 it was introduced to Britain on a consignment of several hundred Flowering Cherry trees from Japan, and these were distributed throughout the country before the presence of the pest was discovered. Change of climate seems to have no ill effect upon the interloper.

The progress of the naturalization of such alien immigrants can often be traced. The Camellia or Cottony Scale (*Pulvinaria floccifera*), a traveller of cosmopolitan experience, found in British glasshouses, cool or warm, has only recently been imported to California, although it is common in the eastern and southern regions of Canada and the United States, and in the western State it has so far obtained foothold only in a single locality near San José.

Scale Insects undoubtedly owe their wide distribution to the smallness and insignificance which enables them to elude observation. Where no system of expert examination of imported plants is in force, the introduction and establishment of injurious Scales may, under the conditions of modern commerce, be absolutely relied upon. Britain, with her lofty disregard of the benefits of a humble science, reverses the old saw, and prefers the cure to prevention. As a result, there has been added to her fauna during the present generation, and this is only one of the items, a long series of foreign scale insects[1] all of them injurious to vegetation and some of them, such as the Mulberry Scale (*Aulacaspis pentagona*) and the Lantana Bug (*Orthezia insignis*) which is steadily extending into fresh districts, containing great possibilities of damage and destruction.

It must not be supposed, however, that trees confine their imports to Scale Insects. The name of one of the most widespread and persistent pests of our orchards—the American Blight—indicates its origin, though it is hard to believe that in a century and a half an alien could have so thoroughly become part and parcel of our fauna. The American Blight or Woolly Apple Aphis (*Eriosoma lanigera*) a species of plant-bug or greenfly, was unknown in England till 1787 when it was traced by Banks to a nursery in Sloane Street, London. Several facts, apart from the tradition of its origin, suggest that America is the true home of the Woolly Aphis; but, much against his will, man has established it in practically all the apple-growing countries of the world.

America has also been generous in her dispersal of a closely related plant-bug—the Grape Phylloxera (*Phylloxera vastatrix*). Introduced into France before 1863, upon rooted vines from America, the Grape Phylloxera gained so secure a footing that by 1884 it had destroyed about two and a half million acres, more than a third of all the vineyards of France, and to-day there is scarcely a vine-growing country of any importance where the Grape Bug has not appeared, to the dread and loss of vine-growers. It is a curious comment on the thoroughness of the distribution of animals through commerce, that this American Vine Bug, which under natural conditions never reached California from

[1] Prof. R. Newstead in 1901 gave the following list of Scale Insects imported on living plants into Britain during the twelve years which closed the past century. (For the sake of uniformity I have substituted the modern synonyms and have added the popular names.) *Aspidiotus alienus*, discovered in 1889; the Mulberry Scale, *Aulacaspis pentagona*, imported in 1898; the Orange Mussel Scale, *Lepidosaphes beckii*, common on oranges and lemons and found also on imported plants; the Coffee Scale, *Ischnaspis longirostris*, now a frequent greenhouse pest; *Gymnaspis æchmeæ*, well established and increasing; *Fiorinia fioriniæ* and *F. kewensis*, on palms; the Screwpine Scale, *Pinnaspis pandani*, a very destructive palm pest; the Lantana Bug, *Orthezia insignis*, steadily extending its area; and the Egyptian Cushion Scale, *Icerya aegyptiacum*, which was discovered and destroyed with its host plants on importation.

its home in the eastern States, should have been introduced there by way of Europe. To California it was brought from France on vine-cuttings imported before 1874; and since then it has spread over the whole State, having destroyed in its progress some 30,000 acres of vineyards.

Moths also have been introduced to new countries upon trees. The Gypsy Moth (*Porthetria dispar*), a common European plague upon shade and fruit trees, was accidentally carried to Massachusetts from Europe in 1869, and is now so widely established and so destructive throughout eastern New England, that the State of Massachussetts has expended over a million dollars in unsuccessful efforts to exterminate it, and the United States Department of Agriculture has instituted a special investigation aiming at the discovery of the best means of control.

In addition to the pests causing surface sores, trees have been known to carry to new countries the canker at their heart. In feeding, the caterpillars of the Leopard Moth (*Zeugera pyrina*), a British and European species, tunnel in the live wood of many fruit and ornamental trees. Probably in such a secret place were hidden the specimens carried unawares to the United States a short time ago, the founders of a colony which now extends throughout the Hudson Valley and along the Atlantic Coast from Massachusetts to New Jersey.

TYPES INTRODUCED WITH PLANTS OF THE FLOWER GARDEN

A strange variety of creatures accompanies the introduced plants of the flower garden. Although they include no pests so serious as some of those which have accompanied the produce of the vegetable garden and of the orchard, they afford in one respect a feature of interest absent from the others. Many of our flowering plants imported from tropical regions bring with them the lesser inhabitants of warmer climes. Under ordinary conditions the aliens, unaccustomed to cold and bitter weather, would soon disappear from our fauna, but the shelter and artificial heat of glasshouses supply them with a climate like their own. So there have arisen in these havens of refuge well established and thriving colonies of insects and other creatures, accidentally introduced, which reproduce on a small and limited scale the minute life of warmer lands.

In these exotic colonies surprising variety of life is represented. Of Spiders, *Theridion tepidariorum* is now very common in most large greenhouses, where it is to be found in all stages of development at all times of the year, and *Hasarius adansonii* has occasionally occurred in the warmer glasshouses. Insects of many orders are numerous: Ants from the tropics, such as *Tetramorium guineense*, and *Technomyrmex albipes*, an inhabitant of India and the islands of the Pacific, have been found in the Royal Botanic Garden in Edinburgh. A common Indian Ant, *Triglyphothrix striatidens*, which has been imported to the United States on tropical plants, has recently been reported to be spreading there. Beetles have come to us in orchid bulbs wherein the grubs tunnel and feed: Dr R. S. MacDougall has found a species of *Xyleborus* attacking *Dendrobium* in an orchid house at Pitlochry, Perthshire; he has also obtained a Beetle, native of the Straits Settlements, *Baridium aterrimus*, in orchid bulbs from Penang, and has collected several specimens of a Burmese Longicorn Beetle *Diaxenes dendrobii*, from orchids growing in Midlothian. This Indian immigrant has of late been found in a number of orchid houses in England and Scotland.

A tiny species of Thrips, *Euthrips orchidaceus*, has recently been found on hothouse orchids in the Royal Botanic Garden, Edinburgh, and in the Glasgow Botanic Garden; and it is more than likely that the Springtail, *Sminthurus igniceps*, which is confined to greenhouses, ranging from Edinburgh and Glasgow, to Germany, Norway, and Sweden, is a foreigner by ancestry.

In another connection, I have mentioned that Cockroaches from America (*Periplaneta americana*) and from Australasia (*P. australasiae*) flourish in many Scottish greenhouses; and isolated examples of other foreign Cockroaches are occasionally captured, such as the specimens of "probably *Lucophaea surinamensis* and *Blabera gigantea*," which, Prof. J. J. F. X. King reports, paid an unexpected visit to the Broomielaw in Glasgow a few years ago. Even more recently two species of Japanese Grasshoppers, *Die strammena marmorata* and *Tachycines asynamorus*, the former of which has been captured in British hothouses, have become naturalized in the greenhouses of Central Europe and have multiplied so greatly as to become pests.

Several other classes of animals have their foreign representatives in our hothouses. An exotic Myriopod, *Paradesmus gracilis*, is well established in several such retreats about Edinburgh, individuals of all ages having been noticed by Mr W. Evans. Even Worms have thus been added to our fauna: a wiry, agile, Indian species of Earthworm, *Perichaeta indica*, which like its congeners possesses unusual activity and power of springing when touched, occasionally accompanies plants, and has become naturalized in the congenial climate of hothouses in Kirkcudbrightshire and Edinburgh. Perichaeta is familiar to gardeners in this country, and though its home is in the East, the genus is now widely distributed in Europe and America—evidence of the ease with which its members, hidden in soil about plant roots, can be transported by man.

It would, however, be matter for surprise that "Leafworms" or Land Planarians, which, like *Placocephalus (Bipalium) kewensis*, may measure 6 to 9 inches in length, and sometimes attain 18 inches at full stretch, should have been introduced unobserved, were it not that they too, like Earthworms, burrow in the soil under conditions of drought. Clearly the species just mentioned is no rare wanderer from its home in the tropics, for it has been observed in several Scottish greenhouses, and has been imported with tropical vegetation and soil to England, Germany, South Africa, and Australia. The transference to foreign and unaccustomed climes has had one curious effect upon the habits of this flattened delicate "Leaf-worm," for it no longer multiplies after the routine manner of its kind through the stages of egg and young, but simply splits into several portions each of which grows into an adult, with a full complement of senses and organs. Plant-browsing animals, such as Snails, are very liable to transportation, and several, such as the West Indian *Bulimus octonus*, and *B. goodalli*, have frequently been found in British greenhouses, while many years ago *Clausilia papillaris* was detected by Joshua Alder amongst exotic plants at Granton, near Edinburgh.

I have already said in effect that the more strange in

appearance an animal is compared with the ordinary natives of a country, the more likely is its appearance in a native fauna to be detected. Hence it comes about that the majority of the stowaways recognized as imported with garden plants, hail from distant and usually tropical lands, although the factor must not be ignored that on account of their very strangeness exotic plants are specially sought after by collectors, and are imported, with their animal associates, in large quantities. This tropical aspect adds a special interest to the transportees of flowering plants, but the peculiar adaptations imposed by the natural habitat of such a fauna have prevented it from obtaining any secure hold at large in Britain. Where they have escaped from the genial warmth and moisture of the greenhouse, the exotic aliens have as a rule, made only a spasmodic appearance in our fauna. Even Mediterranean species, such as the White-keeled Snail (*Helix limbata*) and *Bulimus decolletus*, the former of which was common in hedges around London in 1837, while the latter bred in great numbers in Devon for many successive years about 1826, have generally a short life and not always a merry one. Odd specimens of many other Molluscs, introduced with plants, have appeared in Britain only to disappear again.

Nevertheless, native faunas have been occasionally enriched by plant transportees. There is strong presumption that a number of our familiar types of Earthworms (*Lumbricidae*), which are common to this country and to North America, accompanied the early settlers and their chosen plants from the Old World to the New. Or take that plague of bulbs, the Narcissus Fly (*Merodon equestris*) whose grubs feed upon the juicy scales of the bulbs, not only of Narcissus, but of half a dozen genera besides. A native of southern Europe, the Narcissus Fly has spread to northern Europe, causing serious damage every year amongst the bulbs of Dutch growers, and it was first noticed in Britain in 1869. The later wanderings of infested bulbs have established the pest in Canada, in the United States of America and in New Zealand. It is probable also that the curious "Snail-slug," *Testacella maugei*, which carries its shell on its tail instead of on its back, was introduced with the soil in which it lived. Its native abodes are in south-western Europe, Madeira, the Canary Isles and such southern lands; but it was first noticed in England in a nursery near Bristol between 1812 and 1816, and in 1822 was described as breeding freely in the open. Now it has obtained permanent footing in many of the south-western counties of England and has been found as far north as Cheshire.

From Aberdeen comes a strange instance of a halfway stage to naturalization. Three or four ponds at the Banner Mill in Aberdeen were inhabited for many years by a warm-country Freshwater Snail, *Physa acuta*, which is common in pools in the West Indies and in southern France. With it lived Goldfish and water-plants, as well as the Common Pond Snail, *Limnæa peregra*. There can be little doubt that the foreigner, introduced with either the Goldfish or the plants, was saved from extinction by the fact that the ponds were filled by a flow of warm water discharged from the mill. In these congenial habitations, the visitors flourished and multiplied. If the Common Pond Snail, accustomed to the icy pools of Aberdeenshire, could bear with equanimity the artificial warmth of the Banner Mill overflow, is it not possible that there might also be a halfway stage by which the warm-temperate species might step down from its own to our colder climate? *Physa acuta* has been accidentally introduced and has been found to thrive in the lily tanks of Kew, and in tanks and vessels containing plants in the Royal Botanic Society's Garden in Regent's Park, London.

VIII. 6

TIMBER TRANSPORTEES

TIMBER has been a staple import to Scotland for many centuries, since long before those days in the reign of James VI when the prohibition of Danish timber threatened to stop house-building throughout the country. It is little wonder then that animals brought from Europe in the early days of the trade, finding the new climate congenial, should, having become established, so spread that they are now indistinguishably part and parcel of our fauna.

LONG ESTABLISHED WOOD-WASPS

So, I imagine, it has been with the Wood-Wasps,—the Greater, *Sirex gigas*, and the Steel-Blue, *Sirex noctilio*. They are common on the Continent, and are ever and again turning up in imported wood, for within tree trunks the female lays her eggs, using in the process a long stout sting-like ovipositor, which has earned the group the American name of "horn-tails," and is the cause of much needless alarm among the unsophisticated discoverers of these fine insects. Within the tree the larvae grow and feed for two years, before emerging as adults in the warm days of summer. In many fir woods throughout Scotland, the Greater Wood-Wasp seems to have become permanently established, and the tunnelling larvae do serious damage to the trees, especially of Scots Pine and Silver Fir. The Steel-Blue Wood-Wasp is less common, though it also is held to be destructive amongst conifers, in particular Spruce and Larch. So much at home do the two British species of Wood-Wasps appear to be that some naturalists have regarded them as aboriginal natives of this country. It is next to impossible at this date to prove the former absence of insignificant creatures, where they have been introduced by chance and are long established, but it is tolerably certain that the majority of the Wood-Wasps still captured in Scotland are recent imports in timber. Two examples will enforce the statement. Owing to the long period of two years which the larva spends burrowing in solid wood, it frequently happens that, before the adult insect has emerged, the trees

Fig. 81. Steel-Blue Wood-Wasp (*Sirex noctilio*) (male). Natural size.

Fig. 82. Steel-Blue Wood-Wasp (female). Natural size.

Fig. 83. Steel-Blue Wood-Wasp eating its way out of a pine stem after emerging from pupa. Natural size.

have been cut and the timber put to some use. Thus the appearance of adult Wood-Wasps from finished furniture is no rare occurrence. Dr D. Sharp has recorded that

large numbers of a species of *Sirex* emerged in a house in this country some years after it was built, to the great terror of the inhabitants. The wood in this case was supposed to have been brought from Canada.

And Mr. A. T. Gillanders relates "that one of the best consignments of those insects I ever had sent me was captured by a miner, issuing from props within the coal-pit." Most pit-props are imported from Norway.

TIMBER CARRIES BORING CREATURES

No set of imported animals is more characteristic than the timber transportees. They are almost without exception creatures whose larvae bore in the solid wood of trees, and on this account their kind is mainly limited to Sawflies, such as the Wood-Wasps, and to Beetles, particularly of the boring Longicorn group. In addition, parasites may be carried in the Wood-Wasp or Beetle larvae, and thus foreign Ichneumon flies may emerge on our side of the ocean.

A Few Exceptions

There are at least two striking exceptions to the rule that timber carries only boring creatures. I have seen two specimens of the black and blue, yellow-banded shining lizards known as Blue-tailed Skinks (*Eumeces quinquelineatus*), which a few years ago accompanied a cargo of timber to a Musselburgh timber-yard. The timber had been imported from North America, where, in the United States, the Blue tailed Skink has a wide range. It has a habit of concealing itself under the loose bark of trees, and this no doubt led to its long and involuntary journey.

It is unlikely that the Blue-tailed Skink would have established itself in this country, for even in North America its range is limited; but another rather unexpected introduction with timber has met with greater success.

In 1824, the Linnean Society received the first recorded British specimens of the Zebra Mussel (*Dreissensia polymorpha*), these having been found in abundance attached to shells and timber in the Commercial Docks on the Thames. The Zebra Mussel lives in fresh water, in the Danube and the rivers of Russia, and in northern France, Belgium and Germany. It is supposed to have been originally carried to Britain with cargoes of wood from the Volga, and it has actually been seen attached to Baltic timber ere yet the timber was removed from the ship's hold. The success of the Zebra Mussel as a colonist has been remarkable. It has spread from one locality to another until it has stations in some twenty English counties. In Scotland it is common in the Paisley Canal and in the Forth and Clyde Canal, where it used to be found "in vast abundance." Even in the most out-of-the-way places it has succeeded in obtaining a hold and in making headway; it is a common member of the fauna of water-pipes, and in 1912 a stoppage of the water supply at Hampton-on-Thames led to the discovery that the diameter of the 36-inch main for unfiltered water had been reduced to 9 inches by masses of Zebra Mussels which were growing attached to the inside of the pipe. Ninety tons of the shells are said to have been removed before the main was again put in working order (see Fig. 68, p. 415).

Other Molluscs have been carried with timber, such as the West Indian *Bulimus undatus*, which arrived in Liverpool attached to tropical timber and alive, and is said to have formed a colony near that city. But success seldom greets such casual voyagers.

The characteristic travellers with timber are the boring insects, especially the tunnelling Beetles, whose larvae are long-lived and lie unobserved and secure in their burrows until the time of their transformation and emergence.

Long-horned Invaders

Of invading Beetles the most interesting in appearance are the striking "long-horned" Beetles (*Longicornia*), easily recognized by the extraordinary length of their antennae, which may twice or three times exceed that of the body. Often their wing-cases display rich patterns and colours, and this and their size make them conspicuous visitors. Not many native beetles can rival in appearance the oak-boring Timber Beetles or Capricorns, *Cerambyx heros* and *Cerambyx cerdo*, with their rich brown wing-cases and their fine antennae, in the male almost twice as long as the inch-and-a-half long body. Their distribution in England betrays their origin. By far the greatest number of specimens has been found in the neighbourhood of dockyards, whither clearly they had come as larvae, hidden within beams of oak.

These fine species have not yet been found in Scotland, but in North Britain almost equally handsome relatives have occasionally been captured : *Monochammus dentator* —a North American species, has been found alive in Glasgow,

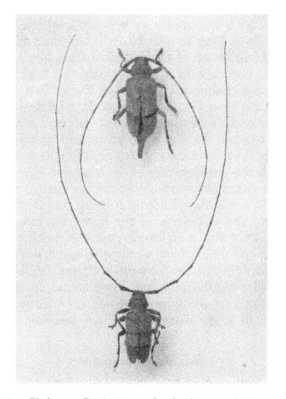

Fig. 84. Timberman Beetles (upper, female; lower, male) imported to Aberdeenshire in Norwegian pine logs. Natural size.

and *Monochammus sutor* in Colinton Dell and Berwick ; the North American *Goes tigrina* has also been found in wood landed in Scotland. A specimen, probably imported, of the large ashy-grey *Saperda carcharias*, a species native to England, was caught in the woods of Strathspey at Cromdale in 1891 ; another English species *Callidium variabile* has been introduced with timber to the Forth area; the lesser

Agelastica alni has been found in Edinburgh, *Pogonochaerus fasciculatus* at Bo'ness Docks, and the North American *Arhopalus speciosus* in the Glasgow district.

None of these Long-horned Beetles has made more than a fleeting visit to our shores, but one interesting form—the Timberman, *Acanthocinus aedilis* (Fig. 84, p. 471)—has made persistent efforts at establishment. No imported Longicorn appears so often as the greyish brown, three quarter inch long Timberman, with extraordinary antennae, which in the male are four times as long as the body. In many parts of the country and in all sorts of curious situations, the Timberman turns up; at one time in a railway carriage near Greenock, at another in a coalpit at Coatbridge, more often near timber yards, as at Bo'ness, Granton, Kilbarchan in the Clyde area, at Berwick and Tweedmouth just beyond the southern border. In the last locality a workman had "frequently seen others [of the Timberman] sticking out of holes in Baltic pine-logs." And the importation of Baltic pine containing the living larvae accounts for most of the specimens captured in Scotland. When a boy, I found a living female Timberman resting upon the doorstep of my home in Aberdeenshire, and on the same day a male was captured not far off. Both, presumably, had come from a stacked cargo of Norwegian pine logs brought to the Inverurie Paper Mills for paper making. The Timberman has been found in most of the Scottish faunal areas from Tweed to Moray, and it is possible that in a few places it may have become established.

Other Beetle Immigrants

Beetles, other than Longicorns, occasionally find their way to Scotland from foreign parts, but they too are of the boring kind. Thus the South European *Bupestris haemorrhoidalis* has been captured in a house in Ayr, and in the Royal Scottish Museum in 1915 *Lyctus brunneus* made its appearance in several cases, the larvae having been contained in wood used by a London taxidermist in mounting some foreign mammals. The destructive Banded Pine Weevil, *Pissodes notatus*, and its relative the Pine-bole Weevil, *Pissodes piniphilus*, are frequently brought with European fir trees to Scotland, and, even if they may not have come thus as new members of the fauna, they at least add their numbers to the somewhat scanty native stock, to the great damage of young pines throughout the land.

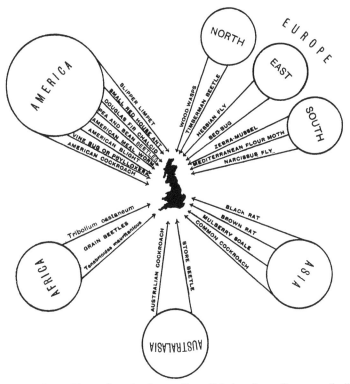

Fig. 85. Some Chance Introductions to Great Britain; shown diagrammatically, associated with the country of their origin.

FINAL REMARKS

The increase and extension of commerce is an end towards which all civilized nations have worked, yet a reading of the summaries given in this chapter will show that the gain in these respects has been no unmixed blessing. "Half the weeds of American agriculture," Professor Riley once said, "and a large proportion of her worst insect pests, have been imported among us from Europe"; and Europe's debt to America stands little in arrears—we think of the Canadian water weed (*Anacharis canadensis*), the tangled and exuberant growth of which has choked ponds and rivers and blocked canals to traffic, of the American Blight upon our apple trees, of the Pea Beetle, and most damaging of all, of the Plant-bug or Phylloxera of the vine (see also introductions from America, in Fig. 85, p. 473).

Many years' experience of the damage caused by unforeseen introductions from other countries, has led progressive nations whose wealth depends largely upon the products of their fields and orchards, to endeavour to check by all means in their power the constant stream of undesirable immigrants amongst animals. Believing the old saw that prevention is better than cure, the agricultural departments of various States, prominent among them California and Hawaii, have set up strict quarantines, so that each intended import passes under the scrutiny of experts and is declared clear before it is admitted to the State.

The monthly reports of the quarantine inspections in California and Hawaii make interesting reading, not only because they demonstrate the value of this method of combating noxious stowaways, but because they show, more strikingly than one could have imagined, the efficiency of commerce as a distributor of animal life. I cannot do better, in closing this chapter on the camp-followers of commerce, than state, in summary form, the insect pests detected during a single month on the point of entering California from abroad. I take one at random from the reports before me; it chances to contain the pests intercepted during November 1916,—a month, it is to be remembered, in the midst of a great war which has upset traffic and the ways of commerce. Nevertheless these are the month's visitors. From Belgium, seven species of Leafminers, Thrips, Scales of azaleas and bay trees; from Holland, the Narcissus Fly in narcissus bulbs, and the Mussel Scale on box shrubs; from China, Scales upon oranges, the Sweet Potato Weevil in sweet potatoes, and the Rice Weevil in rice; from Japan, Weevil larvae, Pine-cone and unidentified Weevils on chestnuts, a Coccid on oranges, and Lepidopterous larvae in Chili peppers; from Hawaii two kinds of Scale Insects on pine-apples, Scales on betel leaves, Red-Spotted Scale and two other kinds on bananas, Fruit-Fly (Trypetid) larvae in string beans, the Scarabee Beetle (*Euscepes batatae*) in sweet potatoes, Lepidopterous larvae on dates, the Purple Scale on oranges; from South America (Colombia), the Coconut Snow Scale on orchids; from Central America, two species of Scale on bananas, and Weevil larvae in avocado seeds; from Costa Rica, the Purple Scale on oranges; from Mexico, Lepidopterous larvae in dates, Long Scale on limes, and this and the Purple Scale on oranges. From various other States in the Union many pests were brought: New Jersey contributed a Chalcid Fly on orchids; Pennsylvania, Scales on gardenia, begonias, coleus, cyclamens, fuschias and spiraea, and Citrus White Fly on citrus; Florida, Scales on pine apple and Black Scale on avocado; Maryland, San José Scale and Codlin Moth with apples; from Mississippi came Citrus White Fly on gardenia; from New York, Coconut Snow Scale and Shield Scale on orchids, San José Scale on apples, Common Mealy Bug on Otaheite orange, Aspidotus Scale on jasmine; Washington was content to send only Rhizoctonia on potatoes.

Close on half a hundred different kinds of Insects caught attempting to run the blockade of the Californian quarantine; and this the fruit of a month's inspection of one tiny stream in the world's great ocean of commerce[1]!

[1] Since these notes were written the United States, having suffered much, have recognized that the unforeseen introduction of pests has become so serious that a "Notice of Quarantine, No. 37" has been issued, totally prohibiting "on and after June 1, 1919, the importation of such nursery stock and other plants and seeds as harbour certain injurious insects and fungous diseases new to or not heretofore widely distributed within and throughout the United States."

Britain has not yet instituted a systematic examination of imports for transported pests!

CONCLUSION

CHAPTER IX

CHAINS OF CIRCUMSTANCE

> The question of animal and vegetable life is too complicated a problem for human intelligence to solve, and we can never know how wide a circle of disturbance we produce in the harmonies of nature when we throw the smallest pebble into the ocean of organic being.
>
> G. P. Marsh.

IN order that he might gain a full and concrete notion of the effects of man's sway over animal life, the reader has been led through a forest whose trees are solid facts, whose undergrowth, thickets of circumstance. There is a danger that in toiling a laborious way through these thickets, he may have lost sight of the wood amongst the trees. Let us turn, then, from the infinite details of the process, in an endeavour to gain a broad view of the results; remembering that the broad view is a limited view, remembering that the human mind is not built on the principle of a fly's eye that sees all ways at once, but that the span of its vision is limited to steps and stages and incidents; remembering most of all that each act of man is no simple deed, done and forgotten, but a complex of actions and interactions whose influences spread and spread like the circles in a disturbed pool, or rather that, like the sound waves impelled from a bursting bomb, reach up and down and all around. To recast the statement on the lines of bare demonstrable fact —no deeds of man permanently affect the existence of any member of a fauna even in a lesser degree, but the shock of the interference is felt amongst the winged creatures of the air, amongst the beasts of the field and even amongst the burrowers beneath the earth's surface. The widening spheres of influence may be traced some distance from the source of their origin, but soon their waves are intersected and lost to view amidst their own reflected wavelets, and amidst new waves and counter-waves from other centres of disturbance.

Granted then that the broad view is still a limited view and neglects all the finer tracery of man's influence, there is still something to be gained from its survey.

IX. 1

A RETROSPECT

MAN'S INFLUENCE A DEVELOPING FACTOR

It is strikingly apparent that the influence of man upon animal life has not been a constant factor throughout. But this is little to be wondered at, when account is made of the changes which have taken place in man's mental attitude towards animals and his consequent relationship to them. Primitive peoples are creatures of the wilds, and fall in with Nature's ways. They kill where food or safety demands, and are killed by the other wild creatures almost on an equal footing. There is yet no enmity between man and the beast; indeed a suggestion of identity with the animal and vegetable kingdoms pervades the simple peoples, so that they see in birds and beasts, in trees and flowers, creatures like unto themselves in essence, which in the myths and folk-tales share their own thoughts and feelings and speech. So when a particular animal comes to be selected to represent their families or tribes, the race of this totem animal becomes as their own race, to be protected in life and avenged in death.

In such conditions man was almost as much a part of wild nature as the beasts themselves; he made clearings in the forest, but the clearings he made for his settlements nature healed in a few years after he had forsaken them; he slew animals, but the thought of destruction other than for his own food or clothing or protection or such simple necessity had not entered his mind.

So we find that the faunas of the older human stages were those least influenced by man. The simple hunters and fishermen of the Old Stone Age left little permanent trace upon their contemporaries of the wood and plain. But with the passing of years man left his place in the woods to seek a place in the sun, and gained in humanity just as he raised his head above wild nature's level. Now began an antagonism to nature uncontrolled, that has grown and intensified with each new stage of culture. The first great step was taken by Neolithic man when he brought creatures of the wild under the yoke, and cleared the ground of its natural plants for the sake of his cultivated grains. A classification expressive of his new relationship to nature was emphasized: plants were useful, to be preserved, or "weeds" to be rooted out; animals likewise fell into the categories of good and bad, and the latter, mainly the beasts and birds of prey, became marked for persecution everywhere and at all seasons.

In spite of this new rivalry, Neolithic man, and even his successors of the Bronze and Iron Ages, left few marks of their passage upon the fauna of Scotland; their weapons were primitive and their needs were simple. It is a remarkable fact that of all the creatures which Neolithic man found inhabiting the forests and moors of Scotland on his settlement there some 7000 years before the opening of the Christian era, only four, so far as we can judge, failed to carry on their race to the dawn of Scottish history in the first century of our era—the Giant Fallow Deer, the Lemming, the Northern Lynx, and the Northern or Rat Vole—and of these it is doubtful if the Giant Fallow Deer (*Megaceros giganteus*) had not died out even before man's arrival, while the remainder are Arctic or northern animals upon which Scotland's fluctuations of climate must have told with special severity.

Nevertheless in domesticating wild animals, Neolithic man set in motion a force which directly and indirectly has been the most potent factor in changing the old order of nature and setting in its place the new order of mankind. In Scotland the intensity of this force increased after the Roman conquest of the Lowlands and the incoming of Christianity. The greater necessity for food and fire for an increasing population led to the spread of agriculture and the destruction of the forest, so that about the tenth century several interesting and long familiar animals had disappeared or were on the verge of extinction, among them the Brown Bear, the Reindeer and the Elk.

The introduction of a new factor—the active encouragement and protection of certain animals by law—cut both ways, for while it helped to make secure the position of certain favoured creatures, it intensified the warfare against others. Yet over all is the influence of the spread of cultivation and of the care of flocks and herds, ever increasing in its power and in its enmity towards wild creatures, till, reinforced by the invention of gunpowder and guns, in the centuries following upon the sixteenth it swept away a great remnant of the original fauna, blotting out from the land such as the Wolf, the Beaver, the Great Bustard, the Crane and the Bittern, the White-tailed Eagle, the Goshawk, the Kite and the Osprey, and driving into corners of the shrinking forests, or to the wilds, such as the Wild Cat, the Pine Marten, the Golden Eagle, the Raven, the Carrion Crow, and other foes of domesticated things.

Lastly the development of the means and ease of travelling, the spread of commerce, and the rapidity of ocean transit have removed almost the last shackles from man's influence, throwing the world at the mercy of a power that heretofore has been limited by nature's barriers.

A CONTRAST—THE WAYS OF NATURE AND MAN

One is tempted to contrast the ways of Nature and Man in their bearing on animal life. A little consideration of the examples given in the preceding pages will show how different in degree are the methods by which natural processes and man's direct intervention reach the same ends.

Nature may indeed destroy by cataclysm, else how can we account for the shoals of dead fish, which died simultaneously in inland waters covering, million of years ago, the spot where Caithness now is, and which became embedded in layers of the Old Red Sandstone Period. In our own day the eruptions of submarine volcanoes work such havoc. But as a rule Nature's method is different. She exterminates an animal: long she seems to ponder over the process, slowly the conditions creep in which render existence more difficult, time gives many opportunities for changing a habit, even for modifying a structure, so that new adaptation may turn aside the threat of extinction—only to incompetence of adjustment does Nature meet out its reward.

But Man is still "Nature's insurgent son," her methods

are too slow for him, he rebels against her deliberation. Man exterminates an animal: he slays rashly, rapidly, remorselessly, and, the deed done, takes time to regret his impetuosity. Nature, it would seem, considers well before she deals the final blow; Man deals the final blow and then considers.

With the introduction of animals, it is as with their extermination—the same coarse handling marks Man's ways. Nature introduces, say, a new molluscan shell-fish to our fresh waters. A single egg, no bigger than a pinhead, or perhaps a dozen mixed with the clay attached to the foot of a migrating duck, form the possible foundation of a colony. But the colony is so small that the newcomer creates no violent revolution in its surroundings. It may die out under competition, leaving no trace; it may increase and in the end choke its neighbour's life, but the latter process is generally a slow one, there is no fierce apparent struggle, and, besides, with the gradual increase of the newcomer, fresh foes develop which tend to keep it in check.

How different when man interferes! He introduces a new mollusc: he dumps basketsfull time and again at the same spot, upsetting the balance of life in the neighbourhood, offering the newcomer, if conditions be suitable, chances of increase, which may cause the countryside to be overrun before a natural enemy arises to keep the alien within bounds.

Even his unforeseen and accidental introductions have the same effect, for the persistence with which the same pest is introduced time and again, in timber or food, with plants or fruit, makes sure that, unless conditions are actively hostile, it will at one time or another set down its foot and become an established guest.

MAIN TRENDS OF MAN'S INFLUENCE

Along three broad lines man's influence has played upon the animal world. On one hand it has tended to reduce the numbers of one set of animals, to limit their range and to exterminate them; on the other hand it has leaned towards increasing the numbers of another set of animals, compelling them to enter fresh areas, which often enough they have added securely to their original territory. Along with these influences, in conjunction with the one or the other, or independently of either, man has both consciously and unconsciously shared in bringing about changes in the physical structures as well as in the habits and temperaments of many wild creatures.

INFLUENCES TENDING TO INCREASE ANIMAL LIFE

Man has enriched the fauna of Scotland in two different directions—in quantity, by encouraging an increase of numbers amongst certain animals, and in quality, by adding new

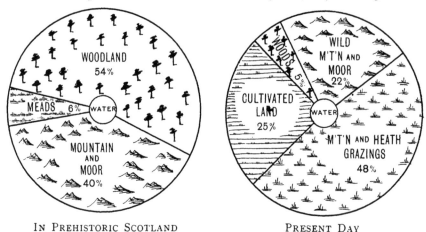

Fig. 86. Comparison of the surface features of Scotland in prehistoric times[1] and at the present day; showing the extent of man's interference with the natural haunts of animals.

creatures from other lands. Both quantitative and qualitative increases involve new demands on the food supply, and these, under natural conditions, entail a decrease in other living things whose food is appropriated by the new-comers. But here again man has stepped in, and has to a certain extent solved the problem of increase without decrease, first and mainly by his cultivation of the soil, and more recently and secondarily by his importation of food stuffs from other lands.

[1] This diagram is based upon the estimated extent of the Upper Forest of the Peat, checked, with allowance for the different physical configuration of the countries, with the present condition of Scandinavia where only 6 per cent. of the land is cultivated.

The importance of cultivation of the soil from the point of view of increasing animal life cannot be over-estimated. The vegetation of even an indifferent pasture is much more abundant than the yield of the same ground untilled, and the value, as feeding stuff, of the produce of an equal field of grain is greater still. Good arable land can be made to yield more and more under intensive methods of cultivation so that, acre for acre, the food-supply has multiplied many times since man turned the first sod in Scotland. When to this increased yield, due wholly to the skill of agriculture, we add the crops of such land—formerly almost unproductive of even wild vegetation—as has been reclaimed from the marshes and sandhills, or stolen from the forest, we gain some vague idea of the increase in vegetable food which has resulted from the cultivation of close on five million acres, one quarter of the total land area of Scotland.

Almost all this increased yield of vegetation has been added to the natural food supply of the inhabitants of Scotland, man and beast, to their enormous multiplication in numbers. Finches and grain-eating birds, Rabbits, Rats, Mice, and vegetarian animals in general, as well as the Stoats, Weasels, and beasts and birds of prey which feed upon them, have benefited by the bounty of the fields; but most important of all, man himself has multiplied, and has been enabled to increase his domesticated stock far beyond the natural bearing capacity of the country.

In another way, man has laboured for the increase in numbers of certain elements in the fauna, by affording them protection from indiscriminate slaughter and by encouraging their multiplication for the sake of sport, for their own intrinsic value, or for the pleasure their presence gives him. This direct protection has been on the whole far less effective from the faunistic point of view than might have been supposed, partly because any success it met with merely stimulated increased slaughter, partly because it was often applied to animals already on the downgrade from causes other than direct destruction, and partly because it tended to clash with agricultural developments and improvements. Yet it has succeeded in preserving for us the Red Deer, which, otherwise, would almost certainly have been exterminated long ago, as were its relatives; it has succeeded in keeping up the stocks of many game-birds which adverse conditions might have brought near the vanishing point, and in recent years it has certainly been instrumental in increasing the numbers of several rare and of many interesting and useful birds.

Apart from increase in sheer weight of numbers, the animal life in Scotland has been varied in kind by the introduction from abroad of creatures which have found climate and food suitable for their firm establishment. Amongst these naturalized aliens, domestic animals are the chief, not only because of their numbers, but also on account of the pressure they have exerted upon the native fauna; for with man the welfare of his stock has always had first consideration. How great this indirect pressure has been we can scarcely realize, but now and again we have caught glimpses of its power, as when the development of sheep pasturage led to the destruction of tracts of forest land in the Highlands as well as in the Lowlands, banishing their tenants and in the latter district finally driving the Roe and the Red Deer to extermination.

How could it be otherwise when new places and new food supplies have had to be found for the millions of domesticated animals man has imposed upon the country? We are apt to forget that by nothing short of a miracle could 8,635,918 foreigners, however desirable, be added to the fauna of a country without disturbing the food and dwelling arrangements of the aboriginals. But this number includes only the larger domestics—Horses, Cattle, Sheep and Pigs[1]—inhabiting Scotland in 1916, and the wonder is that they have found places without entirely dispossessing the old fauna, for, were they equally distributed throughout the land, on barren mountain-top, as well as in fertile plain, they would represent, roughly, an addition of one animal to every two acres. But under natural conditions each Red Deer requires at least 18 acres of moderately good "forest" to keep it in heart! The domestic animals have, it is true, enormously affected the wild fauna; but nevertheless man has accomplished the miracle of heavily overloading the scale without entirely upsetting the balance—of feeding the

[1] The numbers were horses, 207,290; cattle, 1,226,374; sheep, 7,055,864; pigs, 146,390.

multitudes of newcomers on the few loaves and fishes of the aboriginal creatures and yet leaving over more than a few crumbs for the old races, and this miracle has been accomplished through the marvellous art of agriculture which has created fresh food supplies as the domestic animals increased in number and required them.

Although domestic animals still take the lead amongst introduced creatures, other deliberate importations have to be reckoned with in the modern make-up of the fauna. The Squirrel and the Rabbit both play important parts in the economy of vegetation, the interference of the latter, as I shall have occasion to point out later, being particularly effective and widespread. Yet we should be sorry to lose these sprightly guests from our woods and dells, and bleak sand-dunes, just as we should miss the presence of the graceful Fallow Deer or of the brilliant Pheasant now that these have made their homes in wood and covert.

On the other hand we would gladly dispense with the presence of most of the chance introductions which have followed in the trail of commerce. They have added enormously in variety as well as in numbers to the fauna of Scotland, but, as I have already pointed out, their methods of secret insinuation into new countries ensure that a large proportion of these skulkers and stowaways shall prove pests and nuisances. How many anxious housewives, farmers, gardeners, and fruit-growers must bemoan the generous bounty of a commerce which has freely given them the Cockroach, or "Black Beetle" of the kitchen, the loathsome Bed Bug, the Black Rat and his cousin the Brown, the American Blight of apple trees, the Phylloxera of the vine, and scores of other minute but destructive enemies of growing crops, of manufactured goods and indeed, it would seem, of anything man particularly wishes to conserve. Other influences have favoured multiplication and the formation of new resorts for colonization; we think of the inhabitants of houses, of coal-pits, and of water-pipes. But the introduction, fortuitous or deliberate, of new animals, the protection of certain members of the old stock, the multiplication of vegetable feeders and through them of flesh eaters by the arts of land reclamation and cultivation—these are the chief methods by which man has encouraged the increase, in numbers and variety, of the animal inhabitants of Scotland.

Influences Tending to Reduce Animal Life

Two main lines of action have told severely upon the original fauna of Scotland. The first is patent in its directness—the deliberate destruction of creatures, whether for food or profit or because they threatened to destroy the security of man or of his domestic stock. The former lost to us the Great Auk, the latter the Wolf and many beasts and birds of prey. The second group of influences, indirect in their incidence, is associated with the disturbance of the haunts of wild creatures, and can be traced to the needs of cultivation and the demands of advancing civilization (see Fig. 86, p. 484). The former banished the creatures of the moorland when fields were tilled, and the birds and creeping things of the marsh when swamps were drained; both agriculture and civilization, but especially the latter on behalf of its growing industries, contributed to the destruction of the forest and in consequence to the disappearance of the woodland animals; and civilization is responsible for the ill effects of the pollution of rivers and of the obstacles placed in the way of migrant fishes.

It need not happen that the creatures against which man sets his hand, unwittingly or with intent, necessarily become exterminated. Adverse influences are as clearly marked by simple reduction in numbers, or by a gradual drawing in of the outposts of a species, so that its territory becomes less.

It must not be supposed that these influences making for the increase or decrease of a species are simple in their workings. So far from being mutually exclusive they are highly involved, and the final tendency in the history of any creature must be interpreted with due regard to the conflicting influences which have played upon it. Take, for example, the case of the Stoat. On the one hand its numbers have been reduced, directly, by slaughter for the value of its skin, and on account of its depredations on game, and indirectly by the destruction of woods and its safe retreats. On the other hand its numbers have tended to increase because of the killing of the larger beasts and birds of prey which shared the same food supply, and because of the actual increase of its prey, owing to game preservation and to the increase of food through cultivation and the accumulation of garbage. The final effect upon the Stoat's welfare depends upon the tilting of the balance to one side or the other. Other animals are subject to similar diverse and contradictory influences—it is seldom that the influence of man follows an undeviating path.

INFLUENCES TENDING TO MODIFY STRUCTURES AND HABITS

There is a third indication in the history of a species which points to adverse influences—a decline in physique. Where this degeneracy has taken place dependent upon changes in environment caused by man, the structural change may be attributed to his agency. Between prehistoric times and the present day, the weight and complexity of the antlers of Stags have undergone an extraordinary decadence. This I have endeavoured to trace to the adverse influences which have driven Red Deer, from the forest which is their natural home, to the uncongenial bleakness of the moors and mountain sides, where food is less luxuriant and where shelter from hard weather is less perfect. But man destroyed the forest and drove the Deer to the wastes, and to him therefore can be traced the new moulding of the Deer's antlers.

It is not only in the alteration of characteristic features that the influence of man can be traced, for it seems also to be exhibited in an actual reduction in size of body amongst four-footed beasts. In a country such as ours where domestic animals and the acts of man have usurped and destroyed the food supplies of wild creatures, this reduction is probably of a much more general nature than one would suppose, but many careful observations of the refuse of the prehistoric settlements and other deposits must yet be made, before the size relationship between the old and the modern representatives of mammals can be determined generally. This we can say, however, that of the survivors of the race of native Scottish Deer, one, the Scottish Red Deer, has decreased in size by about one third since the days of Neolithic man; while in the other, the Roe, there are suspicions of a reduction. The limb bones of the Mountain or Variable Hare found in the Neolithic deposits of the Inchnadamph Bone Cave are considerably larger than the corresponding bones of a modern individual. The limb bones of a Wild Cat which I examined from the settlements at Dunagoil in Bute showed that the length of limb was considerably greater than that of a modern Wild Cat, which again is larger than that of the domestic race[1]. Indeed the thigh bone (femur) resembled in length that of a modern Fox. A single hint suggests that this latter creature, too, may have been greater in former days, for one of the few records from prehistoric Scotland is that of "a very large fox" found in the "harbour-mound" at Keiss. Remains of the British Brown Bear also indicate proportions greater than those of its modern representatives on the Continent, a fact which Dr J. A. Smith attributed to the greater abundance of food in prehistoric days and to the greater age probably reached by individuals.

More directly man has played upon the physical characters, and temperament as well, of the creatures he has tamed for his own uses. By selecting the animals best suited for his purposes, by using only such animals to perpetuate the race, and by interbreeding the best blood with the selected animals of other lands, the descendants perhaps of a different wild race, he has lengthened the wool of Sheep and bleached its colour, he has increased the bodies and banished the horns of Cattle, he has evolved the strength of the Clydesdale and the Shire, he has trained dogs to hunt by sight, or by scent, and so on. In a hundred ways he has altered

---

[1] The following table indicates these differences:

	Prehistoric Wild Cat	Modern Wild Cat	Domestic Cat
Humerus ...	120 mm.	106 mm.	90 mm.
Ulna	140 mm.	119 mm.	101 mm.
Femur	135 mm.	119 mm.	99 mm.

structures and habits in accord with his own ideals. In many animals, by rigorous restriction of liberty, he has tended to produce bulk rather than brain—developed by activity; in most he has replaced the savage nature of the wild by the docility of domestication: it is typical of his moulding that he has turned the Wild Boar into a common Swine.

IX. 2

SOME FINAL CONCLUSIONS

INCREASE IN NUMBERS

THE most striking result of a broad comparison between the animal inhabitants of Scotland at the time of the coming of man and at the present day, is the discovery that the total number of living things should have so markedly increased. It is true that when we cast back to the animal assemblage of the New Stone Age, with its Reindeer and Elk, its Brown Bear, Lynx and Wolf, its Lemming and Arctic Vole, our first impression is one of present day loss and meagreness; we feel that a once varied and interesting fauna has become thin and dull. The first impression has some justification. Yet as a matter of fact the fauna as a whole has gained both in numbers and variety owing to man's presence and interference.

To what extent did the "original fauna" found in Scotland by man when he arrived some 7000 years before the Christian era, differ in numbers from that of the present day? The question is no simple one to answer, but some general indications may be gathered. The undisturbed natural fauna was a sparse one compared with the fauna of to-day, that is to say, each animal, vegetarian or carnivore, then required on an average a larger extent of territory to support it. In many of their "forests" Red Deer still, to a great extent, live upon the natural produce of the land. How are they distributed? Take Mr Henry Evans's account of the Red Deer of Jura. In one forest of 7000 acres (the Inner District), he found 426 hinds, 80 calves and 250 stags, in another of 9000 acres only 170 hinds, 80 calves and 250 stags; in the former 10 acres supported a deer, but this was in the best breeding ground of the island, in the latter each deer required 18 acres. The whole area of the Jura deer forests, 27,506 acres, carried 2002 Red Deer of all sorts, so that over this extensive region, roughly 14 acres were necessary for the support of a deer. But it is to be remembered that Jura is an island favourable to Deer, and that, even so, when this examination by Mr Henry Evans was made, the Jura forests were, to all appearance, stocked beyond their capacity, for apart from the numbers slain for sport, a ten years' count showed a heavy mortality, averaging fully one hundred each year.

Several factors of unknown weight tell on one side or other of the scale. In the days before man, many natural enemies, beasts and birds of prey, tended to keep down the numbers of Red Deer, and though they were probably no more deadly taken altogether than the sportsman's gun, yet, contrary to the ways of the gun, they discriminated against the more easily captured hinds and calves, a discrimination which must have leant towards keeping the numbers at a low ebb. On the other hand, food was perhaps more abundant in the old days; but then the Deer were larger and would have required more. Probably we should be not far from the truth, if in view of these facts we estimated that in prehistoric times Deer were not more plentiful than, on an average, one to 20 acres, even when the ground was moderately favourable. But in the prehistoric days just before man's arrival a very much larger portion of the country than now was impassable bog and treacherous swamp. I think it likely, therefore, that Scotland could scarcely have carried more than the equivalent of about 700,000 Red Deer.

Yet in 1916, notwithstanding that in Scotland just about one-fourth of the total land surface is cultivated—that is to say, bears crop or permanent grass—the number of domesticated animals alone amounted to 8,635,918, or an average of one animal for every two and one-fifth acres over the whole of Scotland. Read in the light of Red Deer equivalents this number is somewhat inflated, for although each of the 207,290 Horses, and 1,226,374 Cattle eats considerably more than a deer, each of the 7,055,864 Sheep and 146,390 Pigs eats considerably less. Making allowance for these diversities it seems to me that the average population of the larger animals, taken over the whole country, has been increased about eight to ten times over.

Almost all this gain has been made in the cultivated areas, and a good and sufficiently accurate idea of the extraordinary power of agriculture in multiplying numbers can be gained from a very simple comparison of the bearing power of cultivated and uncultivated land at the present day. Contrast the cases of the wild districts of Ardnamurchan in Argyll or Lochaber in Inverness, with the highly cultivated Linlithgow district, taking in each, lest the difference should be over-emphasized, only the areas given over to domestic stock and crops.

Year 1916	Ardnamurchan, Argyll	Lochaber, Inverness	Linlithgow District
Mountain-grazing ...	203,168 acres	402,148 acres	83 acres
Crops & Pasture ...	4,638 ,,	8,802 ,,	28,654 ,,
Horses	202	526	1,261
Cattle	3,229	4,939	4,613
Sheep..................	68,302	143,457	11,189
Pigs	52	149	960

To put these figures in comparative form: The burden of each 1000 acres in the three districts is:

	Ardnamurchan	Lochaber	Linlithgow District
Horses	1	1	45
Cattle	11	12	161
Sheep	329	349	390
Pigs	$\frac{1}{4}$	$\frac{1}{3}$	33

This comparison, graphically represented in the accompanying diagram (Fig. 87), markedly as it shows the influence of agriculture in increasing numbers, minimizes the effects of cultivation, for neither Ardnamurchan nor Lochaber were wholly uncultivated, the former having 4638 and the latter 8802 acres under grass and crops, while even a part of Linlithgow district was unbroken moor. Further it must not be forgotten that in a highly cultivated district a much larger proportion of the produce, such as potatoes, wheat, oats and barley, goes to the feeding of the human population and is lost to the animal inhabitants. This loss is by no means balanced by the import of foreign feeding stuffs. On the other hand, the primitive fertility of the

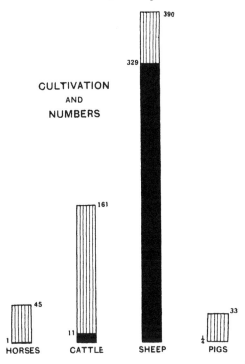

Fig. 87. Comparison between the livestock carried by 1000 acres of cultivated and of uncultivated land. The full height of the columns represents the average numbers of domestic animals on 1000 acres in Linlithgow district (highly cultivated); the black portions those in the same area of Ardnamurchan (almost uncultivated) (see text). The lined portions of columns therefore indicate roughly the increase in numbers due to cultivation. The actual average numbers are given alongside the columns.

district now highly cultivated would probably have been greater.

While an increase in numbers of animals is an unexpected result of the interference of man, it is nevertheless clear that the wild fauna, the animals which are not confined by fences nor tended by human protectors, has suffered serious curtailment. Centres of industry have been planted in its pastures and feeding-grounds, its breeding covers and secluded places have been turned into fields. Even the wastes whither it could retire when hard pressed, have become the grazing grounds of multitudes of Sheep. Throughout this country where the wild beast once roamed free, only one quarter of the surface remains clear of the influences of domestication and civilization, and much of this is barren mountain side. The forests have shrunk, till only one-twentieth of a country which was once half woodland, remains under trees, and under the stress of war this remnant too is disappearing (Fig. 86, p. 484).

But the centres of industry have offered new homes, the crumbs that fall from the table sustain many dependants, and on the multiplied food supplies of the fields new multitudes find a living. So that man seems to have cleared away one race of animals only to make room for others. Have the actual numbers of the wild fauna fallen, since man first put his spoke in the wheel of Nature? I doubt it. Think of the multitudes of Sparrows and other seed-eaters he has created, of the swarms of Rabbits and Rats, Mice, House Flies and other pests which gather about his doors and houses and fields, even of the increasing Earthworms of his gardens. Do their numbers not compensate (as numbers) for the disappearance of a few Reindeer, Elks and Brown Bears, and even of many Wolves, Wild Cats, and other decadents?

INCREASE IN VARIETY

Is then the feeling of meagreness which always arises when we compare the present fauna with the old, due to a new lack of variety? This cannot be, for the fourteen, or so, species of beasts and birds and the few additional insects which have been banished from Scotland since man interfered, are more than replaced, so far as species go, by the creatures he has introduced either deliberately, like the Fallow Deer, the Rabbit, the Squirrels, the Pheasant, and many others, or has brought accidentally with his commerce from foreign lands—the Black and Brown Rats, and a score of parasites and pests which have settled comfortably in the land.

SOME FINAL CONCLUSIONS

THE GREAT CHANGE

How then can we account for our impression of loss in the fauna, since in reality it has gained both in numbers and in variety? The secret lies in the great change. Destruction of feeding grounds and dwelling haunts, as well as deliberate slaughter, makes first for the decrease or extermination of the *larger* animals, whose demands upon covert and food and whose destructiveness are greater. And while man lops off the giants at the head of the scale he adds insignificant pigmies at the bottom—insect marauders which enter unobserved and which are often first noticed, only when they force themselves upon his attention in their myriads.

The wild fauna has not fallen off in numbers nor in variety, but visible numbers and varieties have gone, and their places have been taken by invisible hordes; the smaller things have been added in a bigger ratio than the larger have been lost. The standard of the wild fauna as regards size has fallen and is falling. In spite of statistics and of multitudes of species, we have in effect lost more than we have gained, for how can the increase of Rabbits and Sparrows and Earthworms and Caterpillars, and the addition of millions of Rats and Cockroaches and Crickets and Bugs ever take the place of those fine creatures round the memories of which the glamour of Scotland's past still plays—the Reindeer and the Elk, the Wolf, the Brown Bear, the Lynx, and the Beaver, the Bustard, the Crane, the bumbling Bittern, and many another, lost or disappearing?

IX. 3

SOME INDIRECT RESULTS

ENUMERATION NOT THE WHOLE STORY

But when we have completed this tale of man's influence upon the numbers and character of the fauna, there is yet more to be told. We have focussed our attention mainly upon first causes and first effects; we have, as it were, listened to the bursting of a thunderstorm, and arrested by the first great crash, have failed to notice that sound after sound succeeds, reverberating amongst the hills, echoed back by wood and mountain until in a million eddies and whispers it passes beyond human ken. So it is in the animal world; the chains of circumstance are strong, and the invisible bonds, which link creature with creature and animal life with vegetable, and make each least being a necessary link in the welfare of another, cannot be tampered with without setting loose forces the final workings of which we cannot see:

> Where one step broken, the great scale's destroyed:
> From Nature's chain whatever link you strike,
> Tenth, or ten thousandth breaks the chain alike.

Did the Italians, as they slew wantonly and recklessly their birds of prey, realize that Moles would increase in number and with their burrows riddle those dykes laboriously constructed by man to keep in check the river Po, so that many square miles of fertile land have been flooded? Did the colonists dream of the ultimate influence of the Sparrows and Starlings, the Rabbits and Foxes and Weasels they were so ready to plant in their new lands? Did the early navigators, who in goodwill to their fellowmen, placed Pigs on many of the islands they touched, suppose that history would lay at their door the total extirpation of the Dodo of Mauritius, and possibly of the Solitaire of Rodriguez?

THE COMPLEX OF LIFE

I have already traced in the foregoing pages some examples of the secondary results of man's influence in Scotland: the slaughter of birds of prey and increase of vermin, the increase of cultivated crops which has multiplied graminivorous beasts and birds, which again have benefited their enemies, the beasts and birds of prey, and so on. Here I would add a few more illustrations to emphasize how all-pervading is the complex of life.

It would be easy to invent a series of likely chains of circumstance which ought to happen. It would be easy to argue that the less wild ground there was for the rearing of game the more gamekeepers would be necessary: for cultivation curtails wild ground, but cultivation increases the numbers of insect pests, and this the number of insectivorous birds; these again increase the numbers of hawks and other birds of prey, and so in the long run gamekeepers would have to be multiplied. We are familiar too with Darwin's famous chain of the cats, which, by eating field-mice, and these by eating humble bees, diminish the clover crop, which cannot form ripe seeds in the absence of the humble bee.

But such sequences are frequently little more than purely imaginary, since Nature is not restrained by the limitations of human logic. Yet the succession of events in Nature is no less striking.

RABBITS AND VEGETATION

Man introduced the wild Rabbit to Britain at a not very distant date. It has spread from cultivated field to wild glen and rocky hillside until it has possessed the land. And everywhere it is affecting the vegetation. Mr E. P. Farrow has shown that on moorland a favourite food is the tender stems and shoots of heather (*Calluna vulgaris*), and as a result, heather is eaten bare to the root. Under such conditions, the heather moor rapidly degenerates, and benefiting by the discomfiture of its rival, the sand sedge (*Carex arenaria*) association of plants makes fresh growth and captures the places formerly occupied by heather. This in turn the Rabbits attack, making way for the advance of a "grass-heath" association. So that a simple consequence of the introduction of Rabbits is a tendency on the part of heather moor to disappear and be replaced by grassy moorland. How this change from heather to grass must influence the animals in the area affected the reader can guess.

Influences of Sheep, Goats and Rabbits

When Neolithic man drove his flocks before him across the dry ground of the English Channel, he left them as far as possible to roam at their own free will. Their numbers were greater than those of the animals they displaced, so that cropping of herbage became more intensive, and within the forest and on its margin seedling trees were devoured for their succulent leaves. As under the influence of Rabbits, the general covering of the ground was affected. For many years, on a portion of Scotstown Moor in the neighbourhood of Aberdeen, Professor J. W. H. Trail has told me, Sheep have been annually penned for short periods of ten days or a fortnight, on their autumn migrations from the highlands to the lowland feeding-grounds. There even so short encampments have caused noticeable changes in vegetation, for the areas of the pens stand clear cut to-day in the midst of the moor as areas in which heather has been banished and has been replaced by "grass-heath" associations of plants. Other results, to which I have already alluded (p. 323), followed when the higher lands were cleared of their wild fauna and the custom developed of sending Sheep and Goats to pasture on the upland moors and hills. Here, as food became scarce in winter time, they fed upon the seedlings which formed the advanced posts of the forest and paved the way for new extensions of forest growth. The extension of woodland in all areas open to the inroads of flocks has been prevented.

Further, even within the woodland, Sheep, Goats and Rabbits are exceedingly destructive; young seedlings and even sturdy trees up to three or four feet high are ruined by the demolition of their foliage and leading shoots and by the gnawing of their bark. So that when nature claims an aged and timeworn member of the forest, there is no fresh growth spontaneously to take its place; and the woodland, which was prevented from extending by these aliens, is also prevented by them from recuperating in its old age. Thus there has resulted a general decay of forests bordering the heath and moorland pastures of sheep, the limit of the growth of forest upon the hillsides has been lowered, and woodlands have been forced down the mountain sides into the sheltered valleys where more rapid growth helps to counteract the influence of herbivorous enemies.

Such a consequence of the introduction of new animals is so gradual in development that it is scarcely noticed amidst a confusion of causes and effects. But St Helena offers a vivid presentment of its stages. When it was discovered by the Portuguese in 1502, the island was covered with luxuriant forests and a rich assemblage of peculiar plants. Goats were introduced in 1513 and multiplied rapidly; then native plants and even the tropical forest began to disappear, yet even in 1709, the native ebony was so abundant that it was used as fuel for burning lime. With the enormous increase in numbers of the Goats, however, new forest growth ceased, and when the older trees had been cut down by man, it was realized, in 1810, that the forests were gone, leaving no successors to take their place. The island, which was a precious oasis of tropical plant life set in the midst of the Atlantic, has become a barren and rocky waste.

Results of similar nature, but on a less noticeable scale, have followed the introduction of Sheep, Goats, and Rabbits in Scotland, and the destruction of the forest, for which they were in part responsible, set in motion many trains of circumstance, the origins of some of which have been traced in an earlier chapter.

The Case of Gulls and Moorland

Of necessity these pages have been confined to the main currents of man's influence, but it must not be supposed that his little, insignificant interferences pass unnoticed in the world of nature. Since it has been impossible to illustrate these step by step, it must be taken as sufficient and as typical of many others if I trace a single particular case—that of Gulls and the Moorland, to which my attention was drawn by Mr O. H. Wild.

In the northern confines of Peeblesshire, on the southern slopes of the Pentland Hills, there is a considerable area of peaty land not far from West Linton, known as the White Moss. In 1890 this area was covered with a close growth of heather: it was indeed a typical heather moor with peat and moisture underneath. In 1892 or 1893, according to a former gamekeeper on the estate, a man of intelligence and close observation, a few pairs of Black-headed Gulls came to nest upon the Moss. These were protected and encouraged by the proprietor for the sake of his Pheasants, the eggs of the Gulls being used for feeding the gamebirds. The collection of eggs was made systematically, only the early clutches being taken, while the last clutch was invariably left to hatch. As a result of protection the Gulls increased enormously in numbers, so that in 1897, Mr William Evans regarded the colony as the most populous in the district, and in 1904 came to the conclusion that there could not be fewer than 1500 to 2000 pairs of birds.

With the extension of the gullery, the vegetation gradually underwent a noticeable change. At first the heather (*Calluna*) gradually disappeared and its place was taken by coarse grass, which grew so rankly that the villagers cut from it heavy crops of hay. Then the grass gave way to a dense growth of rushes (*Juncus glomeratus*) amongst which, after a time, sprang up weeds of cultivation, which had before been unnoticed in this peaty ground. Of these some thrived with extraordinary vigour, and a forest of docks (*Rumex*) arose which almost choked the rushes and the disappearing grass.

So crowded did the original gullery become, that an offshoot from it settled upon a part of the White Moss half a mile away, where the surface also was entirely covered with heather (*Calluna*). Here similar changes took place in the vegetation—the heather was successively replaced by rough grass, then by rushes and lastly docks predominated. Nothing could be more striking than to visit these areas in spring before the new year's growth has commenced, and to see, clear cut in the midst of a black heather moor, a circular patch of withered tawny-yellow stems and grass of the past season—the site of the gullery (Fig. 88). Closer inspection shows that the margin of the gullery shades, as one would expect, into the surrounding heather, the borders of which are broken by scattered tussocks of grass becoming

SOME INDIRECT RESULTS 503

Fig. 88. Alteration of moorland by nesting of Gulls, White Moss, West Linton, Spring, 1917.
Site of gullery shown by pale band of withered grass in middle of moor; subsidiary gullery beyond.

Fig. 89. Alteration of moorland by nesting of Gulls; near view of part of above gullery showing merging of tussocks of grass into heather.

more and more frequent until they unite in the solid mass of the gullery itself (Fig. 89, p. 503).

The changes of vegetation which took place in the main and subsidiary gulleries seem to have been due partly to the fertilizing of the soil by the food refuse and excreta of the enormous numbers of birds, partly to the puddling of the surface by their feet and to the surface accumulation of their nests, so that superficial water was retained, and the peat bed with concealed and deep moisture was transformed into a surface marsh.

Such changes did not come alone. No observations were made upon the minute insects and lesser creatures of the moor, although the development of a surface marsh must have closely affected their welfare, their numbers and even their kind. From his practical point of view, however, the keeper noticed that the Grouse ceased to nest and disappeared, and that in their place, attracted by the marsh and the cover of the rushes, there came to feed and nest first a few, then many Teal Ducks. A single flock of these containing as many as seventy birds was seen when the gullery was at its height.

Now after the Gulls had been protected for some fifteen years, they were ousted from the White Moss. The villagers, incensed at the destruction of their hay crop, which had been replaced by docks, made persistent raids on the gullery, and the proprietor, regretful at the disappearance of Grouse, ceased to protect the Gulls for his Pheasants. In the early summer of 1917, scarcely a Gull was to be seen, and not more than thirty pairs nested, where a few years before, there were some 2000 pairs. As a result of the disappearance of the Gulls a gradual reversion towards the old state has been in progress. The docks have almost disappeared, and the rushes are giving way to rough grass and even to heather. The Teal have gone, in 1917 only one pair nested, and the Grouse are returning, so that on a moor where only three brace of Grouse were shot in the heyday of the gullery, now some twenty brace fall to the sportsman's gun.

In the case of the Gulls and the Moorland, an insignificant influence—the protection at first of a few Gulls—set in motion an endless chain of circumstance, of which we have done no more than glance at the most obvious links. And all the changes I have recounted took place in a remarkably short period, the maximum of change being attained just after 15 years, when the Gulls were disturbed and the evolution of the moorland-marsh was checked. In a few years the people had rejoiced at a newly arisen haycrop, which, thanks to the Gulls, had replaced useless heather; and in a few more years, they lamented the growth of rushes and docks, which, malisons upon the Gulls, had ruined their bountiful hay.

The interest of the story of the Gulls and the Moorland lies partly in its suggestiveness. If the natural processes set a-rolling by a tiny and temporary interference of man can be so marked, how can imagination grasp the total effects of man's influence, impressed upon the world of Nature often with great power, and persisted in, not for a few years, nor for a few centuries, but for thousands, nay, even for tens of thousands of years?

IX. 4

THE RECOIL UPON MAN OF HIS INFLUENCE UPON ANIMAL LIFE

There is one last aspect of man's influence upon animals to which I would refer: it might more properly be called the influence of animals upon man, for great as is man's faith in his own independence, he is subject to the recoil of his own actions and influences as surely as the animals he handles.

Man himself would be in a sorry plight indeed were the plaint of the melancholy Jacques the whole truth, that we

> Are mere usurpers, tyrants, and what's worse,
> To fright the animals and to kill them up
> In their assign'd and native dwelling place.

Fortunately there is another side to the matter. Man has not always been the usurper and tyrant, and just in the degree that he has avoided tyranny, he has benefited in a spiritual sense from the recoil of his own influence. On the whole these indirect reactions have been to the advantage of mankind.

A few examples of the material and spiritual influences that have played upon Man and his welfare, as the result of his interference, direct or indirect, with the animal kingdom, will indicate how deeply human life is affected thereby.

RECOILS ON HEALTH

Flies and Disease

Take the case of health. Only a few examples from the hosts available can be given. Man, by his deposition of garbage, stable-manure and rotting organic matter has enormously increased the numbers of the Common House Fly (*Musca domestica*), to which he has furnished suitable breeding-places without number. But the Fly from its very habits is a menace to human life, since it contaminates, with bacteria carried on its feet, food and drink set aside for human use. Thus it has been found to spread cholera, typhoid fever, and tropical dysentery; and a relation, more than suspicious, which has been traced between the numbers of Flies and of human deaths in Manchester, suggests that the House Fly is closely connected with the epidemics of summer or infantile diarrhoea, which in some towns sweep away so many children's lives. It seems obvious too that the Fly may readily help in the dispersal of the "germs" of such diseases as tuberculosis, anthrax, smallpox, ophthalmia and infantile paralysis.

RATS AND DISEASE

Rats have spread vastly under the influence of man. Not only have they been introduced to new countries in ships,—the ancestors of the whole stock of rats in Britain were voyagers brought from across seas,—but in most countries, the increase of offal and of garbage in general, and the development of sewer systems which afford food in abundance and perfect security, have multiplied the numbers of rats beyond belief. This effect of man's civilization results in direct damage to his goods and stored food and buildings, but it is of more account on other grounds. Rats—and the Black Rat and the Brown or Norway Rat are almost equally guilty—nourish undesired colleagues in the many species of Fleas which infest them, and these two allies, rats and rat-fleas, have been responsible for the deaths of upwards of 7,000,000 people in the course of fifteen years (1896–1911) in India alone.

It has been conclusively shown that bubonic plague in man depends entirely upon the disease in the rat, and that the disease is conveyed from one rat to another and from the rat to man solely by one or other of the rat-fleas. Up to the present time eight species of rat-fleas have been shown to be capable of transmitting plague, and of these five are common British species, while the chief sinner, the eastern Plague Flea (*Xenopsylla cheopsis*), is occasionally brought from abroad to our seaports.

There is no need to recall how in past days epidemics of plague flew like messengers of death across the face of Europe, harassing and devastating the great cities, until none were left fit to bury the dead. This infectious and air-borne pneumonic plague is a possible successor of the rat-carried bubonic plague, so that the Great Plagues may well be examples of the recoil upon man's own health of his influence in spreading the Black and the Brown Rats.

Although the danger of great outbreaks in this country seems to have largely disappeared, it is well to remember that in Britain there have been epidemics of plague so recently as 1900 and 1910, and that the disease is almost constantly present in the great seaports of the world. So effective is commerce in its distribution that the epidemic which started in China in 1894 had by 1908 afflicted 51 different countries and had been carried to all the continents of the world.

But the rat is a source of further mischief: it is the chief medium whereby the Round Worm (*Trichinella spiralis*) infects pigs, whence it reaches man through the eating of improperly cooked pork. So widely spread and so serious for man is the resulting disease of trichinosis, due to the lodging of the parasite in muscle or intestine, that most civilized countries have been compelled to institute strict inspections for the detection of infected pork. And rats besides are mechanical distributors of all the ugly contaminations of the sewers—typhoid fever, scarlet fever, diphtheria and many more; for the sewer and the drain are the rat's main thoroughfares.

Fortunately, however, man's influence upon the animal world has not always recoiled upon his own head in new and fell diseases.

Ague in Scotland

Of the ailments which laid hold of our forebears a couple of centuries ago, none was more prevalent or more persistent in its attacks and effects than the Ague. Especially in the spring-time, when trying conditions of weather and reduced physical fitness due to bad housing and poor food and clothing, paved the way for relapses, the ague passed through Scotland like a blight. Contemporary references to the disease show that it prevailed in particular amongst the labouring classes, so that in many districts it was with difficulty that the heavy

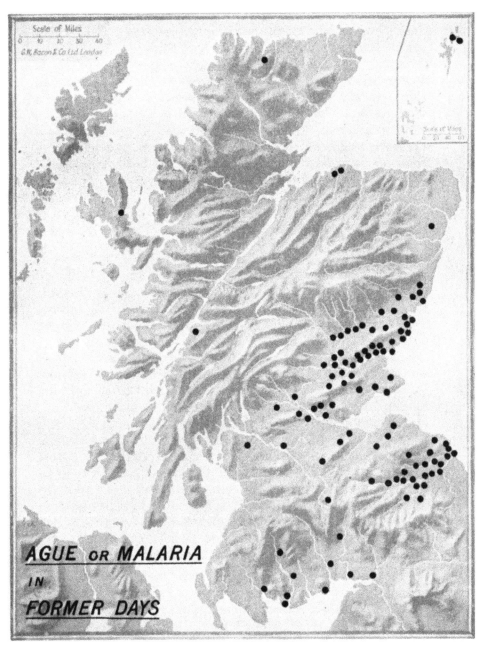

DISTRIBUTION OF AGUE OR MALARIA IN SCOTLAND IN THE 18th CENTURY.
Each dot indicates a parish where Ague was common.

agricultural operations of the spring-time could be performed. The case of the parish of Kirkden in Forfar is typical, where about 1765 the ague was "so general that many farmers found it difficult to sow and harrow their lands, in the proper season, owing to their servants being so much afflicted with it." It is said that in Berwickshire the disease so frequently laid the men aside that the bond women had to leave their lighter labours in the field to take up the heavy work of ploughing. And the writer of the Old Statistical Account of the parish of Abernyte in Perthshire was assured that, about 1760,

if a farmer in the spring wanted four of his cottagers for any piece of work, he generally ordered six, knowing the probability that some of them, before the work could be finished, would be rendered unfit for labour by an attack of the ague.

Through the lowlands and midlands of Scotland, the disease was common; from Berwickshire, Kirkcudbright and Roxburghshire northwards to Forfar and Kincardine, but north of this region, ague seems to have been less prevalent[1]. In the affected areas, the malady reached its height in certain well-defined districts. Of these the Carse of Gowrie was the chief, for scarcely a parish in the region of the flat carse-land but lamented its ravages.

Now the ague gradually disappeared from the infested districts of Scotland, just as it vanished from the fens of Lincolnshire and from other districts in England. About 1734 it was still very common in Linlithgowshire, about 1740 to 1750 in Kincardineshire and Forfar, about 1760 it was "very prevalent" in Perthshire, and in 1780 161 cases occurred in Kelso district alone. Yet by the end of the eighteenth century, ague had almost disappeared from Scotland.

A few cases, typical of many, will illustrate this extraordinary and almost simultaneous disappearance over a wide area. The parish minister of St Vigians in Forfar, writing in 1793 or 1794, says from his own experience that

for many years after 1754, agues were so common in this parish that the incumbent has often seen, in the months of March, April and May, and

---

[1] Since the Map (VIII) was prepared I have found from further investigations of original records, that in the late eighteenth and early nineteenth century ague was very widely distributed, though less common than in the Lowlands, in Aberdeenshire and the neighbouring counties.

sometimes in autumn from fifteen to twenty-five persons in that distemper. He does not remember to have seen a single person in the ague for 20 years past.

A Perthshire record reads: "The ague, which used greatly to prevail here (St Madois parish) as well as in other parts of the Carse of Gowrie, is now hardly known [in 1792]." In Kirkcudbrightshire, it was written in 1744 "agues formerly prevailed very much. There has not, however, been one instance of this disorder for 9 or 10 years past [in the parish of Borgue]." In Berwickshire agues had "almost totally disappeared" in 1792, and the virulence of the disease had much diminished, though, in some districts, it "still returns with such unexpected frequency and force as often baffles all speculation concerning it." And lastly the records of Kelso Dispensary, which I have examined, show that the number of cases treated there declined after 1781, from 161 in that year, to 110 in 1782, after which a gradual fall is noticeable, so that after 1797 they did not exceed 10 in any year, and after 1840 completely disappear.

But what has all this got to do with the influence of man upon animal life? The old writers who have recorded the facts of the prevalence of ague in Scotland and of its disappearance could not have answered the question. To them the miasma of the marsh was a sufficient cause of the disease, though some essayed other speculations. Marshiness, foggy atmosphere, mean houses, defects in cleanliness, and lack of animal food all bore the blame for the mysterious pestilence. Since marsh miasma was the generally assigned cause of ague, the decrease of the latter was generally put down to the drainage of mosses and bogs. Two interesting suggestions of the Scottish chronicles of the late eighteenth century are worth mentioning in view of later discoveries. The Rev. Mr Samuel Smith of Borgue parish in Kirkcudbright held that the disappearance of "intermittent fevers" could not be assigned to the disappearance of bog land since in his district "no mosses or marshes have been drained, of any consequence, for many years past," and he suggests that

when land is deepened and pulverised in consequence of improvements, by lime, shells and marls, it absorbs the rain more quickly and plentifully. Hence less moisture will arise in evaporation; less water will run along the surface, and stagnate in the hollows, which are here to be found in every field.

To these stagnant pools he attributed the "remote cause of intermittent fevers." Even more interesting is the statement of the incumbent of Kirkbean parish in the same county of Kirkcudbright:

Formerly many of the inhabitants went into Lincolnshire [an area long noted for the prevalence of ague] for employment during the harvest, and returned infected with this disease, now they have work sufficient to employ them in the parish and the disease is seldom a complaint.

In recent years science has thrown a new light on the cause of ague, intermittent fever, or malaria, as it is variously termed. The malady is due to the presence of minute parasites of low organization in the human blood. At one stage of their existence these hæmamœbæ (Plasmodium) attack and bore their way into the red blood-corpuscles, in which they undergo a process of multiplication until at last the blood-cells break down and the multitudes of freshly formed parasites are set free in the blood stream. It is at this stage of the disruption of the blood-cell, that the patient has a relapse of fever, the interval between relapses—about forty-eight hours in tertian ague, seventy-two in quartan ague, and an uncertain interval in the irregular fever—depending directly on the length of the life-cycle of the parasite.

Now an interesting and vital connection exists between the tiny Plasmodium parasite of the human blood and the insect world, for the virulence of the parasite in man dwindles if it does not undergo a periodic reincarnation within the body of a Mosquito. The Mosquito becomes infected by stealing a drop of the parasitized blood from a malarial patient, and it passes on the parasites and the infection when it punctures the skin of a healthy subject. The Mosquito therefore is essential to the spread of malaria or ague amongst men.

The habits of Mosquitoes, and by such I mean Anopheline, or, as a rule, "spot-winged" Gnats, are familiar to every naturalist. They like moist, warm, humid places, and they most often frequent marshy and boggy ground, sluggish streams, and stagnant pools, large or small. On the surface of the water, often along the weedy margins of pool or ditch, the eggs are laid and the wriggling larvae and pupae develop beneath the surface, until such time as the fully formed pupa comes to rest at the surface, and from its split skin the winged Mosquito flies into the air, ready, if it be a female, to do its worst to the race of men.

Look again at the Scottish records of ague in the light of the discoveries of Grassi and Laveran, Manson and Ross. Where did it most prevail? In the low-lying and marshy counties bordering the east coast and the Solway, or along the lines of the great rivers, where spring and autumn floods left abundance of stagnant pools; and especially in the Carse of Gowrie, still characterized by its warm humidity: that is to say in just those areas where in the days of unreclaimed marsh we should have expected to find Mosquitoes. It was less common apparently north of the Grampian Hills; but Mosquitoes prefer warmth, and the lower temperature of the northern counties may have been just sufficient to have checked the profusion of multiplication which seems to be necessary for the effective spread of malaria. We know that at the present day, Anopheline Mosquitoes still live in Aberdeenshire, Inverness-shire and as far north as Sutherland, but their numbers are small.

Whom did ague attack? There is scarcely a reference to the upper or artisan classes having suffered, and again and again the farm labourer is singled out for notice on account of his susceptibility to the distemper—just the very man whose work in the fields laid him open to the attentions of the mosquito amazon.

Last link in the chain: In the latter part of the eighteenth century in Scotland agriculture made great strides forward: regular systems of cropping began to replace the primitive and wasteful ways of the "outfields" and the "infields," and the success of the use of marl and lime upon peaty soils led to the reclamation of much marsh and bog-land and to the thorough draining by closed conduits or open ditches of land which previously had lain sodden with moisture throughout most of the year. It never entered man's mind that, by the way, he was destroying Mosquitoes, since he was destroying their breeding-places; he probably never noticed, as the years passed, that he had less often to stop to revenge the prick of the Gnat. Yet unwittingly he had set in motion a chain of circumstances which was to circle and wheel till it returned to him again, when the sweat of his brow was repaid a thousand-fold in the health of his body.

Man drained the marshes and the stagnant pools and so destroyed the Mosquitoes, which could multiply only there, and so he destroyed the only transmitters of the tiny Plasmodium of the human blood, and malaria or ague was no more. Man, Marshes, Mosquitoes, Malaria—steps and stages of a progress which, starting with man, returns through the worlds of inanimate and animate nature, to heap blessings on his head.

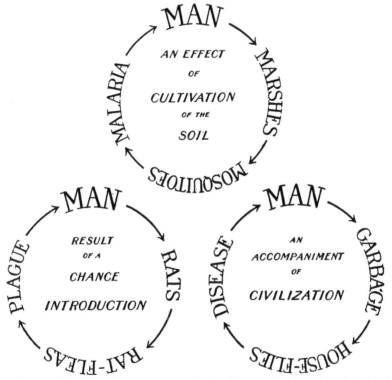

Fig. 90. Diagrammatic representation of Recoils on man's health, due to different types of his interference with animal life.

RECOILS ON MAN'S PROSPERITY

But even where man has not been affected in his own health, the influences he has set a-rolling have recoiled upon his well-being. The draining of the marshes affords a case in point. Man acted unwittingly, for not till long after the chain of circumstances had been completed and had become effective, were the separate links of the chain discovered. Man drained the marshes and strange to say, one of the most serious of the diseases of his domestic stock—the Liver-rot of Sheep,—began to disappear. At its worst, liver-rot caused a mortality of three million sheep in England in a single winter, that of 1879–80, in that of 1830–31, more than two million are said to have succumbed, and Youatt, writing in 1837, considered that the average number of deaths amounted to more than one million sheep and lambs every year. The older writers recognized that the disease was confined to wet seasons, or followed upon pasturing sheep upon ground moist and marshy at all seasons, but who could have guessed the strange stages that led to the reduction of the rot in Britain? Many years of research were necessary before the links in the chain were appreciated.

The rot was caused by a flat-worm—the "Liver-fluke," *Fasciola (Distomum) hepatica*—which choked the bile-ducts, and to the number of 870 has been found in the liver of a single sheep. But the liver-fluke has a curious life-history: the eggs develop in water, and the wriggling larva which hatches from each, dies if it cannot find a Pond Snail (*Limnæa truncatula*), within which it passes through successive stages of development, until, reaching an active stage, it leaves the snail, and forsakes the water to wriggle up any convenient blade of grass. Here it comes to an untimely end if a sheep does not happen to eat the grass; but if a sheep be so unfortunate, the wheel of life is set in motion again, and an adult liver-fluke develops which finds its way to the bile ducts, and gives rise to the deadly liver-rot. In no other way can the liver-fluke come into being, the water and the Pond Snail are as necessary for its development as the sheep itself.

These then are the links in sequence which man unwittingly set in motion: Man drained the marshes, and so doing he destroyed the Pond Snails, and so doing he broke a link in the life-history of the liver-fluke, and destroyed the parasites; so sheep were freed from a scourge which wholly destroyed many flocks, and in a manner threatened, as Youatt says, to extirpate the breed of sheep in Britain.

Yet, to be effective a recoil on man's prosperity need not

follow so tortuous a path. By his cultivation of the soil man increases the number of Earthworms it contains: the earthworms of an intensively cultivated garden are more numerous than those of a cultivated field, and those of a field far outnumber those in uncultivated wilds. Yet earthworms, by penetrating and overturning the soil, render it more porous to moisture and more loose for the penetration of the tiny rootlets of plants, and by dragging leaves and decaying vegetation into their burrows, they add to the fertile humus of the soil itself. Man cultivates the earth for his own good, and so doing increases the number of earthworms, which in their turn add to the fertility of the soil, and to the abundance of man's crops.

Again, man introduced Hive Bees to Britain, or at any rate multiplied their numbers far beyond the stock in natural conditions, but do you think that the benefit he reaps is measured only by the weight of honey he gathers from their summer stores? Do they not by scattering pollen from one flower to another make his orchards fruitful, his garden and his clover fields blossom to some purpose? Darwin found that a hundred heads of red clover (*Trifolium pratense*) exposed to the visits of insects bore 69 grains weight of seeds, or since 40 seeds weigh a grain, over 2700 seeds; but that a hundred heads from which insects were excluded by a net produced not a seed at all. And the Hive Bee is the chief fertilizer of the wild white clover, which in recent years has enormously increased the value of thousands of acres of pasture.

Its work, little though it be evident, is important in the orchard. Mr Harold Bastin records a story of 1907, when, following a cold and wet spring the apple crop was a general failure. In a Huntingdonshire district where there are many orchards, scarcely any fruit was to be found except in three orchards—and these stood close to an apiary of fifty hives. The cold and wet weather restricted the journeys of the bees so that they moved mainly on the trees nearest their hives, and the fruit grower reaped where the bee-keeper had unintentionally sown.

The indirect benefit wrought by the bee is no new or fantastic notion. Fruit growers have accepted it as a matter of course, and the absence of bees due to the ravages of "Isle of Wight" disease, has made necessary other measures for the fertilization of the apple blossom. So it has come about that in several English orchards, in the spring of 1917, girls were employed to dust with powder-puffs the pollen from one flower to another, a labour hitherto performed almost entirely, and who will say less satisfactorily, by the honey bee. Man keeps the bee for its store of sweetness, and the bee, having paid its debt, throws ungrudgingly into the bargain the fruit of a thousand orchards.

THE RECOIL UPON MAN'S CHARACTER AND CIVILIZATION

In other ways man's influence over animals has had a vital power in moulding his own character and in guiding the lines of his civilization. And this is true in particular of the deliberate part he has played in domesticating animals. The attention he has devoted to the fostering of the creatures he has selected for his personal service has given birth to a new feeling of regard for life and living things, a new feeling of proprietorship and of responsibility. And these new feelings of community and of responsibility have together formed the foundations of the complex structure of social life.

It is instructive to trace the steps of early civilization in the light of man's influence upon animals. Primitive man is a child of nature, a simple hunter who claims no country as his own, and has few rights and no comforts save what the strength of his own right arm can bring. But even with the early domestication of such animals as sheep and cattle, man is relieved from the immediate cares of the morrow; for food and raiment lie to his hand, and new responsibilities and interests claim his attention and develop his mind. The herdsman is greater than the hunter, the Eskimo only hunt the Reindeer, the Tunguses of Siberia tame them and live by their herds. The herdsman too has gained in his sense of community, for the herdsman peoples share in the great religions of the world, the Arabs are Moslems, the herdsmen of the steppes of Asia, Buddhists. As flocks and herds increase, the wanderings of the nomad herdsman become more and more pronounced, until the art of the herdsman combines with that of the cultivator, and the family settles into the steady responsibility borne by pastoral tillers of the soil. Now new rights arise, justice comes to replace the rule of might, communities become more permanent, social institutions become fixed and assume greater importance, the earth itself becomes part of man's heritage, and patriotism is born.

It is easy to see how domestication has been a predominating factor in setting the civilization of the settled herdsman-tiller above that of the independent solitary cultivator. Where at first man's own labour tilled a meagre plot or two in the easiest land, his yoke of oxen ploughed far and wide across the plain, and by the labour of his beasts, land, which from its nature lay beyond the power of his arm, was added to his fields. A new wealth arose, more food flowed to his granary, his animals multiplied, and with their multiplication still more fields were added and more wealth accrued. And so came a time when, relieved of the necessity of the daily drudgery of the bread-winner, the choice spirits of the community were free to turn their attention to the development of the arts and sciences, and culture was sped upon its way.

At the heart of the secret of civilization lies the domestication of animals, training man unconsciously in the ways of government and control, adding a new tenderness and regard for life to his nature, guiding him from barbarism to civilization, and impelling him along the ascending pathway of humanity with a force he has been slow to recognize and acknowledge.

So the study of Man's influence upon animals ends where it began—with Man.

INDEX

Abbey Crag, 200
Aberdeen, 100, 116, 222, 250, 415, 428, 443, 447, 466, 500
Aberdeenshire, ague in, 509
  Ant, Small Red House, in, 440
  Buzzard in, 135
  Capercaillie in, 270, 272, 356
  Deer, foreign, in, 254
    Red, in, 211
  Dog, early remains in, 87
  Eagle, Golden, in, 128
  forests of, 311
    destruction, by storm, 324
      for agriculture, 322
  Fox in, 122
  Gulls, change of food habit in, 412
    Kittiwake, as food, 142
  Horses, early remains in, 72
    "wild" in, 80
  Kite in, 133
  Marten in, 161
  Mosquitoes in, 512
  pearl fishing in, 194, 195
  Perch in, 279
  Polecat in, 165
  Rabbit warren in, 250
  Rat, absence from eastern, 426
    Black, in, 427
  Salmon fisheries of, 380, 384
  Squirrel, Red, in, 294, 296, 297
  Tench in, 279
  Trout, "Rainbow," in, 278
  Urus, prehistoric remains in, 51
  Wolf in, 116, 121, 311
  Woodpecker, Great Spotted, in, 358
  wool of, 46
Aberfoyle, 128
Abernethy, 314, 321, 354, 358
Abernyte, parish of, 509
Aboyne, 294
Abraxas grossulariata. See Moths, Magpie
Acanthocinus ædilis, 472
Achatina fulica, 459
Actora æstuum, 395
Adamnan's *Life of St Columba*, 222
Africa, dispersal of pests from and to, 421, 442, 444, 450, 451, 458, 459, 460
  East, 440
  South, 452, 464
Agelæus phœniceus. See Starling, Red-winged
Agelastica alni, 472
Agriculture in Scotland. See Cultivation

Ague in Scotland, 373, 508–513
Ailsa Craig, 130
Airleywight, 346
Alaska, 110, 246, 257, 458
Alauda brachydactyla, 394
Alca impennis. See Auk, Great
Alces alces. See Elk
Alder, Joshua, 464
Aleurobius, 410
Alexandria, 289
Alford, 125, 272
  Vale of, 294
Algeria, 442
Allman, Prof., 386
Allt nan Uamh, Bone Cave of. See Inchnadamph
Alnwick, River, 277
Alosa alabamae. See Shad
Alston, E. R., 186 f.-n., 288
America. See below, *and also* Canada
  dispersal of pests to and from, 435, 444, 447, 457, 473
  introduction of animals and plants to and from, 242, 258
America, North, dispersal of pests to and from, 446, 454, 456, 458, 465, 469, 471, 472. See below, *and also* Canada
  forest, destruction of, 327
America, South, dispersal of pests to and from, 452, 475
America, United States of, animal preserves of, 236, 387
  Colorado Beetle in, 387–388, 397, 409, 411
  counter-pests in, 259, 260
  dispersal of pests to and from, 259, 423, 436, 440, 441, 442, 450, 451, 452, 453, 458, 459, 460, 461, 462, 463, 465, 469
  Earthworms in, 393
  fisheries in, 220, 258, 277
  protective legislation relating to animals, 187, 206, 227
  Rats in, 431
  river barriers affecting salmon fisheries of, 380
  Seal, Fur, protection of, 223
American Blight. See Aphis, Woolly
Indians, 28, 454
Anacharis canadensis, 473
Anderson, Dr James, 47
  Dr Joseph, 58, 87
Angel fish, 19
Anglesey, 423

INDEX

Angus, 356
Animals, acclimatization of, 246
  balance, even, necessary for successful adaptation, 255, 257, 419, 486, 489
  chains of circumstances affecting, 479
  changes in habits, 400-416
  complex of life, 498, 499
  decline in physique, 489
  decrease due to cultivation and civilization, 365-386, 488
  deliberate destruction of, 109-196, 480, 481
  deliberate interference with, 25 et seq.
  development of man's influence upon, 480
  dispersal of, 397-399
  domesticated, numbers added to fauna, 486
  domestication of, 27-107. See Domestication
  extermination of larger and addition of smaller, 497
  fauna, the great change in, 497
  forest, destruction of, affecting, 327-364
  increase, due to cultivation and civilization, 387-396
  in numbers, 492-496
  in variety, 496
  indirect interference with, 303 et seq., 498-505
  influences tending to increase numbers, 387-396, 481-488, 492-496
    to modify character and habits, 489
    to reduce numbers, 488
  introduction of, deliberate, for amenity, 264, 283-300
    for sport, 264-282
    for utility, 246-263
  introduction of, unforeseen, 417-475, 480, 487
  native, on Man's arrival in Scotland, 12 et seq.
  post-glacial, 12
  protection of, 197-240, 481, 485, 487
  recoil upon man of his influence upon, 506-517
  sanctuaries, 236
  selection of domicile by, 400
  transportation of, 278-282
Annan, River, 276
Annandale, 429, 430
Anobium domesticum, 404
  paniceum. See Beetles, Biscuit
Anser anser. See Goose, Grey Lag
Anthonomus pomorum, 392
  signatus, 392
Antilles, the, 421
Ants, as food of pheasants, 268
  introduction in food materials, 440
  Red House, 440
  tropical, introduced into greenhouses, 463
Antwerp, 414

Apanteles glomeratus, 458
Aphis, Cabbage, 458
  Corn, 440
  Turnip, 392
  Woolly (American Blight), 392, 461, 474, 487
Aphodius, increase of, 396
Appin, 352, 398
Apple, introduction of pests with, 449
Apple seed chalcid, 453
Apple trees, American Blight on. See Aphis, Woolly
  Codlin Moth as pest on, 451, 475
  Scale insects on, 450
Aquila chrysaëtus. See Eagle, Golden
Arctic fauna, 12, 13
  flora, 10, 305, 306
Arctocephalus ursinus. See Seal, Fur
Ardea egretta and *candidissima*. See Egrets
Ardnamurchan, 164, 296, 494
Ardrossan shell-mound, 19, 71, 158
Argali, the, 44
Argentina, 262, 459, 460
Argyllshire, Cattle, Highland, in, 61
  Capercaillie in, 270, 273
  Celtic shorthorn in, 57, 58
  Chough in, 178
  Deer, Fallow, in, 287
    Red, in, 31, 334
    Roe, in, 31, 331, 332
  domestic stock in, 494
  forest, destruction in, 319, 320, 354
  Marten in, 161
  Minnow in, 279
  Polecat in, 164
  Rabbit in, 252
  Squirrel in, 294, 296, 352, 353, 354
  Starling, Red-winged, in, 283
  Willow Grouse in, 275
Arhopalus speciosus, 472
Arkaig, Loch, 314
Arran, Capercaillie in, 270, 355
  Chough in, 178
  Deer, Virginian, in, 254, 287
  Hare, Common Brown, in, 280
    Mountain or Variable, in, 281
  Minnow in, 279
Arrochar, 161, 288
Arthur's Seat, 190
Arvicola ratticeps. See Vole, Northern
Ascaris lumbricoides, 422
  megalocephala, 422
Asellus aquaticus, 414
Ashby, Lincolnshire, 139
Ashiestiel, 314
Ashkirk, 336
Ashworth, Prof. J. H., 403
Asia, 421, 442
Askaig, Loch, 134
Asparagus, pests introduced with, 459
Aspidotus alienus, 461 f.-n.
  perniciosus. See Scale Insects, San Jose
Assam, 328

Assynt, 120, 126, 177
Astur palumbarius. See Goshawk
Athalia colibri, 392
Athene noctua. See Owl, Little
Athole, foreign Deer in, 254
  Forest of, 114, 200
    royal hunts in, 119, 210-211, 265, 266, 283, 285, 331, 355
Auk, Great, 17, 148
  destruction of, 110, 138, 142-146, 488
Aulacaspis pentagona. See Scale Insects, Mulberry
Australasia, 463
Australia, Apple Scale in, 450
  Bed-Bug, Tropical, in, 421
  Codlin Moth in, 452
  Corn Weevils in, 442
  Greenbottle Fly in, 423
  Hare and Rabbit in, 241, 253, 255
  Land Planarians in, 464
  Mediterranean Fruit Fly in, 451
  plants and animals introduced into, 241, 242, 245, 255
Awe, Loch, 126, 134, 164, 278, 294, 319, 322, 331
Ayala, Don Pedro de, 41, 149
Ayr, 472
Ayr, River, 276, 277, 293
Ayrshire, Alpine Hare in, 282
  American Brook Trout in, 277
  Beaver, prehistoric remains in, 158
  Boar, Wild, prehistoric remains in, 90
  Capercaillie in, 270, 273
  Cattle, prehistoric remains in, 34, 51, 56, 57
  Chough in, 178
  Deer, Red, prehistoric remains in, 34, 333
  Dog, prehistoric remains in, 34, 87
  Goat, prehistoric remains in, 34
  Grayling in, 276
  Hawks in, 199
  Horse, prehistoric remains in, 34, 70, 71
  Kite in, 133
  Marten in, 161
  pearl fishing in, 195
  Polecat in, 164
  pollution of rivers in, 384
  Rabbit in, 252
  Reindeer, prehistoric remains in, 338, 340, 344
  Sheep, prehistoric remains in, 34
  skin produce of, 156
  Squirrel in, 293
  Trochus lineatus in shell-mounds in, 20
  Urus, prehistoric remains in, 51
  Wolf, prehistoric remains in, 115

Baddingsgill, 402, 403
Badenoch, 211
Badger, denizen of ancient Scottish forests, 308, 329

Badger—*Cont.*
  destruction of, 229, 369
    for skin, 156, 166, 170
    for sport, 184-186
  drowned by floods, 328
  Flea, Human, parasitic on, 420
  importation for sport, 264
  protection through popular favour, 233
Balæna mysticetus. See Whales, Greenland
Balænoptera. See Whales, Rorqual
Balance of animal life, 257, 486, 488, 489
Balanus tintinnabulum. See Barnacle, Common Acorn
Balgone, 69, 89
Ballochbuie Forest, 314
Balquhidder, 161
Baltimore Oriole, 299
Bananas, introductions with, 447, 474
Banchory, 272, 294, 358
Banffshire, "bog butter," discovered in, 58
  Capercaillie in, 272
  Great Spotted Woodpecker in, 35
  Kite in, 133
  Squirrel in, 296
  Wolf in, 120
Banks, Sir J., 461
Barclay, Dr William, 285
Baridium aterrimus, 463
Barnacle, Common Acorn, distribution by ships, 432
Barra, 75, 277
Barrett-Hamilton and Hinton, Messrs, 248, 425, 429
Barry, Rev. Geo., 284
  Sir F. T., 348
Barvas, 280
Bass, Black, 258, 259
  Striped, 380
Bass Rock, 147, 171, 251
Bastin, Harold, 515
Bates, H. W., 171
Bats, 379
Battle Abbey, 401
Bealach na Uidhe, Loch, 279
Beans and peas, pests introduced in, 445, 446
Bear, Brown, denizen of ancient Scottish forests, 308, 329
  effect of destruction of forests upon, 351
  extermination of, 112-115, 481, 496, 497
  prehistoric remains of, 17, 20, 112, 114, 330
  reduction in size, 490
Beasts of prey, destruction of, 109, 111
  destruction of forest to extirpate, 120, 318
  effects of destruction of, 197, 489
Beaufort Castle, 181, 295, 296, 297
Beauly, 273
Beauly Firth, 118
Beauly River, 295
Beaver, in Scotland, 17, 155, 157
  destruction for skin, 155, 157-158, 221, 308, 329, 482, 497

INDEX

Beaver—*Cont.*
  prehistoric remains, 157, 158
  re-introduction for amenity, 242, 298
Bedford, Duke of, 288
Bedfordshire, 288
Bee, Hive, domestication of, 107
  fertilization of flowers by, 515
  indirect effects of Isle of Wight disease of, 515
  introduction to Britain, 515
  to New Zealand, 257
Beetles, Asparagus, European, 459
  Bean, 446
  Biscuit, 410, 444
  Black. *See* Cockroach
  change of food habits of, 409, 410
  Cigarette, 445
  Click, increase of, 392
  "Clock," 239, 416
  Cockchafer, 235, 392
  Colorado, 387-388, 397, 409, 411
  "Death Watch," 404
  Dung, 239, 396
  Furniture, 404
  Grain, 441
  Heather, 206
  houses selected as domicile by, 404
  increase due to dung, 396
  introduction, into greenhouses, 463
    with flour, 443
    with wheat, 440
  Ladybird, 235, 260, 261
  Lentil, 446
  Longicorn, transportation in timber, 470
  Pea, 392, 445, 474
  pine, 329, 472
  Scarabee, 474
  Store, 442
  Timber, or Capricorn, 470
  Timberman, 472
  Tobacco, 410
  Turnip Flea, 392
  Weevils, *q.v.*
Belfast, 436
Belgium, 469, 474
Belhelvie Moss, 51
Bellenden, John, 83, 90, 173, 251
"Ben, Jo.", 173
Ben Lui, 314
Ben More, 126
Ben Venue, 128
Bennachie, Capercaillie in woods of, 272
  Forest of, 311
Bennie, James, 11, 305
Berlese, A., 262
Berner, Dame Juliana, 277
Berriedale, 295
Berry, William, 395
Berwick, North, 432
Berwick-on-Tweed, 226, 471, 472
Berwickshire, ague in, 509, 510
  Beaver in, 157

Berwickshire—*Cont.*
  Boar, Wild, in, 90
  Bustard in, 274
  Chough in, 178
  Deer, Red, in, 333
    Roe, in, 330, 331
  Elk in, 346, 347
  Hare, Mountain or Variable, in, 282
  Polecat in, 164
  Rabbit in, 252
  Squirrel in, 352
  Tench in, 279
  Urus in, 51
  Wild Cat in, 125
Beveridge, F. S., 412
Bignold, Sir Arthur, 288
Bilharzia hæmatobia, 422
Birchwood, 446
"Bird-butter," 148
Birds, carrion, protection as scavengers, 223-226, 395
  change of food habits of, 411
  change of habits induced by houses, 401
  destruction, as pest on fruit-crops, 411
    as vermin, 176-178
    by bird-catchers, 189, 193-194
    by lighthouses and other lights, 377-379
    by Little Owl, 256
    for food, 139
    for plumage, 110, 189
    of eggs, 148
  dispersal of, 399
  effect of destruction of forest upon, 329
  game, destruction for sport, 186-188
  influence of towns upon, 405-408
  insect-eating, increase due to cultivation, 393, 394, 485
  introduction for amenity, 283
  marshland, decrease due to reclamation of swamps, 370, 371, 373, 374, 488
  moorland and plains, decrease due to cultivation, 365
  mortality due to telegraph wires, 379
  of prey, destruction of, 109, 111, 127-137, 217, 489, 493, 498
  Paris Convention for Protection of, 277
  prehistoric remains of, 8, 17
  Protection Acts relating to, 227, 228, 229, 236
  protection by popular favour, 233
    for aesthetic reasons, 229
    for food, 216
    for sport, 183, 197, 202, 203, 204-207, 486
    for utility in agriculture, 227
  Royal Society for the Protection of, 236
  seed-eating, increase of, 389, 485, 496
Birse, Forest of, 80, 81, 311
  parish of, 121 f.-n.
Biscuits, pests infesting, 441, 443, 444
Bishop, A. Henderson, 19, 30, 316
Bison, in Germany, 349

Bittern, in Scotland, 21, 374
   destruction for sport, 187
   extermination due to reclamation of swamps, 374, 482, 497
   prehistoric remains of, 18
   protection for hawking, 203
   use as food, 140
Blabera gigantea, 463
Black Loch, Wigtownshire, 90
Blackbird, as pest, 228, 231
   change of food habit, 411
   increase of, 230, 394
   protection by popular favour, 234
   slaughter for food, 139
Blackgame, 241, 266, 267, 271, 329
   price of, 140
   protection for sport, 203, 205, 206
Blackhall, 274
Blackwater, Slipper Limpet in, 424
Blair Athole, 271, 287, 293, 319
Blair Drummond, 56, 104, 273
Blastophaga grossorum, 257
Blatta orientalis. *See* Cockroach, Common
"Bloomeries," 320 *et seq.*
Boar, Wild, in Scotland, 16, 20, 89
   destruction for sport, 138, 184
   domestication of, 89-95, 491
   prehistoric and early remains of, 30, 33, 89, 90
Boar's Chase, 90
Boarhills, 90
Bobwhite. *See* Quail, Virginian
Boece, Hector, 38, 46, 65, 80, 87, 88, 89, 116, 118, 122, 127, 133, 155, 158, 200, 268, 316, 318, 355, 365, 366, 371, 386
Boelter, Mr, 431
"Bog butter," 58
Bonar Bridge, 295, 398
Bone implements, 7
Bo'ness, 435, 472
"Book worm," 410
Boorde, Andrew, 312
Booth, E. T., 358
Boreray, 146
Borgue parish, 510
Borland Water, 277
Borness Cave, 56, 57, 72, 73, 331, 334
Borough Loch, Edinburgh, 371
Bos taurus longifrons. *See* Cattle, Celtic Shorthorn
Bos taurus primigenius. *See* Urus
Bot-fly, 423
Botanic Garden, Cambridge, 436
   Edinburgh, 436, 463
   Glasgow, 436, 463
   Kew, 436
   Regent's Park, London, 466
Botaurus stellaris. *See* Bittern
Bothriocephalus latus, 421
Braconid, 458
Braemar, 80, 122, 125, 128, 136, 211, 314, 356
Brambling, 389

Brand, Rev. John, 77, 130, 141, 169, 201, 280, 299
Brazil, 421, 460
Breadalbane, 120
   Marquis of, 270
Breaking in of waste lands, 365-370
Bream, Sea, 8, 19, 149
   Black Sea, 19
Brechin, 272, 293
Bressay, 76
Bridge of Frew, 292, 398
Bridges, as means of dispersal of animals, 398
Bristol, 466
British Columbia, 242, 451
Brochs, animal remains of, 52, 54, 57, 67, 72, 76, 87, 90, 112, 114, 115, 144, 331, 333, 337, 340, 341, 348, 350
Brock. *See* Badger
Broom, Loch, 296
Brown, Prof. Hume, 47, 249, 312, 316, 371
Browne, E. T., 433
Bruchus chinensis, 446
   lentis, 446
   obtectus, 446
   pisorum, 445
   rufimanus, 446
Buchan, 426
   Forest of, Aberdeenshire, 311
   Kirkcudbrightshire, 334 f.-n.
Buchan, Rev. A., 145
Buchanan, J. H., 178
Buchat, River, 278
Buckinghamshire, 288
Buffalo, American, 110
Bugs, Bed, introduction of, 421, 487, 497
   houses selected as domiciles by, 404
   Common Mealy, 475
   Lantana, 461
   plant, 459, 474
   Tropical Bed, 421
   Vine, 461, 474, 487
Bulbs, transportation of pests with, 465, 474
Bulimus decolletus, 465
   goodalli, 464
   octonus, 464
   undatus, 470
Bullfinch, 231, 241, 242, 390
Bunawe, 321, 353
Buntings, 140, 394, 413
   Corn, 389
   European, 394
   Snow, 141
Bupestris hæmorrhoidalis, 472
Burbot, in Hamburg water-system, 414
Burghead, 57, 161
Burnham, J. B., 206
Burra, 250
Burt, Capt., 356
Burwick, Broch of, 341
Bustard, Great, 21
   destruction for plumage, 189
   for sport, 187
   extermination, 274, 363, 366, 482, 497

INDEX

Bustard—*Cont.*
  protection for sport, 205
  re-introduction, 274
Bute, Beaver in, 298
  Capercaillie in, 273
  Deer, Red, prehistoric remains of large, 337
    Roe, in, 280
  Dunagoil vitrified fort, 316
    prehistoric animal remains in, 124, 337, 490
  Hare, Common Brown in, 280
  Marten in, 161
  Squirrel in, 293
  Wild Cat, prehistoric remains in, 124, 490
Buteo buteo. *See* Buzzard, Common
Butterflies, Artaxerxes, destruction by collectors, 190, 229
  Cabbage White, increase of, 392, 457
    Small, introduction to Canada, 458
  Comma, 369
  Copper, Large, destruction by collectors, 190, 229
  disappearance due to cultivation, 369
  Orange-tip, 369
  Peacock, 369
  Ringlet, 369
  Skipper, Large and Small, 369
  Speckled Wood, 369
  Wall, 369
Buxton, Sir Fowell, 270
Buzzards, change of habit in selection of nest, 401
  Common, destruction of, 18, 134, 217
  Honey, destruction as pest, 135
  increase during War, 396

Cabbages, pests introduced with, 457, 458, 459
Caccabis petrosa. *See* Partridge, Barbary
  rufa. *See* Partridge, Red-legged
Cairns, horned or chambered, 8, 32, 40, 51, 56, 58, 71, 89
Cairntable, 282
Cairnton of Kemnay, 427
Caithness, Bear, Brown, in, 17, 112, 114
  Boar, early remains in, 89, 90
  Celtic Shorthorn in, 56
  Deer, Red, in, 187, 333
    Roe, 331
  Dog, early remains in, 87
  Elk in, 348
  Falcon in, 200
  forest, 310
  fossil fishes in, 482
  Fox, prehistoric remains of large size in, 490
  Goose, Grey Lag, in, 104
  Great Auk in, 144
  Horse, early remains in, 71, 72
  Pigeons in, 100
  Rabbit in, 252
  Rats in, 426, 427, 432

Caithness—*Cont.*
  Reindeer in, 14, 187, 341, 342, 345
  Sheep, Peat or Turbary, in, 40
  Snow Bunting in, 141
  Squirrel in, 295
  Urus in, 50, 51
Calandra granaria, 441, 445 f.-n.
  oryzae, 445 f.-n.
Calder Wood, 312, 314
Calderwood, W. L., 151
Caledonian Canal, 397
  Forest, 114, 312, 314
    Wild Cattle in, 65
California, Fig Insect introduced into, 257
  fishes introduced into, 258
  "Jack-Rabbit" in, 369
  Mediterranean Fruit Fly in, 451
  Phylloxera in, 461, 462
  quarantine inspection in, 474
  Scale insects and their counterpests in, 260
Callander, 273
Callidium variabile, 471
Calliphora erythrocephala. *See* Flies, Blow
  vomitoria. *See* Flies, Blow
Cambrensis, Giraldus, 158, 202, 376
Camden, W., 114
Campbell, Dr John, 27, 68, 186, 214, 221, 222
  Bruce, 98
  J. F., 114
  J. G., 239
  Mr, Kingussie, 153
Campsie, 292
  Fells, 178, 273
Camster, 51, 56
Canada, Asparagus Beetle in, 459
  Cabbage Aphis in, 459
    Butterfly in, 458
    Fly in, 458
  dispersal of animals by water transport in, 397
  introduction of plants and animals into, 242
  Moth, Angoumois Grain, in, 442
    Codlin, in, 452
    Mediterranean Flour, in, 443
  Narcissus Fly in, 465
  Scale insect, Apple or Mussel, in, 450
    Camellia or Cottony, in, 460
    Mulberry, in, 460
  Wood-Wasp (*Sirex*) from, 468
Canadian Water Weed, 473
Canals as means of dispersal of animals, 397
Canary, the, 299
Canary Islands, 246, 466
Cannabina linota. *See* Linnet
Canobie parish, 293
Canoes, Neolithic, 9, 51, 56
Cape of Boars, 90
Cape of Good Hope, 451
Capello, Venetian Ambassador, 225

INDEX

Capercaillie, as food, 266
    denizen of Scottish forests, 329
    extermination of, 354-357
    introduction of, 268-274
    prehistoric remains of, 18
Capplach, Mount, 118
Capreolus capreolus. See Deer, Roe
Caprimulga europæus. See Nightjar
Cara, 178
Cardan, Dr Jerome, 41, 76
Cardiff, 414
Cardross, 199
Cariacus virginianus. See Deer, Virginian
Caribou, 329, 345
Carlisle, 292
Carluke, 299
Carmel River, 277
Carmichael of Skirling, Lord, 284
Carncruinch, 358
Carp, introduction for sport, 277
    into California, 258
Carrbridge, 314
Carron Iron Works, 321, 322
Carron River, 312
Carse clays, prehistoric remains in, 9, 10, 20
Carse of Gowrie. See Gowrie
    of Stirling, 9, 10, 20
Castle Coole, 104
Castle Grant, 358
Castor fiber. See Beaver
Cat, domestication of, 106
    Wild, destruction of, 124-126, 176, 177, 184, 229, 329, 482, 496
    effect of destruction of forests upon, 351, 362
    prehistoric remains, 124, 490
    reduction in size, 490
Caterpillars. See Butterflies *and* Moths
Cattle, 20, 49-67
    Aberdeen-Angus, 28, 60, 62
    Ayrshire, 28, 61, 64
    Celtic Shorthorn, 54, 55-60, 67
    colour of, 62
    development of milk supply, 64
    domestication of, 55
    early fattening of, 64
    effect of grazing on forest growth, 323
    Galloway, 29, 60, 62, 63, 64, 247
        Belted or Sheeted, 62
    Highland Kyloes, 60, 62, 64
    hornless, 62, 490
    modern, 60-64
    native wild, 49-52
    numbers added to fauna, 486, 493
    Orkney and Shetland, 60
    parasites introduced with, 422, 423
    prehistoric and early remains, 14, 31, 32, 33, 51, 55, 56, 57, 58, 73
    Shorthorn, 60, 62, 64
    South American Wild, 110
    Urus. See Urus
    "Wild" White, 54, 65

Cave dwellings, animal remains in, 32, 34, 87, 89, 111, 112, 115, 331, 334, 337, 340, 344, 373
Cawdor, 125, 296, 297
Centipedes, increase of, 393
Cerambyx cerdo, 470
    heros, 470
Ceratitis capitata, 448, 450
Cercyon, increase of, 396
Cervus dama. See Deer, Fallow
    sika. See Deer, Japanese
Cessnock River, 277
Ceylon, 459
Chad. See Shad
Chaffinch, 389, 390
Chalcids, 456
    Apple Seed, 453
    Douglas Fir Seed, 454, 473
    Rose Seed, 456
Champion, E. C., 446
Changes in fauna, v
Char, destruction by pollution in Lake Coniston, 385
    destruction for food, 152
    disappearance from Loch Leven, 372
    in prehistoric Scotland, 19
    introduction for sport, 277
    transportation of, 279
Chatauqua Lake, 258
Cheimatobia brumata, 392
Cheltenham, 405
Cherry trees, Scale insect on, 460
Cheshire, 466
Chesters, Roman Camp at, 112
Cheviot Hills, 128, 160, 292
Chichester, 431
Chili, 450
Chillingham Forest, 310
    White Cattle in, 65
Chilocorus similis, 261
China, 423
Chironomus, 415
Chloropisca ornata, 403
Chortophila (Phorbia) brassicæ, 458
Chough in Scotland, destruction as pest, 177
Christian monuments, early, 76, 342
Chrysomphalus aonidum, 448
Chrysophanus dispar. See Butterflies, Large Copper
Cimex hemiptera. See Bugs, Tropical Bed
    lectularius. See Bugs, Bed
Cinclus cinclus. See Dipper
Cinn Trölla, Broch of, 87, 331, 341
Circus cyaneus, 134
Civilization, change in habits of animals induced by, 400-416
    influences on animal life, 363, 377-417, 488, 516
Clackmannanshire, Capercaillie in, 273
    Deer, Roe, in, 332
    Rabbit in, 252
    Squirrel in, 292
Clams, destruction for food, 153

Clarke, Dr W. Eagle, 141, 375, 378, 399, 401, 428
Clausilia, 329
  papillaris, 464
Cleadon Hills, 144
Cleaves Cove, Dalry, 158
Clelland, Prof., 34, 87
Climate of Scotland, changes in, 10, 305, 306, 308, 481
  destruction of forest due to, 308
  effect of destruction of forest upon, 327
Clova Hills, 178
Clover, red, in New Zealand, 257
Clusius, 225
Clyde, River, 69, 178, 276, 277, 385
  area of, 51, 312, 472
Clythe, 51, 56, 71
Coal, use of, 315
Coal-pits, Cockroaches in, 435
  fauna of, 415-416
  Wood Wasps in, 469
Coalecken iron-furnace, 322
Coatbridge, 472
Coccid. *See* Scale Insects
Coccothraustes coccothraustes, 230
Cockle-beds of Forth, 386
Cockroaches, American or Ship, 436
  as pest, 487
  Australian, 436
  Common, 435
  German, 438
  houses selected as domiciles, 405
  in coal-pits, 435
  introduced in dry food materials, 435-438
  introduced with bananas, 448
  introduction into greenhouses, 463
  Snowy Tree, 447
"Cod, Murray," 259
Codlin Moth, 451, 475
Cœlopa frigida, 395
Colchester, Essex, 112
Coldingham, 330, 346
Coldstone parish, 324
Colinton Dell, 471
Colintraive, 161
Colinus virginianus. *See* Quail, Virginian
Coll, 280
"Collectors," destruction of animal life by, 189
Collembola, introduced. *See* Springtails
Collessie, 58
Coln, River, 424
Colombia, 475
Colonsay, Chough in, 178
  prehistoric remains in, 31, 32, 33, 34, 74, 89
Columba, St, 222
Columba livia. *See* Rock Dove
Columbia Red Wood, 454
Commerce, introduction of animals unawares by, 416-475, 487
Condor, destruction for plumage, 189
Conger-eel, 8

Coniston Lake, 385
Connecticut River, 380
Constantine's Cave, 173
Contarinia (*Cecidomyia*) *tritici*, 440
  (*Diplopsis*) *pyrivora*, 451
Cony, origin of name, 248. *See* Rabbit
Coot, 18
Coralline, Oat-pipe, 433
Cormorant, 8, 17, 141
Corncrake or Landrail, decrease in numbers, 368
  protection for sport, 205
Cornwall, 411
Corra Linn Fall, 178
Corrour, 399, 401
  Forest of, 337
Corse Burn, 294
Corstorphine, 10, 233, 254, 289, 428
Cortachy, 272
Corvus corax. *See* Raven
Costa Rica, 475
Coturnix coturnix. *See* Quail
Counter-pests, introduction of, 259-263, 419
Cowden Burn, 50
Crab, destruction for food, 154
  prehistoric remains of, 19
  protection for food, 220
Craig Green, 340
  Phaidrich, 331
Craigcrook, 346
Craigleith, island of, 432
Crail, 207, 249, 250
Cramond, 346
Crane in Scotland, 21, 375
  as food, 140, 266
  destruction for plumage, 189
  extermination due to reclamation of swamps, 375, 482, 497
  in Ireland, 376
  introduced species, 299
  prehistoric remains, 18
  price of, 140
Cranefly, increase of, 392
Crannich, Forest of the, 358
Crannogs, 40, 57, 71, 89
Crawfordjohn, 178
Cree, River, 293, 347
Crepidula fornicata, 424
Crex crex. *See* Corncrake
Cricket, House, houses selected as domiciles by, 405
  increase of, 497
  introduced in dry food materials, 434
Criffel Moor, 282
Crinan Canal, 273
Crioteris asparagi, 459
Croftamie, 338
Crofthead, 50, 68, 69
Cromarty, 31, 211, 440
  Firth, 295
Cromdale, 471
Crossbill, 329

Crouch, River, 424
Crow, Carrion, destruction as vermin, 136, 177, 217, 482
  increase as scavengers, 395
  increase during War, 396
  Hooded, destruction of, 137
  prehistoric remains of, 18
  protection as scavengers, 224, 225, 226
Cryptopleurum, increase of, 396
Cuckoo, 234
Culdross, 289
Cultivation, breaking in of waste lands, 365, 481, 485
  change of habit induced in animals by, 400-416
  decrease of animal life due to, 365-386, 488
  destruction of forest for, 322
  disappearance of ague due to, 512
  increase of animal life due to, 387-396, 484, 485, 486, 487, 493-496
  influences on animal life, 363-417
  proportion of land in, 484, 493
Cumberland, 293
Cumbrae, 280
"Cunningares." *See* Rabbit-warrens
Cunningham, J. T., 150
Cupar, 447
  Abbey, 250, 376
Curle, A. O., 214, 316, 426
  James, 72, 79
Curlew, 266, 363
  decline due to reclamation of swamps, 376, 377
  price, 140
Currie, 55, 86
Cydia (Carpocarpsa) pomonella. *See* Moth, Codlin
Cyprinus carpio. *See* Carp

"Daddy-long-legs," increase of, 392
Dalkeith, 292, 293, 435
  Palace, 286, 292
Dalmeny, 290
Dalmuir, 10
Dalnaglar, 126
Dalnaspidal, 399
Dalry, 158
  Forest of, 126
Dalzell, Sir J., 259
Damon, Mr, 154
Danube, River, 469
Daphnia pulex, 373
Darenth Wood, Kent, 446
Dark Forest, the, 314
Darnaway, Forest of, 133, 334, 358
Dartford, 446
Darwin, Charles, 95, 98, 323
Dawkins, Prof. Boyd, 56
Day Poaching Act, 205
Decoys, destruction of wild-fowl by, 139
Dee, River, 272, 293, 294
Deeside, 165

Deer, conclusions regarding the race of, 349
  Elk, *q.v.*
  export of skins, 155, 156
  Fallow, introduction for amenity, 284-287, 496
    protection of, 213, 286
  foreign, failure to acclimatize, 254
  Giant Fallow, 14, 50, 52 f.-n., 68, 481
  "Irish Elk." *See* Giant Fallow Deer *above*
  Japanese, introduction of, 287, 288
  Red, banishment from lowlands owing to pasturage of sheep flocks, 42, 331, 334, 486
    denizen of ancient Scottish forests, 308, 322, 329
    destruction as pest, 181, 212
    destruction for sport, 138, 187, 207, 213
    effect of destruction of forest upon, 333-338, 350
    fewer numbers in modern Scotland, 493
    importation from Germany, 264
    in Jura, 212, 492
    in Lowlands, 207
    introduced into New Zealand, 242
    physical degeneracy, 16, 335, 489
    prehistoric and early remains, 16, 30, 31, 34, 284, 331, 333, 336, 337
    protection for food, 217
    protection for sport, 207-213, 286, 485
    restriction of range, 208, 333, 334, 486
  Reindeer, *q.v.*
  Roe, denizen of ancient Scottish forests, 308, 322, 329
    effect of destruction of forest upon, 330-333
    physical degeneracy of, 332, 490
    prehistoric and early remains of, 16, 31, 284, 330, 331
    protection for food, 217
    protection for sport, 213
    restriction of range, 330
    transportation of, 280
  royal hunts in Atholl, 119, 210-211, 265, 283, 285, 331, 355
  Virginian, introduction of, 254, 287
  Wapiti, introduction of, 254
Deer forests, 207
  of Jura, 212, 492
Deerness, 250
Denmark, 431
Dermacentor reticulatus, 423
Dermanyssus gallinae, 423
Derwent, River, 276
Destruction of animals, as vermin and pests, 176-183, 212
  deliberate, 108-196, 488
  due to cultivation and civilization, 365-386, 488
  for food, 138-154
  for pleasure or luxury, 189-196

INDEX

Destruction of animals—*Cont.*
  for safety of man and his domestic stock, 109, 111-137, 488
  for skins and oil, 155-175
  for sport, 184-188
  indirect results of, 137, 488
Deveron, River, 272, 279
Devon, 465
Diaxenes dendrobii, 463
Diestrammena marmorata, 463
Dipodomys, 389
Dipper, destruction as pest, 181, 182
Dispersal of animals due to transport, 397-399
Dochart, River, 273
Dochfour, 296
Dochmius (Ancylostomum) duodenale, 422
Dodo, extermination of, 498
Dog, domestication of, 29, 86-89, 490
  early remains of, 32, 33, 34, 86
  introduction to Australia, 243
  parasites introduced with, 422
  Scottish varieties of, 88
Dog-fish, 8
  Spiny, 19, 149
Dolphin, 8, 19
Domestic animals, 21, 25, 35-108
  introduction of, 246
    into colonies, 242
  numbers added to fauna, 486
  parasites introduced with, 420, 422
  sequence of introduction, 32
Domestication of animals, 27-108
  beginnings of, 28, 29
  distinctive Scottish breeds, 28
  general effects of, 27
  potent factor in faunal changes, 481
  recoil upon man, 506-517
Domiciles, selection by animals of, 400
Don, River, 195, 278, 279, 294, 380, 381, 384
  area of, 125, 165, 272
Dordogne, Caves of, 344
Dormouse, 233
Dornoch, 287
  Firth, 126
Doryphora (Leptinotarsa) decem-lineata. See Beetle, Colorado
Dotterel, 140
Douglas Fir Seed Chalcid, 454
Dovecots, 97, 98
Doves. *See also* Pigeons
  introduction of exotic, 283
  Rock. *See* Rock Dove
  Turtle, 234, 241, 242, 266
Dowally, 252
Dowalton, Loch of, 57
Drake, Sir Francis, 435
Dreissensia polymorpha, 414, 469
Drumblade, 296
Drumlanrig, 287
Drummelzier, parish of, 323
Drumoak, forest of, 311
Drymen, 289

Ducks, domesticated, 106
  Eider, 18
  Golden Eye, 104, 203, 266
  Shoveller, 203
  Teal, *q.v.*
  Wigeon, *q.v.*
  Wild, change of food habits in, 411, 412
    destruction of, 139, 188
    domestication of, 106
    prehistoric remains, 17
    price of, 140, 265
    protection for food, 17
    protection for hawking, 203
Duddingston Loch, 346, 347
Dulnan Valley, 314
Dumbarton, 105
Dumbartonshire, Capercaillie in, 273
  Falcon eyries in, 199
  Marten in, 161
  Polecat in, 164
  Reindeer, prehistoric remains in, 338
  Squirrel, Grey, in, 289
    Red, in, 292, 352
Dumfries, 282, 293, 434, 447
  Fur Fair, 156, 162, 165, 167, 169, 170, 180, 252, 390
Dumfriesshire, Bear, Brown, prehistoric remains in, 17, 112
  Beaver, prehistoric remains in, 157
  "bloomeries" in, 320
  Chalcid flies imported, 456
  Deer, Fallow, in, 287
    Red, prehistoric remains in, 333
    Roe, prehistoric remains in, 16, 330
  forests of, 310
  Fox in, 122
  fur produce of, 156
  Grayling in, 276
  Osprey in, 133
  Pigeon, 100
  Polecat in, 164
  Rabbit in, 168
  Reindeer in, 14, 340
  Squirrel in, 292, 293
  Wolf in, 121 f.-n.
Dunagoil, Bute, vitrified fort of, 316
  animal remains in, 124, 337, 490
Dundee, 174
Dunfermline, 161, 172, 442
Dunipace, 199
Dunkeld, 252, 254, 270, 293, 360
Dunnet, parish of, 427
Dunottar, 293
Dunrobin, 126, 178, 295
Duns, 347
  Castle, 292
Duntroon, Bone Cave at, 31
Durham, 144
Durness, 126, 278
Durrant and Beveridge, Messrs, 445
Durris, 454, 455
Duthil, 358
Dyce, Forest of, 311

Eagle, destruction for protection of food animals, 217
  Golden, destruction of, 127-128
    increase of, 230
    prehistoric remains of, 18
  White-tailed, or Sea, extermination of, 130-132, 482
  Skua as enemy of, in Shetland, 228
Earn, River, 272, 333, 381
Earthworms, exotic, in hot-houses, 464
  in coal-pits, 416
  increase due to cultivation, 392-393, 496, 497
  effects of, 515
  transportation of, 464, 465
East Anglia, 275, 376
East Grange, 256
East Lothian. *See* Haddingtonshire
Edderton, 126
Eddystone Lighthouse, 378
Ederachillis, 119
Edinburgh, animals as scavengers in streets of, 224, 395
  Ant, Small Red House in, 440
  Ants, tropical, in Botanic Gardens, 463
  Arctic fauna of district, 14
    flora of district, 10
  Artaxerxes Butterfly on Arthur's Seat, 190
  Badger near, 233
  Beetles, Longicorn, in, 472
  Celtic Shorthorn, prehistoric remains in Nor' Loch, 55
  Cockroach, American, in, 436
    German, in, 438
  Cricket in, 434
  Death's Head Moth in beehives near, 409
  Deer, Red, prehistoric remains of large size in, 336
  Dove, primitive type, in, 96
  Flies, swarms of, in, 403
  Owls in, 406
  Perichæta indica in, 464
  Rat, Black, in, 427
  Raven in, 136, 395
  Reindeer, prehistoric remains near, 340
    in Zoological Garden, 254
  rookeries in, 405
  Snake, Smooth or Ringed as "escape" in, 299
  Thrips, in Botanic Garden, 463
  Tree-Frog transported in bananas to, 447
  Wagtail, Pied, in, 406
  Water Flea in pond near, 373
  Wild fowl in, 371
  wooden houses of, 316
Edrom, 157, 331
Edward, Robert, 356
Edwards, J., 446
Eel, Common, in waterworks, 413
  migrations of, 278
  Conger, 149

Egew, Schir Robert, 46, 251
Eggs, birds', destruction for food, 148
  destruction of bird-life by collectors of, 191
Egrets, destruction for plumage, 110, 189
Egypt, 441, 443, 446
Eil, Loch, 296
Einig, Glen, 296
Eird, or Picts' Houses, animal remains of, 40, 52, 57, 72, 74, 87, 331, 333
Elgin, 279, 371
Elginshire. *See* Morayshire
Elibank, 314
Elk, as denizen of Scottish forests, 14, 329
  extermination of, 207, 213, 346-349, 481, 496, 497
  effect of destruction of forest upon, 346-350
"Elk, Irish." *See* Deer, Giant Fallow
Ellis, Sir H., 401
Ellon, 161
Elwes, H. J., 39
Emberiza hortulana. *See* Ortolan
Emu, destruction for plumage, 189
English Channel, the, 12, 13
Enniskillen, 104
Entomostraca, dispersal by canals, 397
Ephestia kühniella, 443, 445 f.-n.
Erie Canal, 397
Eriosoma lanigera. *See* Aphis, Woolly
Eriska, Isle of, 57
Ermine. *See* Stoat
Erne. *See* Eagle, White-tailed
Ernle, Lord, 391
"Escapes," 298
Esk, North, River, 272, 293
Eskdale, 117, 199, 331
Esmore, 320
Esox lucius. *See* Pike
Essex, 330
Ettrick, River, 347
  Forest, 310, 313, 314, 334
    Deer, Red, in, 334
    royal chase in, 119, 208
    Sheep in, 41
    Wolf in, 119
Eumeces quinquelineatus, 469
Euxoa (Agrotis) segetum. *See* Moth, Diamond-backed
Evans, Henry, 493, 494
Evans, William, 190, 369, 378, 428, 429 f.-n., 435, 464, 492, 502
Ewart, Prof. J. Cossar, 37, 39, 40, 44, 51, 62, 69, 72, 75, 79

Fair Isle, 97, 141
Fala Hill, 164
Falco peregrinus. *See* Falcon
Falcon, Peregrine, change of habit in nesting, 401
  destruction of, 134
  protection for sport, 199-203
Falkirk, 322

INDEX

Falkland, 164
 Islands, 175
Farnham, 323
Farquharson, Rev. J., 324
Farrow, E. P., 253, 499
Fasciola (*Distomum*) *hepatica*. *See* Liver fluke
Fauna, changes affecting, v
Ferenze, forest of, 207
Ferret, 107
Fertilizing agents, introduction of, 515
Feugh, River, 294
Fiddich, 384
Fieldfare, 266, 394
Fife, Capercaillie in, 270, 272, 273
 Celtic Shorthorn, early remains in, 57
 Deer, Red, in, 207
  early remains of large size, in, 337
  Roe, in, 332
 Dog, prehistoric remains in, 87
 Flies, swarms of, in, 403
 forests, destruction in, 312, 318
 lochs and swamps in, 371
 Marten in, 161
 Owl, Little, in, 256
 Partridge, Red-legged, in, 275
 Pigeon caves and dovecots in, 97, 100
 Polecat in, 164
 Rabbit warrens in, 249, 250, 252
 Rat, Brown, in, 395
 Squirrel in, 292
Fifty-foot Beach, 9, 10, 11, 19, 20, 56, 70
Fig Insect, introduction to California, 257
Finches, 256, 413
 increase of, 230, 485
Findhorn, River, 328
Finnart, 288
Fiorinia fioriniae, 461 f.-n.
 kewensis, 461 f.-n.
Fire, the Great, 316
Fisheries. *See also* Shell-Fisheries
 protection of, 217, 220
 yield of, 149
Fishes, decrease due to pollution, 382-386
 decrease due to river barriers, 379-382
 destruction for food, 149, 150, 151
 dispersal by waterways, 397
 food of early inhabitants, 8, 19, 149
 fossil, in Caithness, 482
 in waterworks, 413, 414
 introduced, 258, 276, 284
 introduction for amenity, 284
 introduction for sport, 276-278
 prehistoric remains of, 8, 19
 protection for food, 217-220
 transportation of, 278-279
Fish-hatcheries, 219
Fish-hawk. *See* Osprey
Fittis, R. S., 116
Fleas, Brown Rat as carrier, 423, 507
 Human, in houses, 404
  increase during War, 396

Fleas—*Cont.*
 introduction unawares, 420
 parasitic on Badger, 420
Plague, introduction unawares, 420, 507
"Shore," 395
"Water," 373, 397
Flies, Apple Seed Chalcid, 453
 Bot-, 423
 Blow-, or Bluebottle, 404
  increase during War, 396
 Cabbage, 458
 Chalcid, 453, 454, 456, 473, 475
 Citrus White, 475.
 Fruit, 474
 Green-bottle, 423
 Hessian, 439
 House, 403, 404
  and disease, 506
  increase due to garbage, 396, 496, 506
 Ichneumon, 469
 in coal-pit, 416
 Mediterranean Fruit, 448, 450
 Narcissus, 465, 474
 Pear Midge, 451
 relation to disease, 506
 "Sheep-tick," 423
 "Spider," parasitic on horses, 423
 "Tick," parasitic on horses, 423
 Warble, 423
 Wheat Midge, 440
 Yellow Dung, 396
Flora, post-glacial, 10
Florida, 436
Flounder, in waterworks, 414
Flour, animals introduced with, 443
Flower garden, introductions with plants of, 462
"Flukes," transportation of, 421
 Liver, of sheep, 373, 422, 514
Food, habits, changes due to cultivation and civilization, 408-413
 materials, animals introduced in, 434-446
 protection of animals for, 216-220
 scarcity and high prices in Scotland, 216, 217
 supplies, increase affecting fauna, 387
Forbes, Prof. Edward, 386
Forests, curtailment of growth by sheep, goats and rabbits, 323, 500
 destruction of, 303, 304-364, 488, 496
  effects on animal life, 327-364
  effects on climate, 327
  final results of, 325-326, 360, 362
 for agriculture, 322-323
 for extirpation of beasts of prey, 120, 318
 for fuel, 315
 for housebuilding, 316
 for iron-smelting, 319
 for safety in travelling, 318
 for sheep pasturage. *See under* Sheep
 for shipbuilding, 319
 incidental to conquest, 317

Forests, destruction of—*Cont.*
    influence on Bear, Brown, 351
        on Capercaillie, 268, 274, 354-357
        on Deer, 330-338
        on Elk, 346-350
        on Fox, 351
        on Great Spotted Woodpecker, 357-360, 361
        on Grouse, 361
        on Reindeer, 338-346, 350
        on Squirrel, 290, 351-354
        on Urus, 350
        on Wolf, 120, 318, 351
    physical changes wrought by, 327
    wind and fire as agents of, 324
  early historical, 309-314
  laws relating to, 309, 310, 311
  Lower Peat, 305
  post-glacial, 10, 12
  prehistoric, 305
  relations of Scottish fauna to, 12, 13, 328
  Upper Peat, 306
Forfar, 90, 272
  Loch of, 90, 372
Forfarshire, Capercaillie in, 272, 274
  Chough in, 178
  Elk, prehistoric remains in, 346
  Horse, prehistoric remains in, 69
  Marten in, 161
  Polecat in, 164
  Rat, Black, in, 427
  Wolf in, 120
Forres, 126, 199
"Forrest-shire," 313
Fort Augustus, 296
Forth, Firth of, decrease of fauna in, 386
  pollution of, 385
  shell fisheries of, 153, 386
Fotheringham, 274
Foula, 228
Foumart. *See* Polecat
Fowls. *See* Poultry
Fox, Blue, slaughter for skins, 110
  Common, 17, 329, 365
    destruction of, 121-124
      for skin, 155, 156, 166, 168, 221
      for sport, 184, 264
    effect of destruction of forest upon, 351
    introduction to colonies, 243, 498
    early remains of, 69, 490
    reduction in size, 490
France, 431, 453, 461, 462, 466, 469
Franck, R., 426
Frankfield Loch, 298
Frew, Bridge of, 292, 398
Frogs, decrease due to reclamation of swamps, 372, 373
  Green Tree-, 447
  prehistoric remains of, 373
Fruit, introduction of animals with, 447
Fruit-crops, birds as pests of, 411
Fruit Fly, Mediterranean, 448, 450

Fulmar Petrel, 145
  destruction for oil, 171
  increase of, 230
  slaughter for food, 147
Fulmart. *See* Polecat
Fulmarus glacialis. *See* Fulmar Petrel
Fur-bearing animals, destruction of, 155-174
  protection of, 221-223
Furnace, 321, 353
Fyne, Loch, 294, 321
Fyvie, 161, 296

Gairsey, 282
Gala, River, 384
Galloper Lightship, 378
Galloway, 46, 156, 239
  Cattle. *See under* Cattle
Game Laws, 180, 198-215, 481
  special Orders during Great War, 205, 212, 215
Gamhna, Loch, 193
Gannet, 8, 17, 145
  destruction for food, 146
  for oil, 171
  increase of, 230
Ganvich, Loch na, 279
Garbage, animals as scavengers of, 223-226
  dispersal of rats due to, 432
  increase of flies due to, 396, 506
  influence on increase of animal life, 395-396
Garden, 164
Garden pests, increase of, 392
  plants, introductions with, 462
Garefowl. *See* Auk, Great
Gare Loch, 289
Garelochhead, 289
Garnett, T., 134
Garry, River, 271
Garrywhin, 71, 87
Garson, Dr J. G., 87
Gartmore, 10
Gartmorin Dam, 279
Gastrophilus equi, 423
Georgia, South, 110, 170, 175
Germany, 431, 463, 464, 469
  Black Rat in, 425
  Capercaillie in, 269
  Elk in, 349
  Red Deer in, 264, 337
Gesner, Konrad v., 401
Giga, 178
Gilbey, Sir Walter, 83
Gillanders, A. T., 469
Gilpin, W., 313, 325
Glacial period, 2, 6, 10, 11, 12, 49, 279
Glamis, 293
  Castle, 272
Glasgow, 69, 427, 434, 438, 441, 463, 471, 472
Glass parish, 272
Gled. *See* Kite
Glenalmond, 272, 293

INDEX

Glenapp, 273
Glen Brierachan, 133
Glen Dochart, 126
Glendye, 125
Glen Falloch, 273, 314
Glen Feshie, 314
Glengarden, 136
Glengarry, 165
Glen Isla, 272
Glenkindie, 294
Glenlee, 164
Glen Lui, 358
Glen Lyon, 178, 273
Glenmore, 119, 122, 126
    Forest, 321
Glen Moriston, 269, 296, 322, 356
Glenmuick, 136
Glen Nevis, 314
Glen of Rothes, 296
Glen Orchy, 126, 294
Glenshee, 126, 272, 348
    "Spittal" of, 120
Glen Strath-Farar, 211, 323
Glen Tanar, 125, 294, 297
Glen Tilt, 285
Glen Urquhart, 296
Globicephalus melas. See Whale, Pilot
Glycyphagus, 445
Gnats. *See also* Mosquitoes
    larvae in waterworks, 415
Goat, domestication of, 107
    influence on forest growth and vegetation, 323, 500, 501
    prehistoric remains of, 34
Goatfield, 321, 353, 354
Godwit, Blacktailed, 191
Goes tigrina, 471
Goldfinch, destruction by bird-catchers, 193
    increase of, 230
Goldfishes, introduction of, 284, 466
"Goldings," 203, 266
Golspie, 122
Goosander, the, 231
Goose, Barnacle, 104, 370
    Brent, 104
    death caused by artificial manures, 370
    destruction for sport, 188
    domestication of, 34, 103–106
    hawking of, 203
    Grey Lag, 103, 104
        change of food habit of, 412
        destruction for food, 139, 266
        price, 140
    Merganser, 231
    ornamental, 299
    Pink-footed, 104, 370
    prehistoric remains of, 17
    Solan. *See* Gannet
Gordiidae, 415
Gordon, Duke of, 321
    Rev. Geo., 353
    Sir Robert, 120, 351, 356, 375

Goshawk, 200, 401
    extermination of, 133, 182
Gourdas, Fyvie, 161
Gowrie, Carse of, ague in, 509, 510, 512
    Squirrel in, 293
Graham, Dr Patrick, 128, 134
    Lord, 115
Grain, animals and insects introduced with, 434 *et seq.*
Grant, Col. Sir Arthur, 254, 272, 274
    Sir James, 321
Granton, 472
Grantown-on-Spey, 297, 342
Grape Phylloxera, 461, 474, 487
Grasshoppers, Japanese, introduction of, 463
Gravesend, 446
Gray, Robert, 128, 132
Grayling, introduction for sport, 276
Grebe, Great Crested, 18
    increase of, 230
    influence of towns upon, 406
Greenburn, River, 276
Greenfinch, 241, 389
Greenfly, increase of, 392. *See* Aphis
Greenhouses, introductions with tropical plants to, 404, 436, 460, 461 f.-n., 462, 463
Greenock, 298, 427, 472
Grieve, Symington, 31, 32, 33, 74
Grillus Focarius, 435
Grilse, destruction of, 151
    protection for food, 219
Grimshaw, P. H., 206
Grocers' Itch, 445
Grosbeak, Rose-breasted, 411
Grouse or Moorfowl, 206, 266, 267, 329, 504
    influence of destruction of forest upon, 361
    protection for hawking, 203
    transportation of, 280
    Willow, introduction for sport, 275
Grus grus. See Crane
Gryffe Water, 276
Gryllus domesticus. See Cricket, House
Guillemot, 8, 17
Guinea Fowl, 106
Guiraca ludoviciana, 411
Gulls, as food, 140, 142
    as pests, 228, 231
    Blackheaded, 228, 231, 412
    change of food habits, 412
    Common, 412
    effect on vegetation of moorland, 501–505
    increase due to cultivation, 394
    influence of towns upon, 406
    Kittiwake, 142
    protection for eggs as pheasants' food, 207, 502
Gunpowder and guns, effects of invention of, 109, 482
Gymnaspis aechmeae, 461 f.-n.

Habits of animals, changes induced by cultivation and civilization, 400–416
Haddington, 92, 224
Haddingtonshire, Boar, Wild, prehistoric remains in, 89
  Celtic Shorthorn, early remains in, 56
  Deer, Red, prehistoric remains of large size, in, 333, 337
  Hare, Mountain, in, 282
  Horse, prehistoric remains in, 69
  Polecat in, 164
  Rats in, 430
  Squirrel in, 292
  Tench in, 279
  Traprain Law, animal remains at. *See* Traprain
Haddo House, 296
Hadfield Moss, 317
Hair worms in water-pipes, 415
Haliaëtus albicilla. *See* Eagle, White-tailed
Halichærus grypus. *See* Seal, Grey
Hamburg, fauna of waterworks of, 413
Hamilton, 310
Hampshire, 446
Hampstead, 288
Hampton-on-Thames, 415, 470
Handa, Isle of, 119
Hare, Alpine, Mountain or Variable, 14
  reduction in size, 490
  transportation of, 280
  Californian, 369
  Common Brown, 14, 365
    destruction as pest, 179–180
    destruction for skins, 156, 166–168
    destruction for sport, 184
    increase of, 391
    protection of, 203, 213–215, 217
    relations to Rabbits, 168, 253
    roads as means of dispersal, 398
    transportation of, 280
      to colonies, 253, 365
Harmer, Dr S. F., 415
Harrier, Hen, 134
Harris, Carp in, 277
  Celtic Shorthorn, prehistoric remains in, 57
  Dog, prehistoric remains in, 87
  Eagle, Sea, in, 130
  Goose, Grey Lag, in, 104
  Hare, Alpine, in, 282
  Horse, prehistoric remains in, 72, 74
  Marten in, 160
  Rat, Black, as plague in, 427
Harting, J. E., 65, 116
Harvie-Brown, Dr J. A., 124, 160, 164, 186 f.-n., 270, 271, 279, 291, 325, 352, 357, 359, 360, 398
Hasarius adansonii, 463
Haskeir, 141, 172, 173
Hawaii, 458, 459, 460, 474
Hawfinch, 230
Hawick, 292
Hawking, 199–203

Hawks, Buzzard, *q.v.*
  destruction of, 132–136, 217
  Fish. *See* Osprey
  Goshawk, *q.v.*
  Hen Harrier, 134
  increase during War, 136
  Kestrel, 136, 182
  Kite, *q.v.*
  Merlin, 136
  protection for sport, 199–203
  Sparrow-, 136, 200, 202
Hawskeir. *See* Haskeir
Health, man's influence on animal life in relation to, 506 *et seq.*
Heather, extermination by rabbits, 499
  influence of gulls upon, 502
Heather beetle, 206
Hebridean Islands, Eagle, Golden, in, 128
  White-tailed, in, 130
  Goose, Grey Lag, in, 104
  Horses of, 73, 74–76
  prehistoric forests of, 305, 308
  Rabbit in, 251
  Rock Dove in, 95, 141
Hebrides, Inner, Chough in, 178
  Hare, Alpine, in, 282
  Common Brown, in, 280
Hebrides, Outer, Carp in, 277
  Celtic Shorthorn in, 57
  Eagle, Golden, in, 127
  Great Auk in, 144
  Hare, Alpine, in, 282
  Common Brown, in, 280
  Marten in, 160
  Rat, Black, in, 427
  Raven in, 226
  Rock Dove in, 141
  Seals in, 172, 173
  whaling stations of, 175
Heiskir, Island of. *See* Haskeir
Helensburgh, 289
Helix, 329
  limbata, 465
  nemoralis, 154
  pomatia, 154
Hemichionaspis minor, 448
Hendrick, Rev. J., 355
Hensen, von, 392
Heron, as food, 266
  as pest, 140, 231
  confusion of identity with Crane, 376
  decline due to reclamation of marshes, 376
  price, 140, 376
  protection for sport of hawking, 202
Herring, abundance of, 149, 150
Hessian Fly, 439
Hill forts, animal remains of, 57, 124, 331, 333, 337, 490
Hill of Bruan, 56, 71
Hillswick, 331
Hilston, Cornwall, 411

INDEX

Hippobosca equina, 423
Hirsel Lough, 279
Holinshed, 184, 207
Holland, 474
Hopetoun, 287
Horne, Dr John, 17, 31, 111, 340
Horses, in Scotland, 68-85
    breeding and interbreeding, 81
    Celtic pony, 70, 76
    Clydesdale, 28, 83, 490
    "desert" or "plateau" types, 69, 70
    domestication of, 34, 67
    Great or War Horse, 83
    Hebridean, 73, 74-76
    Highland garron, 80
    influences modifying native race, 73-80
    in prehistoric and early times, 71-73
    modern Scottish breeds, 83
    native, 68
    Norse influence on, 76, 79-80, 81, 85
    numbers added to fauna, 486, 493
    of Scottish mainland, 78
    parasites introduced with, 422, 423
    prehistoric remains of, 32, 33, 50, 68, 70, 71, 72, 73, 76, 109
    Shetland pony, 73, 76-78
    Shire, 83, 490
    "Wild," 14, 20, 80-82, 365
Houghton House, 292, 293
Houses, fauna of, 403, 487
    influence on animal life, 401
Hoy, island of, 93, 128, 280, 282
Hudson Valley, 462
Hugues, M. A., 394
Huntingdonshire, 190, 515
Huntly, 296
Hybernia defoliaria, 392
Hydrochelidon nigra, 191
Hydrophilidae, increase of, 396
Hydrozoa, distribution by ships, 432, 433
Hypoderma, 423
Hypothenemus eruditus, 410

Ice Age in Scotland. *See* Glacial period
Icerya ægyptiacum, 461 f.-n.
    purchasi, 260
Ichneumons, transportation of, 469
Imports, quarantine inspection of, 474
Inch Galbraith, 134
Inchkeith, 173
Inchmahoma, 134
Inchmarlo, 272
Inchmurrin, 115
Inchnadamph Bone Cave, 14, 17, 31, 111, 112, 165, 340, 373, 490
India, effect of destruction of forest in, 328
    transportation of Grain Beetles to, 441
Indian Corn, Angoumois Grain Moth in, 442
Industries, reduction of animal life due to, 488, 495
Innerarity, 118
Innerwick, 379

Innes, Prof. Cosmo, 285
Insects, decrease due to cultivation, 369
    destruction by lighthouses and other lights, 378
    hordes in collected sea-weed, 395
    increase due to cultivation, 391
    introduced, as fertilizing agents, 515
    as parasites on livestock, 423
    in dry food materials, 434-446
    into greenhouses, 463
    with fruit, 447-456
    with plants and vegetables, 457-466
    with timber, 467-473
    pests, transportation of, 473
    detected in Californian quarantine, 474
    prehistoric remains of, 18
Scale, introduction of, 474. *See* Scale insects
Insh, Loch, Char in, 153
Introduction of new animals, deliberate, 241-300, 480, 487
    even balance necessary for successful adaptation, 255, 257, 419, 480, 486, 489
    for amenity, 283-300
    for sport, 264-282
    for utility, 246-263
Introduction of animals unawares, 417-475, 480, 487
    amongst fruit, 447
    by ships, 425
    conveyed by plants and vegetables, 457
    in dry food materials, 434
    in timber, 467
    parasitic on Man and his domestic stock, 420
Inverary, 294
Inverbeg, 289
Invercauld forest, 358
Inveresk, 56, 248
Invergarry, 321, 354
Inverness, 155, 296
Inverness-shire, Beaver in, 158
    "bog butter," in, 58
    Capercaillie in, 269, 273, 355, 356
    Char in, 153
    Deer, Red, in, 211, 333
    Roe, 331, 332
    forest, destruction in, 319, 322
    by sheep, goats and rabbits, 323
    Goshawk in, 133
    iron furnaces in, 354
    Mosquitoes in, 512
    Osprey in, 134, 192
    Polecat in, 165, 166
    Rabbit in, 401
    Scottish Crested Tit in, 230
    Sparrow in, 399
    Squirrel in, 181, 295, 296, 297, 354
    Wild Cat in, 126
    Wolf in, 319
    Woodpecker, Great Spotted in, 358

Invershin Bridge, 295, 398
Inverurie, 472
Iona, Chough in, 178
  Dog, remains in, 87
  Horse, remains in, 72
  Seals in, 171, 222
Ireland, Crane in, 376
  destruction of shellfish in, 154
  protection of falcons in, 202
Iron-smelting industry, 309, 319-322, 353, 354
Irvine, 70
  River, 277, 385
Ischnaspis longirostris, 461 f.-n.
Isla, River, 195, 271, 272
Islay, 98, 132, 178, 280, 282, 287
Italians, slaughter of birds of prey by, and indirect results, 498
Italy, Scales and counterpests in, 262
Ixodes ricinus, 423

Jackdaw, 177
"Jack-Rabbit" in California, 369
Jamaica, 448
Japan, 261, 450, 456, 459, 460, 474
Jarlshof, Sumburgh, 76
Jay, 230, 329
Jefferies, Richard, 328
Jones, H., 225
Jura, Deer, Red, in, 212, 492
  Roe, in, 280
  deer forests of, 212, 492
  Hare, Common Brown in, 280

Kea, 410
Keiss, animal remains in Broch and Mound at, 52, 72, 87, 112, 114, 144, 341, 345, 348, 490
Kellogg, Prof. V. L., 436
Kelly Burn, 194
Kelso, ague in, 509, 510
Kelvin, River, 340
Kemna, Dr A., 414
Kemnay, 427
Kenmore, 276
Kent, 446
Kesserloch, 344
Kestrel, 136, 182
Kettleburn, 87, 331, 341
Kilbarchan, 472
Kilbrandon, 178
Kilchattan, 178
  Bay, 298
Kildrummy, 72, 87, 294
Killearn, 292
Killiecrankie, 120
Killin, Forest of, 287
Kilmarnock River, 277
Kilmaurs, 331, 338, 344
Kilmuir Castle, 295
Kimmerghame, 157
Kincardine-on-Forth, 273, 292

Kincardineshire, ague in, 509
  Capercaillie in, 272, 274
  Douglas Fir Seed Chalcid in, 454, 455
  Rabbit in, 252
  Squirrel in, 293
  Wild Cat in, 125
King, Colonel, 98
  Prof. J. J. F. X., 463
Kingfisher, increase of, 230
  Smyrnian, destruction for plumage, 189
Kingussie, 153
Kinloch, 17, 157, 346
Kinneil, 164
Kinross-shire, Capercaillie in, 272
  Polecat in, 164
  Rabbit in, 252
Kintore, forest of, 311
Kintrawell, Sutherland, 52
Kirkcudbrightshire, ague in, 509, 510
  Celtic Shorthorn, remains in, 56, 57
  Deer, Red in, 334
    early remains in, 331
  Roe, early remains in, 331
  Horse, remains in, 72, 73
  Marten in, 160
  Polecat in, 164
  Squirrel in, 293
Kirkden, parish of, 509
Kirkgunzeon, 282
Kirkmichael, Banffshire, 120
Kirkoswald, 87
Kirkurd, 346
Kirkwall, 57, 130
Kitchen middens, animal remains of, 8, 16, 17, 19, 29, 31, 32, 34, 56, 57, 72, 76, 87, 89, 115, 158, 247, 331, 333, 355
Kite, as scavenger, 225, 395
  destruction of, 133, 135, 136, 191, 217, 482
Kittiwake, as food, 142
Kiwi, the, 245
Knighton, H., 317
Knot, 140
Koebele, A., 260
Kyle of Sutherland, 279
Kyloe, Highland. *See under* Cattle

Ladybird beetles, protection by popular favour, 235
  as counterpests, 260, 261
Lagopus lagopus. *See* Grouse, Willow
Laing, J., 132
Lairg, 165
Lake deposits, animal remains in, 331, 333, 340, 346
  dwellings. *See* Crannogs
Lakes, Great Canadian, dispersal of animals by, 397
Lamington parish, 292
Lammermoor Hills, 282, 292
Lanarkshire, fur produce of, 156
  Hare, Alpine, in, 282
  Polecat in, 164

Lanarkshire—*Cont.*
   Reindeer, prehistoric remains in, 338
   Snake, Ringed, as "escape" in, 299
Landrail. *See* Corncrake
Langwell and Sandside estates, destruction of pests and vermin on, 126, 128, 136, 137, 177, 182
Lantana bug, 461
Lantz, Prof. D. E., 431
Lapwings, as insect-eaters, 394
   decrease in numbers due to cultivation, 368
   destruction of eggs for food, 148
   transportation to colonies, 241, 242
Larch, Japanese, 456
Lark, Crested, 394
   Short-toed, 394
   Sky, destruction for food, 139, 266
   price of, 140
Lasioderma serricorne, 445
   testacea, 410
Lathirsk, 270
Lauder, Sir Thomas Dick, 325
Lauderdale, 282
Laws, Game, 180, 198-215, 481
Leadenhall market, 139
Leaf-mining insects, 474
"Leaf-worms," in Scottish greenhouses, 464
"Leather-jackets," increase of, 392
Leeches, 259
   in waterworks, 414
Leeds, 438
Leges forestarum, 59, 92, 309
Leith, ale of, 313
   birds of carrion as scavengers in, 224, 226, 395
   Cockroach, American, in, 436
   Oyster beds near, 153, 386
   Rat, Alexandrine, in, 428
      Black, in, 427
   Raven in, 136
   shipbuilding at, 319
   skins and furs exported from, 155
   transportations in fruit to, 447
Lemming, Arctic, 14, 481
Lentils, pests introduced with, 445, 446
Leochel Burn, 294
Lepidosaphes beckii, 461 f.-n.
   ulmi. *See* Scale Insects, Mussel
Lepisma, 404
Leptyphantes nebulosus, 404
Lepus europæus. *See* Hare, Common Brown
   texianus, 369
   timidus. *See* Hare, Alpine or Variable
Leslie, Bishop Jhone, 6, 35, 39, 61, 62, 66, 88, 105, 114, 119, 140, 149, 155, 184, 194, 208, 248, 266, 312, 326, 355, 376, 426
Lessertia dentichelis, 416
Letterfearn MS., 320
Leucotermes flavipes, 440
Leven, Loch, bird life of, 372
   Char, disappearance from, 372

Leven, Loch—*Cont.*
   Leech, medicinal in, 259
   Trout, transportation of, 241, 278
Lewis, Carp in, 277
   Hare, Alpine, in, 282
      Common Brown, in, 280
   Hive bee in, 107
   Horse in, 74
Lewis, Francis J., 305
Liberton parish, 316
Lice, Clothes, 421
   Crab, 421
   distribution by man, 421
   Head, 421
   increase during War, 396
   plant, 392
Liddesdale, 164, 427
Lighthouses and lights, destruction of birds by, 377-379
Lilford, Lord, 256
Limax maximus, 416
Limnæa peregra, 466. *See also* Snails, Pond
   stagnalis and *glabra*, 298
   truncatula, 373, 514
Limnophora septem-notata, 403
Limosa limosa, 191
Limpet, Slipper, 424
Lincolnshire, 371, 509
Lindsay, Alexander, 224
   Robert, of Pitscottie, 41, 66, 83, 208, 210, 266
Linhope, 164
Linlithgow, 118, 121 f.-n.
Linlithgowshire, ague in, 509
   Deer, Red, prehistoric remains in, 333
   Roe, in, 332
   numbers of domestic animals in cultivated area of, 494
   Polecat in, 164
Linnean Society, 469
Linnets, 193, 230, 242, 394
Linnhe, Loch, 178, 314
Linton Loch, 157, 336
Liothrix luteus, 283
Liponyssus bursa, 420
Lismore and Appin, 352, 398
   Island, 178
Lithgow, William, 46
Littlewood, 165
Liver-fluke of sheep, 373, 422, 514
Liver-rot, disappearance due to reclamation of swamps, 373, 514
Liverpool, 470
Livestock, parasites introduced with, 422
Lizards, transportation with bananas, 448
   with timber, 469
Lobster, destruction for food, 154
   protection for food, 220
Lochaber, Beaver in, 158
   Capercaillie in, 355
   forest, destruction in, 312, 319
   numbers of domestic stock in, 494
   Wolf in, 119, 319

INDEX

Loch-an-Eilan, 134, 192, 193
Lochan Coire an Lochan, 401
Lochawe, 126, 134, 164, 278, 294, 319, 322, 331
Lochbrune, 318
Lochbuie lochs, 278
Locheil Old Forest, 314
Lochgilphead, 294
Lochlee, 57, 71, 90, 331, 340
Lochmæa suturalis, 206
Lochnagar, 314
Lochrosque, 288
Lochspouts, 57, 331
Lochy, Loch, 397
Locustella luscinioides, 191
Locusts, transported with bananas, 448
  transported with cauliflowers, 457
Lomond, Loch, 115, 134, 273, 277, 289, 314
London, Ant, Red House, in, 440
  bird's nest and eggs transported with bananas to, 448
  birds of, 284, 405, 406
  Cockroaches in, 435, 438
  Grain Mite in, 443
  Kite in, 225, 395
  Physa acuta in, 466
  site on reclaimed swamps, 370
  Snail, White-keeled, near, 465
  timber houses of, 316
  wild-fowl on site of, 371
  Woolly Aphis, arrival in, 461
Long Island, 178
  Loch, 288
Longicornia, transportation in timber, 470
Lossie, River, 279
Lothian, Marquis of, 276
  East. *See* Haddingtonshire
  West. *See* Linlithgowshire
Lothians, the, 273
Louse. *See* Lice
Louth Market, 139
Lovat, Lord, 323
Low, Prof. D., 43, 93, 130, 173, 228, 276
  Rev. G., 93
Lowther, C., 105, 153, 372
  Hills, 282, 293
Lucas, Sir Charles, 371
Lucilia sericata. *See* Flies, Green-bottle
Lucophæa surinamensis, 463
Lugar, River, 276
Luing, Island of, 331
Lundie, Loch, 81
Lundy Island, 147, 249
Luss, 289, 292
Lutra vulgaris. *See* Otter
Lyctus brunneus, 472
Lyne, River, 430
Lynx, Northern, denizen of Scottish forests, 20, 308, 329
  extermination of, 111, 481
  prehistoric remains, 17

Macadam, Prof. W. Ivison, 320, 321, 322, 354
MacArthur Cave, Oban, 8, 34, 56
Macaulay, Rev. Kenneth, 74, 141, 146, 147, 173
MacDougall, Dr R. S., 440, 442, 446, 454
Macgillivray, Prof., 98, 160, 173, 226, 280, 358
Mackaile, 107, 395
M'Kenzie, Sir George, 144
J. H. Munro, 75
McWilliam, Rev. J. M., 412
Macrosiphum granarium, 440
Madeira, 465
Magdalenian cave pictures, 30, 344
Magnusson, E., 341
Magpie, in Scotland, 329
  destruction as pest, 177
  in French vineyards, 394
  increase of, 230
  protection of, 226
Malaria in Scotland. *See* Ague
Mallard, changed food habit of, 411, 412
  destruction for sport, 188
  domestication of, 106
  price of, 265
  protection for food, 217
  protection for hawking, 203
Malloch, P. D., 328, 385
Malmesbury, William of, 236, 371
Mammoth, 108, 338
Man, arrival in Scotland, 2, 6
  influence on animal life a developing factor, 480
  main trends of his influence, 4, 483
  methods contrasted with Nature's, 482
  recoil upon himself of his influence on animal life, 480, 506-517
Manar, 294
Manchester, 507
Mann, Ludovic McL., 30, 316
Manor parish, 282
Mar and Kelly charter, 47, 156, 222
Mar Forest, 254, 311, 358. *See also* Marr, Brea of, and Braemar Lodge, 270
Maree, Loch, 134, 314
Marlee Loch, 157
"Marr, Brea of," 80, 356
  deer-hunts in, 211
Marsh, Professor G. P., 304, 327, 393, 397, 479
Marshall, Dr J. N., 316
Marshes, reclamation of, 370-377
  disappearance of ague due to, 373, 510
  disappearance of liver-rot in sheep due to, 373, 514
  extermination of Bittern due to, 374, 482, 497
  extermination of Crane due to, 375, 482, 497
  reduction of animal life due to, 370-377, 488

INDEX

Marten, destruction as pest, 126, 127, 176, 229, 353, 362, 482
    destruction for fur, 155, 156, 159-161, 166
    effect of destruction of forest upon, 350
Martes martes. See Marten
Martin, House, habit of selecting a domicile, 401, 402
    protection for hawking, 203
    protection through superstition, 239
Martin, Martin, 74, 127, 130, 144, 145, 146, 148, 160, 172, 175, 212, 427
Maryland, 475
Massachussetts, 462
Master of Game, The, 248
Mauchline, 293
Mauritius, destruction of Mayer's Pigeon by Macaque monkeys in, 245
    extermination of Dodo in, 498
Maxwell, Sir Herbert, 191, 293
May Island, 249
    lighthouse, 378
Maybole, 161
Mayetiola (Cecidomyia) destructor, 439
Meadows, the, Edinburgh, prehistoric remains of deer in, 336
    wild fowl in, 371
"Meal Worms," 410, 443
Megaceros giganteus. See Deer, Giant Fallow
Megasternum, increase of, 396
Megastigmus spermotrophus, 455
Meles meles. See Badger
Melophagus ovinus, 423
Melrose, 156, 331, 381
Menteith, Lake, 134
Merganser, Redbreasted, 17
Merlin, 136
Merodon equestris. See Narcissus Fly
Merse, the, 366
Mertoun, 347
    Loch, 347
Mertrick. See Marten
Methlick, 272
Methven, 293
Mexico, 436
"Michael, the Great," 319
Michelet, Jules, 227
Micropterus dolomieu. See Bass, Black
Midge, Pear, 451
Midlothian, Deer, Red, prehistoric remains in, 333
    Dog, prehistoric remains in, 86
    Elk, prehistoric remains in, 346
    forest in, 314
    Polecat in, 164
    Squirrel, Red, in, 292
    tropical Longicorn Beetles transported to, 463
Migdale Loch, 279
Miliaria europæa, 394
Millais, J. G., 274, 332
Millipedes, increase of, 392

Milvus milvus. See Kite
Minard House, 294
Minnoch Water, 161
Minnow, transportation of, 279
Minto, 292, 293
Miramichi, forest fire of, 324
Misty Law Hills, 282
Mites, Cheese, 410
    Chicken, 423
    Flour, 410
    Fowl, 420
    Grain, 443
    introduced as parasites on livestock, 423
    in dry food materials, 434
    Itch, 421
    Red, 392
    Sugar, 445
Mole, destruction as pest, 179
    increase, due to cultivation, 393
    in Italy, 498
Mollusca, in water systems, 414, 415
Monkey, Macaque, in Mauritius, 245
Monnair, Forest of, 211
Monochammus dentator, 471
    sutor, 471
Monomorium pharaonis, 440
Monro, Don., 74, 160, 165, 251, 318
Monteviot, 276
Montrose, 356
Monymusk, 254, 272, 274
Moorfoot Hills, 282, 292
Moorfowl. See Grouse
Moorland, influence of Gulls on, 501-505
    of Rabbits on, 499
    of Sheep and Goats on, 500
    reclamation of, destruction of animal life due to, 365 *et seq.*, 488
Moose. See Elk
    in America, 346
    in New Zealand, 242
Moravia, 109
Moray, Right Hon. the Earl of, 214, 426
    Firth, 125, 188, 296, 297, 334, 358
Morayshire, "bog butter" in, 58
    Capercaillie in, 273, 356
    Deer, Red, in, 211
    Goshawk in, 133
    Rat, Black, in, 427
    Squirrel in, 296, 352
    Swans on Loch Spynie in, 372
    Tit, Scottish Crested, in, 230
Morbhaich Moor, 86
Morlich, Loch, 193
Mortlake, 421
Morven, 295
Moryson, Fynes, 312
Mosquitoes, disappearance due to reclamation of swamps, 373, 510
    relation to ague, 511
Moths, Angoumois Grain, 442
    Burnet, 448
    Clothes, 410
    Codlin, 451, 475

INDEX

Moths—*Cont.*
  Death's Head Hawk, 408, 409
  destruction by lighthouses and other lights, 378
  Diamond-backed, 378, 392, 394
  distribution by biscuits, 444, 445
  European Grain, or Wolf, 443
  Gypsy, 462
  introduction upon trees, 462
  Leopard, 462
  Magpie, 392
  Meal Snout, 443
  Mediterranean Flour, 443, 445 f.-n.
  Turnip, 392
  Umber, 392
  Winter, 392
Moufet, 421, 435
Mouflon, 35, 36, 39, 43
Mount Rainier National Park, 236
Mouse, Field and Harvest, 391, 485
  House, houses selected as domiciles, 405
  in coal-pits, 416
  increase of, 391, 485, 496
  railways as means of dispersal of, 399
Mouswald, 310
Muck, Isle of, 432
Muckross, 90
Mull, Island of, Chough in, 178
  Deer, Fallow, in, 287
    Roe, in, 280
  Fox, disappearance from, 124
  Hare, Alpine, in, 282
    Common Brown, in, 280
  Horse in, 75
  Rabbit in, 251
  Trout, American Brook, in, 277
    Lake Geneva, in, 277
    Rainbow, in, 278
Munro, Dr Robert, 340
Murchison, Sir Roderick, 237
Mus (Epimys) norvegicus. See Rat, Common or Brown
  (Epimys) rattus. See Rat, Black
  (Epimys) rattus alexandrinus. See Rat, Alexandrine
Musca domestica. See Flies, House
Muskelunges, 259
Mussel, beds in River Forth, 386
  destruction for food, 153
  Freshwater, in waterworks, 414
    pearl fishery, 194-196
  protection of, 220
  Zebra, in water-pipes, 414
    transportation of, 469
Mussel Scale Insect. See Scale Insects, Mussel
Musselburgh, 56, 248, 447, 469
Mustela martes. See Marten
  putorius. See Polecat
Muthil, 346
Mycetozoon, in coal-pit, 416
Mygalid spider, 447
Myodes torquatus. See Lemming, Arctic

Myriopod, exotic, in greenhouses, 464
Mytilaspis pomorum. See Scale Insects, Mussel

Nairnshire, Capercaillie in, 273
  forest, destruction in, 354
  Scottish Crested Tit in, 230
  Squirrel in, 296, 354
  Wild Cat in, 126
Narcissus Fly, 465, 474
Nature, her methods contrasted with Man's, 482
Neolithic man in Scotland, 6, 7, 20, 51, 333, 340, 345, 347, 417, 481
Nesœnas mayeri, 245
Ness, Loch, 158, 296, 322
Nestor notabilis, 410
Netherlands, the, 423
New Forest, 236, 446
New Jersey, 462, 475
New South Wales, Hare in, 253
  Rabbit in, 253
New Zealand, animals introduced for pleasure into, 241, 242
  Apple Scale pest in, 450
  Codlin Moth in, 452
  domestic stock introduced to, 246
  Honey Bee in, 257
  Kea in, 410
  Mediterranean Fruit Fly in, 451
  Narcissus Fly in, 465
  Rabbit in, 241, 255
  Scottish Red Deer in, 242, 338 f.-n.
  Sparrow, House, in, 255
  Starling in, 256
  Stoat and Weasel introduced into, 245, 255
  Trout, Loch Leven, transported to, 278
Newfoundland, 241
Newhaven, 319
Newlands parish, 430
Newstead, Prof. R., 262, 393, 461
  Roman station at, 37, 51, 54, 56, 62, 72, 79, 87, 90, 248, 265, 331, 348, 375
Night Poaching Act, 205
Night-Reeler, Red, 191
Nightingale, 234, 266
Nightjar, 394
Nilsson, Prof., 52, 58
Niptus crenatus, 442
  hololeucus, 404, 442
Nisibost, Harris, 72, 74, 87
Nith, River, 276, 384
Nor' Loch, Edinburgh, 55, 371
Norfolk, 191
Normans, probable introduction of Rabbit by, 248, 249
North Berwick, 69, 348
North Sea, the, 12, 13
Northamptonshire, 256
Norway, Brown Rat introduced from, 429, 430
  Codlin Moth in, 452

INDEX

Norway—*Cont.*
    Longicorn Beetles in timber from, 472
    Springtails in, 463
    Wood-Wasps transported in timber from, 469
Nuthatch, transportation of, 284
Nyassaland, 442

Oakwood, 347
Oban, 8, 34, 56, 87, 294
Ochil Hills, 121 f.-n., 178, 200, 272
Oecanthus (Panchlora) niveus, 447
Ogilvie, Mrs D., 119
Oich, Loch, 397
Oil, destruction of animals for, 170 *et seq.*
Old Cambus, 125
Old Deer parish, 296
Old Stone Age, 138, 480
Oligorus macquariensis, 259
Orchard pests, increase of, 392
Orchestia gammarellus, 395
Orchids, transportation of pests with, 463, 475
Orkney Islands, Bee, Hive, introduction into, 107
    birds' eggs, collection for food, in, 148
    Boar, Wild, early remains in, 89, 90
    Cattle in, 60
    Celtic Shorthorn, prehistoric remains in, 56, 57, 60
    Death's Head Hawk Moth caterpillars in, 409
    Deer, Fallow, prehistoric remains in, 284
        Red, prehistoric and early remains in, 16, 284, 333, 337
    Dog, early remains in, 87
    Eagle, White-tailed, in, 130
    fur produce of, 155
    Grayling in, 276
    Great Auk in, 144, 146
    Grouse in, 280
    Gulls, change of food habits in, 412
    Hare, Alpine in, 282
    Hawks in, 134, 200
    Hen Harrier in, 134
    Horse, early remains in, 71, 72, 76
    Otter in, 169
    Peregrine Falcons of, 200
    Pig, primitive, in, 93
    Ptinus tectus in bakehouse at Stromness in, 442
    Rabbit in, 250, 251
    Rat, Black, in, 427
    Reindeer, attempts to acclimatize foreign, in, 254
        prehistoric remains in, 14, 338, 340, 341
    Rock Dove in, 95, 98
    Seals in, 169, 173
    Snail, Banded Garden, in, 154
    Starlings feeding upon insects in seaware in, 396
    Stoat, disappearance from, 299
    Urus, prehistoric remains in, 52

Orkneyinga Saga, 187, 310, 341
Ormiegill, 56, 71, 87
Ornithodoros megnini, 421
Oronsay, Hares and Fulmarts in, 165
    prehistoric huts in, 316
    shell-mound in, 19, 30, 56, 144, 149
Orthezia insignis, 461
Ortolan, 140, 394
Osborn, Prof. H. F., 344, 351
Osprey, extermination of, 133, 192, 482
Otiorrhynchus picipes, 392
Otis tarda. *See* Bustard
Otter, in prehistoric Scotland, 8, 17, 308, 373
    destruction for depredations on fishing-streams, 169
    destruction for skin, 156, 166, 168-170, 221
    prehistoric remains, 30
    protection by popular favour, 233
Ouzel, Water. *See* Dipper
    Ring, decrease in numbers, 370
Ovis aries palustris. *See* Sheep, Turbary or Peat
Owen, Prof. Richard, 345
Owl, Brown, in Edinburgh, 406
    destruction as pest, 181, 182
    Little, introduction of, 256
    Snowy, 18
Ox. *See* Cattle
Oxfordshire, 330
Oxyuris vermicularis, 422
Oykell, River, 295, 314
Oyster, 20
    association of Slipper Limpet with, 424
    beds, 153, 154, 386
    destruction for food, 153
    protection of, 220
Oystercatcher, 140

Pabbay, Island of, 226
Paisley, 10, 207, 299, 312, 427, 434, 436, 438
    Canal, 470
Palaeolithic animal remains, 108, 109
    man, 138, 344, 480
    sculptures, 30, 344
Pallas, P. S., 428
Paludicella articulata, 415
Pamphila linea, 369
    silvanus, 369
Panama Canal, 397
Pandion haliaëtus. *See* Osprey
Pappa Westray, 146, 250
Paradesmus gracilis, 464
Pararge megæra, 369
Parasites, of livestock, 422-424
    of man, 420-422
Paris, Convention for Bird Protection, 227
    fauna of water-pipes of, 414
Partridge, Barbary, introduction for sport, 275
    destruction for food, 140, 267

Partridge—*Cont.*
    Hungarian, introduction for sport, 275
    protection for food, 217, 266
        for hawking, 203
        for sport, 204, 205, 206
    Red-legged, introduction for sport, 275
    transportation to colonies, 241, 242
Parus cristatus scoticus. See Tit, Scottish Crested
    major. See Tit, Great
Paterson, J., 288
Peach, Dr B. N., 9, 17, 31, 111, 226, 282, 340
Peacock, as food, 140, 266
    introduction of, 265, 283
Pear midge, 451
Pearl fishing, 194–196
Peas and beans, pests introduced in, 445
Peat, formation of, 11, 309
    lower forest layer in, 305
    upper forest layer in, 306, 326
Pediculoides ventricosus, 443
Pediculus. See Lice
Peebles, 430
Peeblesshire, Deer, Red, prehistoric remains in, 333
    Douglas Fir Seed Chalcid as pest in, 456
    Elk, prehistoric remains in, 346
    fur produce of, 156
    Gulls on moorland, effect in, 501
    Hare, Alpine, in, 282
    Martins' nests on house in, 402
    Rat, Brown, in, 430
    Squirrel in, 292
Pelican, 18
Penang, 463
Penguins, destruction for oil, 171
Penicuik, 332
Pennant, Thos., 114, 142, 195, 269, 285, 352, 356
Pennsylvania, 475
Pentland Hills, Celtic Shorthorn, prehistoric remains on, 55
    Deer, Red, on, 208, 334
    Roe, on, 332
    Fox, prehistoric remains on, 69
    Gulls, effect on moorland on, 501
    Hare, Alpine, on, 282
    Horse, prehistoric remains on, 69
    Martins' nests on house on, 402
    Reindeer, prehistoric remains on, 69, 340
    Ring Ouzel, disappearance from, 370
    Wolf, prehistoric remains on, 69, 115, 121 f.-n.
Perca fluviatilis. See Perch
Perch, Californian, 258
    transportation of, 279
Peregrine Falcon, change of habit in selection of nesting site, 401
    protection for sport, 200, 202
Perichaeta indica, 464
Periplaneta americana, 436, 463
    australasiae, 436, 463

Periwinkles, use as food, 153
Perkins, Principal, 323
Pernis apivorus. See Buzzard, Honey
Perodipus, 389
Perth, 9, 276
Perthshire, ague in, 509, 510
    Bear in, 114
    Beaver in, 17
        prehistoric remains in, 157
    Capercaillie in, 270, 271, 272, 355, 356
    Chough in, 178
    Deer, Fallow, in, 285, 287
        Red, in, 334
        Roe, in, 331, 332
    Eagle, Golden, in, 128
    Elk, prehistoric remains, in, 346, 348
    forest, destruction in, 319, 361
    Kite in, 133
    Marten in, 161
    Osprey in, 134
    pearl-fishing in, 195
    Polecat in, 164
    Rabbit in, 252, 391
    Squirrel in, 293, 398
    Urus, prehistoric remains in, 51
    Wild Cat in, 126
    Wolf in, 120, 121 f.-n., 319
    Woodpecker, Great Spotted, in, 360
Pests, Californian quarantine inspections for, 474
    counter-pests introduced to combat, 259–263, 419
    destruction of, 176–183
    increase due to cultivation, 387
    introduction of, 21, 255, 387–396, 417–475, 496
Peterhead, 164, 174
Petrel, Fulmar. See Fulmar
Phalarope, Red-necked, 230
Pheasant, as food, 140
    change of food habit in, 410
    Chinese, or Ring-necked, 268
        in America, 206
    food of, 206
    Gulls protected for, 206, 502
    introduction for sport, 264–268, 496
    Japanese, 268
    Mongolian, 268
    protection for sport, 205
    Reeve's, 268
    sensitiveness to sound, 379
    transportation to colonies, 241
Philip, George, 394
Phoca vitulina. See Seal, Common
Phora rufipes, 416
Phoxinus phoxinus. See Minnow
Phthirus pubis. See Lice, Crab
Phyllodromia germanica, 438
Phyllotreta nemorum, 392
Phylloxera vastatrix, 461, 474, 487
Physa acuta, 466
Physeter macrocephalus. See Whale, Sperm
"Pictish Towers." See Brochs

INDEX

"Picts' houses." *See* Eird houses
Picus major. See Woodpecker, Great Spotted
Pieris brassicae and *rapae. See* Butterflies, Cabbage
Piette, M., 30
Pig. *See also* Boar, Wild
   as street scavenger, 224
   domestication of, 89, 92-95, 491
   extermination of Dodo of Mauritius by, 498
   in New Zealand, 245
   numbers added to fauna, 486
   parasites introduced by, 422
   prehistoric remains of, 32, 33, 34
   primitive, in Highlands and Islands, 92, 93
Pigeon, 34, 95-100, 267. *See also* Rock Dove
   Crowned, destruction for plumage, 189
   houses, evolution and legislation, 97-100
   introduction of exotic, 283
   Mayer's, 245
   protection for food, 98, 217
   Wood, 389, 390
      as pest, 231
      in London, 406
Pike, transportation of, 279
   in California, 258
Pine forests, 308
Pines, weevils transported with, 472, 474
Pinnaspis pandani, 461
Pipit, Tree, 399
Pissodes notatus, 472
   piniphilus, 472
Pitfour House, 296
   Ponds, 279
Pitlochry, 133, 360
Placocephalus (Bipalium) kewensis, 464
Plague, Bubonic, 420, 423
   relation of Fleas to, 420, 507
   of Rats to, 423, 507
Plaice, destruction of, 150
Plains, animals of, 12, 13
   destruction owing to cultivation, 365, 366
Planarians, Land, introduction to Scottish greenhouses, 464
Plant-bugs, 459, 474. *See also* Aphis
Plants, introduction of animals with, 457-466, 474, 475
   introduction with animals of, 418
Pliny the Younger, 45
Plover. *See also* Lapwing
   abundance of, 266, 356
   destruction of eggs for food, 148
   Golden, mortality caused by telegraph wires, 379
   price, 140
   protection for food, 217
Plumatella emarginata, 414
Plutella maculipennis. See Moth, Diamond-backed

Plymouth Sound, 433
Poaching Acts, 205
Podiceps cristatus. See Grebe, Great Crested
Pogonochaerus fasciculatus, 472
Polecat, in Scotland, 107, 329
   destruction as vermin, 127, 176, 177
   destruction for skin, 156, 161-166, 221
   effect of destruction of forest upon, 350
Pollution of rivers, 382-386
Polyommatus agrestis var. *artaxerxes. See* Butterflies, Artaxerxes
Polyzoa, in water-pipes, 414
Pony. *See under* Horse
Poplar, 414
Population, increase of, 25, 394, 481, 485
Port Elphinstone, 294
Porthetria dispar, 462
Possil Marsh, 298
Potato, birds as pests on, 411, 412
   Colorado Beetle as pest upon, 387
   pests transported with, 475
   plants as food of Death's Head Hawk Moth caterpillars, 408
   Sweet, pests transported with, 474
Poultry, domestication of, 34, 101-106
   parasites introduced with, 423
Pratincola rubicola. See Stonechat
Prêdmost, 109
Pressmennan Lake, 279
Prestonpans, 154
Pribilof Islands, 110, 223, 257
Prospaltella berlesei, 262
Protection of animals, 197-240, 481, 485, 487
   as scavengers, 216
   by law, 198, 478
   by popular favour, 198, 233-237
   by superstition, 198, 237
   for aesthetic reasons, 198, 229-232
   for food, 216
   for fur and skins, 216, 221
   for special services to agriculture, 216-227
   for sport, 197, 198, 199-215
   for utility, 216-228
Prothero, R., 391
Psychoda humeralis, 416
Ptarmigan, as food, 267, 356
   prehistoric remains, 17
   protection for hawking, 203, 205
Pteronus ribesii, 392
Ptilinus pecticornis, 404
Ptinus tectus, 442, 445 f.-n.
Puffin, rats in burrows of, 432
Puffin Island, 432
Pulex irritans. See Fleas, Human
Pulvinaria floccifera, 460
Punjab, 328
Punt-gunning, 188
Pupa, 329
Putney, Surrey, 446
Pyralis farinalis, 443

Quail, as food, 266
  Button-, introduction of, 275
  decrease in number, 367
  destruction for food, 139
  protection for sport, 205
  Virginian, introduction of, 275
Quarantine inspections of imports, 474
Quebec, 458
Quedius mesomelinus, 416
Queensberry Hill, 282

Raasay, 178, 282
Rabbit, change of habit in selection of domicile, 400
  destruction as pest, 21, 137, 179-180, 252, 253, 390, 391
  destruction for skin, 156, 166-168, 222, 252, 390
  dispersal of, 398
  domestication of, 106
  effects of introduction of, 253-254, 499, 500
  increase due to cultivation, 390, 485, 496, 497
  influence on forest growth, 323
  on vegetation, 487, 499-500
  introduction and spread in Scotland, 247-253, 400, 496
  "Jack," 369
  pest in the colonies, 241, 243, 245, 255, 498
  protection for food, 205, 217, 221, 250, 251
  for fur, 221-222, 249, 250
  warrens, 249 *et seq.*
Raehills, 287
Raesgill, 338
Railways, decrease of animal life due to, 379
  dispersal of animals by, 398
Rangifer tarandus. See Reindeer
Rannoch, 119, 126, 319
  Black Wood of, 314
  Loch, 271
Rat, absence from certain areas, 426
  Alexandrine, 428
  as pest in Harris, 427
  bird-colonies infested by, 432
  Black, or "Ship," 425-428, 487, 496
  Common or Brown, colonization of Europe by, 428-432, 487, 496
  damage done by, 430, 431
  dispersal of, 398
  fleas infesting, 423, 507
  in coal-pit, 416
  in sewers, 413
  increase of, 137, 391, 395, 432, 497, 507
    during the War, 396
  introduction by ships, 425 *et seq.*
  Kangaroo, 389
  "Norway." *See* Common or Brown
  parasites introduced by, 420, 423, 507
  railways as means of dispersal of, 399
  relation to disease, 507-508

Raven, in Scotland, 17
  as scavenger, 224, 225, 226, 395, 405
  destruction of 136, 482
  increase of, 230
  influence of towns upon, 405
"Ravenois foullis," 136
Ray, Thornback, 8, 19
Razorbill, 8, 17
"Red Maggot," 440
Redpolls, 193, 329
Redshank, 140
Reeves, 140
Regent's Park, American Cockroach in, 436
  Grey Squirrel in, 288
Reid, Clement, 11, 305
Reindeer, in Scotland, 14, 20, 308, 497
  destruction for sport, 187
  effect of destruction of forest upon, 338-346, 350
  extermination of, 207, 213, 481, 496, 497
  failure to re-acclimatize, 242, 254
  in Pribilof Islands, 257
  Magdalenian sculpture of, 30
  prehistoric and early remains of, 31, 69, 330, 338, 340, 344
  racial character of Scottish, 343-346
  Woodland, 329, 343
Renfrewshire, Capercaillie in, 273
  Deer, Giant Fallow, prehistoric remains in, 68
  Red, in 207
  Grayling in, 276
  Hare, Alpine, in, 282
  Horse, prehistoric remains in, 68
  Polecat in, 164
  Trout, American Brook, in, 277
  Urus, prehistoric remains in, 50, 68
Reyce, *Breviary of Suffolk*, 248
Rhidorroch Forest, 314
Rhinocola dianthi, 392
Rhizoctonia, 475
Rhyssa tridactyla. See Gulls, Kittiwake
Rhytina stelleri, 138
Richardson, Dr, 402
Richmond, Yorkshire, 112
  Park, 288
Riley, Prof., 458, 473
Rivers, barriers causing decrease of fisheries, in, 379-382
  pollution causing decrease of animal life in, 382-386, 488
Roads, dispersal of animals by, 398
Robbers, destruction of forests for, 318
Robert Bruce, King, 66, 82, 199, 207, 310, 311
Robin, association with man, 406
  increase due to cultivation, 394
  introduction to New Zealand, 241, 242
  "Pekin," introduced, 283
  protection by popular favour, 234
  by superstition, 237

INDEX

Rock Dove, domestication of, 95–98
  slaughter for food, 141
Rodriguez, 498
Roeding, Mr, 257
Rogers, Prof., 200, 249
Romans in Scotland, 56, 309
  animal remains in settlements, 40, 51, 54, 56, 72, 79, 87, 90, 112, 113, 248, 331, 333, 348, 395
  destruction of forests by, 317
  horses of, 72, 79
  introduction of Peacock by, 283
  of Pheasant by, 265
  protection of Raven by, 225
  reclamation of swamps by, 370
Ronaldshay, North, 154, 412
  South, 427
Rooks, as pest, 177–178, 217, 228, 231, 390
  as scavengers, 224, 225, 226, 395
  dispersal by railways, 399
  increase due to cultivation, 390, 394
  influence of towns upon, 405, 406
Rose-seed Chalcid, 456
Rosehall, 314
Ross, animal sanctuary in, 237
  Buzzard, Honey, in, 135
  Capercaillie in, 273, 274, 355
  Celtic Shorthorn, early remains in, 55
  Deer, Japanese, in, 288
    Red, in, 211, 264
    Roe, in, 332
  Dog, prehistoric remains in, 86
  Goose, Grey Lag, in, 104
  Oyster-beds off coast of, 154
  pearl-fisheries in, 195
  Polecat in, 165, 166
  Rat, absence from, 426
  Reindeer, prehistoric remains in, 340
  Scottish Crested Tit, in, 230
  Squirrel in, 295
    as pest in, 181
  Wild Cat in, 126
Rothiemurchus, Forest of, 133, 290, 297, 314, 321, 358
Rougemont Gardens, Exeter, 288
Rousay, 14, 280, 340
Row, 289
Roxburghshire, ague in, 509
  Ant, Small Red House, in, 440
  Beaver, prehistoric remains in, 157
  Celtic Shorthorn, prehistoric remains in, 55
  Deer, Red, prehistoric remains in, 333, 336
  Elk, prehistoric remains in, 347
  fur produce of, 156
  Pigeon, price in, 100
  Polecat in, 164
  Rabbit in, 252
  Squirrel in, 292, 352
  Wild Cat in, 125
  Wolf in, 121 f.-n.

Ruff. See Reeves
Rum, Isle of, 178
Russia, 469
Rutherglen, 56
Ryper. See Grouse, Willow

St Abbs, Chough at, 178
  pigeon-houses at, 97
  Rock Dove at, 95, 97
St Andrews, 90
St Cyrus, 87
St Helena, influence of Goats on forest-growth and vegetation in, 501
  introduction of plants to, 418
St Kilda, destruction of Gannet and Fulmar in, 146
  destruction of Great Auk in, 144–166
  egg-gathering on, 148
  ponies of, 74
  sheep of, 37, 38
  Wren, 190, 229
St Marcel, Indre, sculptures at, 30, 344
St Medan's Cave, 57
St Ninian's Cave, 57, 87
St Osyth, 424
St Vigians, 509
Salmo irideus. See Trout, Rainbow
  lemanus. See Trout, Lake Geneva
  levenensis. See Trout, Loch Leven
  trutta. See Trout, Brown
Salmon in Scotland, 19
  destruction in Tweed, 150–152
  interference with, by barriers, 380
  pollution causing decrease of, 384–386
  protection for food, 217–219
  spawn destroyed by flood, 328
Salt manufacture, 319
Salvelinus alpinus. See Char
San Francisco, 436
San José, 460
  Scale Insect. See *under* Scale Insects
Sanctuaries, animal, 236
Sanday, 250, 251, 280
Sandwick, 72, 250, 251, 341
Saperda carcharias, 471
Sarcoptes scabiei, 421
Sawflies, Gooseberry, 392
  transportation in timber, 469
  Turnip, 392
Saxicola œnanthe. See Wheatear
Saxony, Elk in, 349
Scale Insects, Apple. See *below* Mussel
  Aspidiotus, 461 f.-n., 475
  Black, 475
  Camellia or Cottony, 460
  Coconut Snow, 475
  Coffee, 461 f.-n.
  Egyptian Cushion, 461 f.-n.
  Fluted or Cottony Cushion, 260
  importation into Britain, 448, 461
  Ladybirds as counter-pests of, 260
  Long, 475
  Mulberry, 261, 460, 461

INDEX

Scale Insects—*Cont.*
  Mussel, 450, 460, 474
  Orange Mussel, 461 f.-n.
  Purple, 475
  Red-Spotted, 474
  San José, 261, 475
  Screw-pine, 461 f.-n.
  Shield, 475
Scandinavia, 458, 484 f.-n.
Scatophaga stercoraria, 396
Scavengers, protection of beasts and birds as, 223–226, 395
Sciurus carolinensis. *See* Squirrel, Grey
  vulgaris. *See* Squirrel, Common Red
Scotland, advantages of fauna for study, 2 *et seq.*
  arrival of man in, 2, 6
  early climate and vegetation, 10
  early inhabitants, 8
  early physical condition, 9
  original fauna of, 12, 492
  surface features on man's arrival, 9, 484
Scots, Fir, 308
  Grey fowls, 101, 102
Scotstown Moor, 500
Scott, Dr Thomas, 373
Scottish Dumpie fowl, 102
Sculptured stones, 76, 342
Sea fisheries, protection of. *See* Fisheries
  prehistoric inhabitants and, 19
Sea-gulls. *See* Gulls
Sea-weed, animals breeding in, 395
Seacliff, 87
Seal, Common, in Scotland, 19
    destruction for food, 138, 172, 222
      for oil, 171–175, 222
    flesh as food of early inhabitants, 8
    in Iona, 171, 222
  Fur, destruction of, 170
    protection of, 222–223
    slaughter on Pribilof Islands, 110
  Grey, destruction of, 169, 172, 173, 174, 223
    early remains of, 19, 173
    protection of, 223
Sedge, Sand-, association and rabbits, 499
Seeds, transportation of pests in, 454
Selborne, 435
Selkirk, 430
  Forest, of, 208. *See* Ettrick Forest
Selkirkshire, Deer, Red, prehistoric remains in, 333
  royal hunts in, 119, 208, 211
  Elk, prehistoric remains in, 346, 347
  forest, destruction in, 313, 314. *See* Ettrick Forest
  fur produce of, 156
  Rat, Brown, in, 430
  Squirrel, Red, in, 292
  Urus, prehistoric remains in, 51
Service, R., 162
Settle, 112
Seychelles, the, 460

Shad, in Thames, 384
  introduction to California, 258
Shag, 8, 17
Shapinsay, 280
Sharp, Dr D., 468
Shaw, Rev. Lachlan, 352, 356
Shaws, Dumfriesshire, animal remains at, 16, 112, 330, 340
Shebley, W. H., 258
Sheep in Scotland, 35–48, 68
  Blackfaced, 29, 43, 44, 45, 48
  Cheviot, 29, 43, 44, 45, 48
  Deer, banishment from Lowlands due to sheep flocks, 42, 208, 331, 334, 486
  destruction of forest for pasturage, 322, 331, 334, 486, 496
  flocks in Tweeddale, 41, 42
  four-horned, 41, 42
  influence on vegetation, 323, 500
  liver-rot in, 373, 422, 514
  man's influence on characteristics, 39, 42–48
  numbers added to fauna, 486, 493
  parasites introduced with, 422, 423
  Peat or Turbary, 35, 40–43
  prehistoric and early remains of, 31, 32, 33, 34, 40, 73
  quality of rapid fattening, 45
  Shetland, 35, 40–43, 47
  Soay, 35, 37–40, 46, 48
  Ticks of, 423
  wild ancestors of, 36, 44
  wool, characteristic, 39, 43, 45–48
    improvement of, 45–48, 490
Sheldrake, protection for hawking, 203
Shell-fish, in water-systems, 414, 415
  use as food in times of scarcity, 216
Shell-fisheries, depletion for food, 153, 154
  destruction caused by pollution, 386
  protection for food, 220
Shell-mounds, 9, 19, 30, 31, 32, 33, 56, 57, 71, 72, 87, 89, 158
Shells, Land. *See under* Snails
Shetland Islands, Birds' eggs, collection for food in, 148
  Cattle of, 60
  Death's Head Hawk Moth caterpillars in, 409
  Deer, Red, prehistoric remains in, 16, 333
  Roe, 16, 331, 332
  Eagle, White-tailed, in, 130
  Falcons and Hawks of, 200, 201
  forest, prehistoric, in, 305, 308
  Grouse in, 280
  Pig, primitive, of, 93
  Ponies of, 73, 76–78
  Quail in, 368
  Ravens on whale carcases in, 226
  Rock Dove in, 95, 98
  Seal, Grey, destruction in, 173
  Sheep of, 35, 40–43, 47

Shetland Islands—*Cont.*
  Skua, protection in, as enemy of Sea-Eagle, 228
  Weasel (Stoat), introduction by Falconer to, 299
  Whaling industry in, 175, 226
Shin, Loch, 165
Ship-worm, 432
Shipbuilding industry in Scotland, 319
Shipping as means of transport and dispersal of animals, 397, 425 *et seq.*
Ships, stowaways on, 425-433
Shore-flea, 395
Shrimp, freshwater, 383
  in water-pipes, 414, 415
Sibbald, Sir Robert, 142, 144, 351, 435
Siegfried, 52, 349
Silchester, Roman settlement at, 225, 265
Silesia, Elk in, 349
"Silver-fish" insect, 404.
Sirex gigas, 467-469
  noctilio, 467
Siskin, 193, 329
Sitotrepa (Anobium) panicea, 410, 444
Sitotroga cerealella, 442
Skara, Orkney, 52, 57
Skaw lighthouse, 378
Skeen, Loch, 134
Skink, Blue-tailed, 469
Skins. *See* Fur-bearing animals
Skua, Great, protection in Shetland, 228
Skye, "Bog butter" in, 58
  Chough in, 178
  Eagle, Golden, in, 128
    White-tailed, or Sea, in, 132
  Hare, Alpine, in, 282
    Common Brown, in, 280
  Hebridean ponies of, 75
Slag-furnaces, 320
Slateford, 435
Slater, Freshwater, in waterworks, 414
Slugs, in coal-pits, 416
Smeaton-Hepburn, 337
Sminthurus igniceps, 463
Smith, Dr John Alexander, 58, 347, 490
  Major Hamilton, 53
  Rev. Samuel, 510
Smoo Cave, 278
"Snail-slug," transportation of, 465
Snails, Banded Garden, 154
  Land, destruction for food, 154
    effect of destruction of forest upon, 329
    introduced in cabbages to Ceylon, 459
    transportation of, 464, 465
  Mediterranean species, transportation of, 465
  Pond, disappearance due to reclamation of swamps, 373
    in waterworks, 414, 415
    naturalization of, 466
    relation to liver-rot, 514
    transportation of, 298

Snails—*Cont.*
  Roman, destruction for food, 154
  White-keeled, 465
Snake, Smooth or Ringed, as "escape," 299
Snipe, as food, 266
  formerly shot on site of London, 371
  price of, 140, 266, 367
  protection for sport, 205
Soay, Gannets on, 146
  Sheep. *See under* Sheep
Solenopsis molesta, 440
Solitaire, extermination of, 498
Solutré, 109, 138
Solway Firth, punt-gunning on, 188
  district of, ague in, 512
  Deer, Red, prehistoric remains in, 334
  Grebe, Great Crested, in, 230
  Hare, Alpine, in, 282
  Rat, arrival in, 429
Spain, 366, 400, 401
Sparrow, House, destruction as pest, 179, 389
  habit of selecting a domicile, 403
  increase of, 389, 496, 497
  influence of towns upon, 405, 407
  introduced into the colonies, 255, 498
  local migrations for food, 412
  railways as means of dispersal of, 399
  White-throated, 299
Sparrow Clubs, 179, 389
Sparrow-hawk, increase during War, 136
  protection for sport, 200, 202
Spey, River, Char in, 153
  Minnow in, 279
  pollution affecting fish of, 384
  Water Ouzel, destruction on, 182
Speyside, Beetles, Longicorn, in 471
  Capercaillie in, 356
  forests, destruction in, 314, 321
  iron furnaces in, 321, 354
  Squirrel in, 296, 297, 354
  Woodpecker, Great Spotted in, 358, 360
Sphæridium, increase of, 396
Sphærium lacustre, 298
Spiders, Bird-eating, 447
  Cave, in coal-pits, 416
  exotic, in greenhouses, 404, 463
  protection through superstition, 239
  selection of houses as domicile, 404
Sponges, freshwater, in waterworks, 414, 415
Spongilla lacustris, 414, 415
Sport, destruction of animals for, 184-188
  introduction of animals for, 264-282
  protection of animals for, 197, 199-215
  transportation of animals for, 278-282
Spring-tails, in coal-pits, 416
  introduction into greenhouses, 463
Spynie, Loch, 279, 371, 372

Squirrel, Common Red, in Scotland, 290–297, 351–354
  as swimmer, 295
  bridges as means of dispersal, 295, 398
  change of food habits in, 411
  decline of, 351–354, 359
  destruction as pest, 181, 244, 297
  dispersal of, 398
  effect of destruction of forest upon, 290, 351–354
  introduction and spread of, 290–297, 361, 496
  Grey, introduction and spread of, 288–290
Stanley, 287
Stannergate, shell-mound at, 31
Starling, as pest, 228, 231
  change of food habit in, 409, 410, 411
  increase due to cultivation, 393, 394, 396
  influence of towns upon, 406, 407
  introduction into colonies, 256, 498
  introduction to Australia of Fowl-mite with, 420
  Red-winged, introduction of, 283
  slaughter for food, 139
  Woodpecker, Great Spotted, relations to, 359
Stejneger, L., 80
Steller's Sea Cow, destruction of, 138
Stennis, Loch, 56, 71, 87
Sticklebacks in waterworks, 414
Stirling, Carse of, 10, 20, 31
  Deer, Fallow, in park of, 285
  fox-hunter at, 121
  Peregrine falcons near, 200
  Whales, remains of, near, 9, 20, 70
  Wolf-hunting near, 117, 120
Stirlingshire, Deer, Fallow, in, 285
  Roe, in 332
  Falcon in, 199, 200
  forest, destruction for iron-smelting industry, 322
  Hawks in, 200
  Kite in, 133
  Marten in, 161
  Polecat in, 164
  Rabbit in, 252
  Squirrel, Red, in, 292, 398
  Wild Ca in, 125
  Wolf in, 117, 120
Stoat in Scotland, 17, 329
  decrease in numbers, 488
  destruction as pest, 127, 181
  for skin, 155, 221, 488
  in Shetland, 299
  increase in numbers, 485, 488
  introduction to New Zealand, 245, 255
Stocket, forest of, 116, 311
Stoddart, Thos. T., 150, 277
Stonechat, in vineyards of France, 394
  protection by superstition, 238
Storer, Rev. John, 65

Stork, in Edinburgh, 377
  protection by popular favour, 239, 401
  rarity in Scotland, 266
Stornoway, 280
Straits Settlements, 463
Stratford-on-Avon, 402
Strathardle, 272
Strathbeg Loch, 279
Strathblane, 125
Strathdee, 161
Strathdon, 161. *See also* Don, area of
Strathglaish, 211
Strathglass, 269, 356
Strath-Halladale, 16, 347
Strathmore, 90, 273, 293
Strathnairn, 273
Strath Oykell, 296
Strathspey. *See* Speyside
Stromness, 252, 442
Structures of animals, modification of, 489. *Also* Domestication
Stuart, John Sobieski and Charles Edward, 323, 325, 334, 348, 358
Suffolk, 298
Sugar, introduction of mites with, 445
Sula bassana. *See* Gannet
Suleskerry, 173
Sumburgh, 76
Sunart, 164
Surrey, 288, 323, 446
Sussex, 275, 320
Sutherland, Bear, prehistoric remains in, 17, 112
  Bittern in, 375
  "Bog butter" in, 58
  Capercaillie in, 356
  Chough in, 178
  Crows, Rooks, Magpies and Jackdaws in, 177
  Deer, Fallow, in, 287
  Red, in, 187, 334
  prehistoric remains in, 31, 333
  Roe, early remains in, 331
  destruction of vermin and pests in, 124, 126, 128, 136, 137, 170, 177, 182
  Dog, early remains in, 87
  Eagle, Golden, in, 128
  Elk, prehistoric remains in, 16, 347
  forest, destruction by burning in, 325
  prehistoric, 308
  Fox in, 122, 124
  Frogs and Toads, prehistoric remains in, 373
  Goose, Grey Lag in, 104
  Hawks in, 136
  Horses, Wild, in, 80, 81
  Kites in, 136
  Lemming, Arctic, prehistoric remains in, 14
  Lynx, prehistoric remains in, 17, 111
  Marten in, 161, 176
  Mosquito in, 512
  Osprey in, 134

Sutherland—*Cont.*
  Otter in, 170
  Ouzel, Water, in, 182
  Pike, in, 279
  Polecat in, 165, 166, 177
  Rabbit in, 252
  Rats, absence from, 426
  Raven in, 137
  Reindeer in, 31, 187, 340, 341
  Squirrel in, 181, 295, 297, 351, 398
  Trout, transported to lochs of, 279
  Urus, early remains in, 52
  Wild Cat, in, 126, 177
  Wolf in, 119, 120
Sutherland, Kyle of, 279
Swallows, habit of selecting a domicile, 401, 402
  protection through superstition, 239
  telegraph wires as roosting places, 379
Swallows, Cliff, 402
Swamps, reclamation of, 370-377, 488, 510
Swan, Wild, as food, 140, 266
  destruction for sport, 188
  on Loch Spynie, 372
  prehistoric remains of, 18
  price of, 140, 376
Sweden, 463
Swift, Dean, 404
Swifts, food of, 403
  habit of selection of domicile, 401
Swine, as city scavengers, 224. *See* Boar, Wild. *Also* Pigs
Swinton, 90
Swinwood, 90
Switzerland, 219, 344, 453
Sylvius, Aeneas, 312, 315
Syntomaspis druparum, 453
Syntomidae, 448

Taenia nana, 421
  saginata, 421, 422
  solium, 421, 422
Talitrus saltator, 395
Taradale, animal sanctuary at, 237
Tarbat House, 295
Tarbert, 288
Tarbolton, 71
Tarland Burn, 294
Tarnaway Forest. *See* Darnaway F.
Tasmania, 442, 452
Tay, Loch, 271, 273
Tay, River, Grayling in, 276
  pearl fishing in, 195
  Perch in, 279
  pollution of, 385
  Salmon spawn destroyed by floods in, 328
  Tench in, 279
Tay valley, Capercaillie in, 272
  Celtic Shorthorn, prehistoric remains in, 56
  Deer, Fallow, in 287
    Red, in, 333
    Roe, in, 332

Tay valley—*Cont.*
  Squirrel in, 293
  Urus, prehistoric remains in, 51
Tayfield, Fife, 395
Taylor, John, the Water Poet, 80, 120, 211, 267, 356
Taymouth Castle, 270, 271, 273
Taynuilt, 321, 353, 354
Teal, as food, 140
  on White Moss at West Linton, 504
  protection of, 266
  for hawking, 203
Technomyrmex albipes, 463
Tegenaria derhamii, 404
Teith, valley of, 273
Telegraph wires, as roosting-places of birds, 379, 399
  mortality among birds due to, 379
Tench, transportation of, 279
Tenebrio molitor, 410, 443
  obscurus, 443
Tenebrioides mauritanicus, 441, 445 f.-n.
Teredo, introduction by ships, 432
Termite, 440
Terns, 17
  Black, 191
  White, destruction for plumage, 189
Testacella maugei, 465
Tetramorium guineense, 463
Tetranychus telarius, 392
Tetrao urogallus. *See* Capercaillie
Teviot, River, 276, 384
Thainstone, 294
Thames, the, 384, 469
  Estuary, 424
Thanasimus formicarius, 416
Thayngen, 344
Theridion tepidariorum, 404, 463
Thick-knee, 401
Thornton, Colonel, 115
Thortorwald, 310
Thrips, foreign species introduced into greenhouses, 463
  transportation of, 474
Thrumster, 331
Thrushes, as pests, 231, 411
  as prey of Little Owl, 256
  change of food habits, 411
  destruction for food, 139
  increase due to cultivation, 394
    due to protection, 230
  influence of towns upon, 406, 407
  protection by popular favour, 234
Thymallus thymallus. *See* Grayling
Ticks, Asiatic, 423
  introduced as parasites on livestock, 423
  "Sheep-," 423
  Sheep, true, 423
  Spiny ear-, 421
Timber, animals introduced with, 467-473
Tinca tinca. *See* Tench
Tinea granella, 443
Tineidae, 410

Tinwald, 310
Tiree, Celtic Shorthorn, prehistoric remains in, 57
    Chough in, 178
    Hare, Common Brown, in, 280
    Hebridean ponies in, 75
Tits, Great, as insect-eaters, 393
    increase due to cultivation, 394
    influence of towns upon, 406
    Scottish Crested, 230
Toads, decrease due to reclamation of swamps, 373
    Natterjack and Common, prehistoric remains in Sutherland, 373
Tomocerus minor, 416
Tongue, 325
Top shell, 20
Tope, 19
Tor Wood, 114, 312
    Wild White Cattle of, 66
Torquay, fauna of water-supply of, 414
Torrey, 427
Towns, influence on habits of animals, 405–408
    influence on song and nesting of birds, 407
Trail, Prof. J. W. H., 500
Transport affecting dispersal of animals, 397–399. *See also* Chapter VIII
Transportations of animals, 278–282
Trap-netting, 139
Traprain, 56, 248, 316
Traquair, 430
Tree-Creeper, 329
Trees, creatures introduced with, 459
Treig, Loch, 399
Treves, Sir Frederick, 290
Tribolium castaneum, 441, 445 f.-n.
    confusum, 441, 445 f.-n.
    ferrugineum, 441
Trichinella spiralis, 508
Trichinosis, 508
Trichocephalus dispar, 422
Triglyphothrix striatidens, 463
Trochus lineatus, 20
Troglodytes hirtensis. *See* Wren, St Kilda
Tropical plants, introductions with, 462
Troqueer, 164
Troup Head, 296
Trout, American Brook, or "Rainbow," 277, 383
    Brown, transportation of, 278
    Lake Geneva, introduction of, 277
    Loch Leven, transportation of, 241, 278
    protection for food, 217, 219
    Rainbow (*Salmo iridens*), 278
    spawning beds destroyed by floods, 328
    transportation of, 241, 242, 278
Trypetid, 474
Tubularia crocea, 433
Tulliallan, 270, 273
Tulloch, 130
Tummel, River, 271
Tunis, 451
Tunisia, forest destruction in, 323
Turkey, the, 106, 266
Turner, Sir William, 52
Turnip crops, birds as pests on, 412
    increase of pests of, 392
Turnix sylvatica. *See* Quail, Button-
Turriff, 296
Turtle, destruction of eggs for oil, 171
Tweed, River, decline of fisheries, 151
    Grayling in, 276
    Salmon run in, 381
    spearing of, in, 150
Tweed valley, Elk, prehistoric remains in, 16, 346
    Great Crested Grebe in, 230
    Mallard in, 411
    Rat, Brown, in, 430
    Sheep, destruction of woods by, 323
        flocks in, 41, 42
        four-horned, in, 41
    Timberman Beetle in, 472
    Urus, prehistoric remains in, 51
Tweedmouth, 472
Twite, 193
Tyree. *See* Tiree
Tyroglyphus, 410

Ubaldini, Pet., 149
Uisg, Loch, 278
Uist, North and South, 74, 75, 128, 172, 277, 412
Ulbster, 51, 56, 71
Ullapool, 314
Ulverston, 320
Unapool, Loch, 279
Urial, Asiatic, 36, 40, 42, 43
Uruguay, counter-pests in, 262
Urus in Scotland, 49–55, 60, 62, 63, 67, 68, 308
    characteristics of, 52
    effect of destruction of forest upon, 350
    resistance to domestication, 54

Vanessa c-album, 369
Vedalia cardinalis, 260
Vegetables, introduction of animals with, 457–466
Vegetation, influence of rabbits etc. on, 253, 499, 500
    post-glacial, 10
Vemantry, 132
Venus Shell, Warty, 154
Vermin, destruction of animals as, 176–183
    increase of, 137
    during War, 109, 182
Vienna, 438
Vine, Phylloxera of, 461, 474, 487
Vineyards, French, birds of, 394
Virginia, West, bird protection in, 227
Voles, destruction as pest, 179
    plague of, 137, 391
    Northern, 14
        extermination of, 14, 481

INDEX

Volga, River, 469
Vultures, increase during War in Eastern Europe, 396

Wagtails, Pied, in Edinburgh, 406
    railway embankments as nesting-place of, 399
Wainfleet, 139
Walker, Prof. John, 352, 429
Wallace, Prof. Robert, 43, 80
Walls, island of, 280
Walrus, destruction for oil, 171
    disappearance from Scotland, 171 f.-n.
Waltham Abbey, 265
Wangford, 298
War, the Great, destruction of birds as food during, 139
    distribution of Beetles and Moths in dry food materials during, 444
    effects of, on animal life, 109, 136, 137, 205, 212, 214, 361, 396, 402
Warble, Ox, 423
Warblers, 256
    as pest, 394
    increase due to cultivation, 230, 394
    Savi's, 191
    Sedge, increase of, 230
    Wood, increase of, 230
Wardlaw Chronicles, 149, 211, 287
Washington, 475
Wasps, change of food habits, 411
    Wood. See Sirex
Waste land, effects of breaking in, 364-370
Water-fleas, in waterworks, 414, 415
    influence of drought upon, 373
Water-pipes, fauna of, 413-415, 470, 487
Water-rail, 17
Waterston, Rev. James, 404
Waterways, dispersal of animals by, 397
Waterworks, fauna of, 413-415, 470, 487
Weasel, in Scotland, 17, 329
    destruction as pest, 127, 181
        for skin, 221
    increase of, 485
    introduction to New Zealand, 245, 255
        to Shetland (Stoats), 299
Weeds, transportation of, 418, 473
Weevils, 475
    Apple blossom, 392
    Corn or Granary, 441, 445 f.-n.
    increase due to cultivation, 392
    Pine, 472, 474
    Raspberry, 392
    Rice, 474
    Strawberry, 392
    Sweet Potato, 474
Wedderburn, Lieut.-Col., 370
Wedderburn's Accompt Book, 136, 224, 395
Wedel, Baron von, 225, 251
Weisdale Voe, 226, 280
Weldon, Sir Anthony, 312
Wemyss Caves, 57, 87, 97

West Indies, 450, 460, 464, 466, 470
West Kilbride, 34, 87
West Linton, 406, 502
Whale, destruction for oil, 174-175
    Finner. See Rorqual *below*
    food of early inhabitants, 8
    Greenland, Right or Whalebone, 174, 175
    in Firth of Forth, 386
    Pilot, or Ca'ing, 175
    prehistoric remains of, 9, 20, 70
    Ravens feasting on, 226
    Rorqual, or Finner, 20, 175
    slaughter at S. Georgia, 110, 175
    Sperm, 174, 175
    Whalebone. See Greenland *above*
Wheat, animals introduced with, 438-442
    Midge, 440
Wheatear, 140, 394
White, Gilbert, 435
    Rev. Thomas, 316
White Moss, West Linton, 502
Whitewreath, 165
Whitmuir Hill, 51
Whitret. See Weasel
Whitrig Bog, 347
Whittlesea Meer, 190
Wick, 341
Wigeon, change of food habits, 412
    protection for hawking, 203
Wigtownshire, Boar, Wild, prehistoric remains in, 90
    Capercaillie in, 273
    Celtic Shorthorn in, 57
    Chough in, 178
    Deer, Red, prehistoric remains in, 333
    Dog, early remains in, 87
    Elk, prehistoric remains in, 16, 347
    forest, prehistoric, in, 308
    Squirrel, Red, in, 293, 297
    Urus, prehistoric remains in, 50
Wild, O. H., 179, 402, 403, 412, 432, 501
Wild-fowl, destruction by decoy, 139
    by punt-gunning, 188
    protection for food, 217
        for hawking, 203
    reclamation of marsh affecting, 370 *et seq.*
William of Malmesbury, 236, 371
Williestruther Loch, 347
Wilson, Dr R. M., 296
Wireworms, increase of, 392
Woburn, 288
Wolfelee, 121 f.-n., 292
Wolves in Scotland, 17, 20, 308, 329, 365, 497
    destruction of, 115-121, 482, 488, 496
        for sport, 184
    effect of destruction of forests upon, 351
    forests, destruction of, to extirpate, 120, 318
    increase during war, 396
    in Poland, 109
    legislation relating to, 116 *et seq.*
    place names as evidence of, 121 f.-n.
    prehistoric remains of, 69, 115

Wood-Wasps, introduction of, 467-469
Woodburn, H. A., 277
Woodcock, 266
 price of, 140
 protection for sport, 205
Woodpecker, Great Spotted, influence of destruction of forest upon, 329, 357-360, 361
Wool of sheep, characteristic, 39, 43, 45-48
 improvement of, 45-48, 490
 manufacture in Scotland, 46
Worms. *See also* Earthworms
 exotic, in hothouses, 464
 in waterworks, 414
 Liver-flukes. *See* Flukes
 Maw, 422
 Miners', 422
 Pork Thread-, 422, 508
 Round, 422, 508
  of Horse, 422
  transmission by man, 421, 422
 Ship, 432
 Tape, Beef, 421, 422
  Broad, 421
  Dwarf, 421
  of Dog, 422
  Pork, 421

Worms—*Cont.*
 Tape, transmission by man, 421
 Whip, 422
Wrasse, 8, 19, 149
Wren, increase due to cultivation, 394
 protection through superstition, 238
 St Kilda, 190, 229
Wright, Dr Strethill, 386
Wyntoun, Andrew, 90

Xenopsylla cheopsis, 420, 507
Xestobium tesselatum, 404
Xyleborus, 463

Yak, Thibetan, 247
Yarhouse, 341
Yell, Island of, 58, 280
Yellowhammer, 389
Yellowstone Park, 236, 387
Yorkshire, 112, 317
Youatt, Wm, 43, 514
Ypres, fauna of water pipes at, 414
Ythan, River, 194, 296

Zanzibar, 460
Zoological Gardens, London, 288, 436
 Park, Edinburgh, 254, 289, 339, 428
Zygaenidae, 448

For EU product safety concerns, contact us at Calle de José Abascal, 56–1°,
28003 Madrid, Spain or eugpsr@cambridge.org.

www.ingramcontent.com/pod-product-compliance
Ingram Content Group UK Ltd.
Pitfield, Milton Keynes, MK11 3LW, UK
UKHW030857150625
459647UK00021B/2763